OCEANOGRAPHIC APPLICATIONS *of*
REMOTE SENSING

Edited by
Motoyoshi Ikeda
Frederic W. Dobson

CRC Press
Boca Raton   New York   London   Tokyo

Cover photo courtesy of William J. Emery.

Library of Congress Cataloging-in-Publication Data

Catalog record is available from the Library of Congress

Foreword

Most existing books dealing with remote sensing have been written by remote sensing specialists, and hence the emphasis has been placed on sensor characteristics, the mechanisms of imaging, and data calibration. But the use of remote sensing data by physical oceanographers in solving oceanographic problems is steadily increasing. In spite of the obvious need for a suitable reference book, the only existing ones are not kind enough for the typical oceanographer trying to "find a way" into the field. Oceanographers who have not used remote sensing data usually begin with a particular physical, chemical, or biological problem, and then look for a useful method and some relevant data to help solve the problem. They will be best served by a reference source which is organized by processes rather than sensors: this is why this book is organized by processes. The important contents of the book include: comments on the reliability of the geophysical information retrieved from remote sensing data, descriptions of the information that has been used for particular process studies, which information is potentially useful in studying a given process, and descriptions of some of the more important mistakes made by experienced users of remotely sensed data. Technical information on each of the sensors is provided in the appendices.

The editors gratefully thank all the contributors, who first agreed on objectives, and then devoted their valuable time to writing articles, absorbing along the way a wide variety of requests from the editors. This acknowledgment should also include their efforts in reviewing the articles submitted by the other contributors. The reviewers involved in this cross-checking system were W. Emery, K. Katsaros, P.Y. LeTraon, M. Lewis, A. Liu, P. Minnett, G. Mitchum, B. Topliss, K. Raney, K. Steffen and P.Y. Vachon. The book would not be as authoritative without careful reviews by world-class experts on the individual topics from outside the contributor group. The outside reviewers were F. Carsey, R. Cheney, W. Dierking, J. Gower, B. Holt, K. Kawamura, C. Koblinsky, C. Mutlow, D. Reynolds, A. Stoffelen, I. Young and V. Zlotnicki.

The Editors

Motoyoshi Ikeda, Ph.D., was working as a research scientist at Physical and Chemical Science Branch, Department of Fisheries and Oceans, in Dartmouth, Nova Scotia, Canada when this book was compiled. He earned his Ph.D in fluid dynamics at the University of Tokyo in 1974. Since he changed his field from aeronautical fluid dynamics to physical oceanography in 1978, he has been working on numerical modelling of mesoscale ocean variability, a coupled sea ice-ocean system and climate change as well as application of remote sensing to physical oceanography. He has published nearly 50 scientific papers in *Journal of Physical Oceanography, Journal of Geophysical Research* etc. He is currently a faculty member at the Graduate School of Environmental Earth Science, Hokkaido University, Sapporo, Japan.

Frederic W. Dobson, Ph.D., is a research scientist at Physical and Chemical Science Branch, Department of Fisheries and Oceans, in Dartmouth, Nova Scotia, Canada. He earned his Ph.D in physical oceanography from the University of British Columbia in 1969. Since that time he has been working mostly on the fluid dynamics of air-sea interaction, in particular, wave generation — the physics of climate change, experimental physical oceanography and applications of remote sensing techniques to oceanography and meteorology. He is the co-author of two books: *Air-Sea Interaction Instruments and Methods* with L. Hasse and R. L. Davis and *Introductory Physics of the Atmosphere and Ocean* with L. Hasse, and numerous journal articles on his subjects of interest.

Contributors

John R. Apel, Ph.D.
Applied Physics Laboratory
The Johns Hopkins University
Laurel, Maryland

Alain Cavanié, Ph.D.
Oceanographic Spatiale
Centre IFREMER
Plouzané, France

Josefino C. Comiso, Ph.D.
Laboratory for Hydrospheric Processes
NASA/Goddard Space Flight Center
Greenbelt, Maryland

Ella B. Dobson, Ph.D.
Applied Physics Laboratory
The Johns Hopkins University
Laurel, Maryland

Frederic W. Dobson, Ph.D.
Ocean Circulation Division
Bedford Institute of Oceanography
Dartmouth, Nova Scotia, Canada

Mark R. Drinkwater, Ph.D.
Jet Propulsion Laboratory
California Institute of Technology
Pasadena, California

Åsmund Drottning, Ph.D.
Nansen Environmental and Remote
 Sensing Center
Bergen, Norway

William J. Emery, Ph.D.
CCAR
University of Colorado
Boulder, Colorado

Charles W. Fowler, Ph.D.
CCAR
University of Colorado
Boulder, Colorado

Catherine Gautier, Ph.D.
ICESS
University of California
Santa Barbara, California

Francis Gohin, Ph.D.
Oceanographic Spatiale
Centre IFREMER
Plouzané, France

T.H. Guymer, M.Sc.
James Rennell Center for Ocean Circulation
Chilworth Research Park
Southhampton, United Kingdom

Guoqi Han, Ph.D.
Ocean Circulation Division
Bedford Institute of Oceanongraphy
Dartmouth, Nova Scotia, Canada

Motoyoshi Ikeda, Ph.D.
Graduate School of Environmental Earth
 Science
Hokkaido University
Sapporo, Japan

Johnny André Johannessen, Ph.D.
European Space Agency
   ESTEC
Noordwijk, The Netherlands

Ola M. Johannessen, Ph.D.
Nansen Environmental and Remote
   Sensing Center
University of Bergen
Bergen, Norway

Kristina B. Katsaros, Ph.D.
Centre Ifremer
Plouzané, France

Bernard J. Kilonsky, Ph.D.
UH Sea Level Center
University of Hawaii
Honolulu, Hawaii

Harald E. Krogstrad, Ph.D.
SINTEF Industrial Mathematics
Trondheim, Norway

Masahisa Kubota, Ph.D.
School of Marine Science and Technology
Tokai University
Shizuoka, Japan

Pierre-Yves Le Traon, Ph.D.
CLS Space Oceanography Group
Toulouse, France

Marlon R. Lewis, Ph.D.
Department of Oceanography
Dalhousie Unviersity
Halifax, Nova Scotia, Canada

Anthony K. Liu, Ph.D.
Oceans and Ice Branch
NASA/Goddard Space Flight Center
Greenbelt, Maryland

W. Timothy Liu, Ph.D.
Jet Propulsion Laboratory
California Institute of Technology
Pasadena, California

J.A. Maslanik, Ph.D.
CCAR
University of Colorado
Bouolder, Colorado

Peter J. Minnett, Ph.D.
Oceanographic and Atmospheric Division
Brookhaven National Laboratory
Upton, New York

Gary Mitchum, Ph.D.
UH Sea Level Center
University of Hawaii
Honolulu, Hawaii

Frank M. Monaldo, Ph.D.
Applied Physics Laboratory
The Johns Hopkins University
Laurel, Maryland

Lasse H. Pettersson, Ph.D.
Nansen Environmental and Remote
   Sensing Center
Bergen, Norway

R. Keith Raney, Ph.D.
Applied Physics Laboratory
The Johns Hopkins University
Laurel, Maryland

Sei-ichi Saitoh, Ph.D.
Faculty of Fisheries
Hokkaido University
Hakodate, Japan

Paul Samuel, Ph.D.
Nansen Environmental and Remote
   Sensing Center
Bergen, Norway

P. Schluessel, Ph.D.
Meteorologisches Institut
Universität Hamburg
Hamburg, Germany

Robert A. Shuchman, Ph.D.
Environmental Research Institute of Michigan
Ann Arbor, Michigan

Konrad Steffen, Ph.D.
Cooperative Institute for Research
  in Environmental Sciences
University of Colorado
Boulder, Colorado

B.J. Topliss, Ph.D.
Coastal Oceanography
Bedford Institute of Oceanography
Dartmouth, Nova Scotia, Canada

Paris W. Vachon, Ph.D.
Canada Center for Remote Sensing
Ottawa, Ontario, Canada

Gary A. Wick, Ph.D.
CCAR
University of Colorado
Boulder, Colorado

· Contents

Part II   Water Properties

Part III   Wind, Wind Waves, and the Marine Boundary Layer

Section C   Marine Boundary Layer

Part IV   Application to Sea Ice

Section A   Ice Surface Condition

Section B   Ice Movement

Section C   Waves in Ice-Covered Oceans

Part V Appendices

Part I

Ocean Circulation Dynamics

1

Mesoscale Variability Revealed with Sea Surface Temperature Imaged by AVHRR on NOAA Satellites

Motoyoshi Ikeda

In this chapter, the usage of sea surface temperature (SST) data collected on Advanced Very High-Resolution Radiometer (AVHRR) by the National Oceanic and Atmospheric Administration (NOAA) is reviewed for revealing mesoscale variability of ocean circulation. Detection of fronts associated with meandering ocean currents as well as eddies separated from the currents is first presented in Section 1.2 as a remarkable finding in early times, along with field observations to verify the remotely sensed mesoscale variability in Section 1.3. Following such qualitative remote sensing observations, more objective methods of quantitative velocity measurements are discussed next in Section 1.4 as recent developments. Numerical and laboratory experiments useful for understanding and predicting

0-8493-4525-1/95/$0.00+$.50

3

observed mesoscale variability are referred to in Section 1.5, recommendations are given in Section 1.6.

1.1 Historical Summary

The greatest advantage gained from satellite remote sensing is to measure geophysical fields with wide coverage at once so that we can display horizontal structures of SST and other parameters, and observe their evolutions. In addition to the primary objective of observing cloud cover, meteorological satellites are capable of exhibiting sea surface thermal features, which clearly show the meandering axes of the major current systems such as the Gulf Stream and Kuroshio. Following the Geostationary Operational Environmental Satellites (GOES) in early times, AVHRR on NOAA has continuously supplied data of SST for nearly two decades.

Although we had known the mesoscale (a few tens to a few hundreds of kilometers) variability in the Gulf Stream from hydrographic data, satellite infrared imagery gave us confidence on the meandering Gulf Stream and enabled us to observe its evolution.[1] Even in a relatively less energetic current such as the California Current, Bernstein et al.[2] showed clear thermal features implying mesoscale variability. A cold band near the coast due to wind-driven coastal upwelling becomes wavy caused by the undulating California Current. Legeckis[3] collected SST structures related to fronts in various regions and showed that mesoscale variability was common in the world ocean. Apel[4] gave a summary of satellite remote sensing used for studies of dynamic oceanography in the early days.

Satellite infrared imagery soon came to play a major role in oceanographic observation, in particular, a study of mesoscale variability because of its advantage of seeing a large area at once. The infrared imagery has been used in forecast as well because of its quick accessibility. For near-real time forecast experiments in the Gulf Stream, Robinson et al.[5] used SST data in NOAA infrared images to initialize a numerical model. The NOAA images were essential in determining the Gulf Stream axes and positions of cold and warm core rings generated from the stream. Note that the mesoscale variability near the Gulf Stream is appropriately captured with high horizontal resolution of AVHRR, while altimeter data have marginally high resolution (Chapter 2).

1.2 Detection of Current Meanders and Eddies

Sea surface thermal features in the major current systems infer their dynamic structures for the following reason: the currents match the boundaries between two different water masses (one is warm, and the other is cold); e.g., the Gulf Stream is the boundary between the subtropical and the subpolar water masses. The current axis with a maximum speed exists near the steepest horizontal temperature (and density) gradient, because vertical shear is related to a horizontal pressure gradient with a geostrophic balance, which is dynamically similar to thermal wind relationship in the atmosphere.

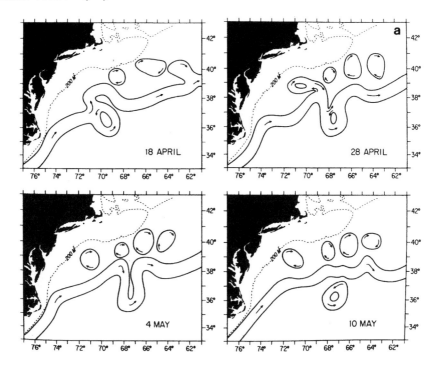

Figure 1.1 The schematic diagram of the Gulf Stream path and rings based on NOAA infrared images. Two rings with warm water existed north of the Gulf Stream, and a cold-core ring was formed southward during this time series. (From Richardson, P.L., *J. Phys. Oceanogr.*, 10, 90, 1980. With permission.)

Using NOAA infrared imagery, Apel[4] detected wavelengths (~300 km) and phase speeds (10 ~ 20 cm/sec) of meanders in the Gulf Stream east of Cape Hatteras. It was clearly shown that the meanders grew in time, and the stream became more perturbed as it went eastward (downstream). Richardson[6] drew the Gulf Stream path and separated rings using NOAA images as shown in Figure 1.1. Numerous articles have been published on both cold-core and warm-core rings: their separation from the stream and drift and coalescence with the stream. Doblar and Cheney[7] captured an instance of cold-core ring separation from the Gulf Stream southward in NOAA images and confirmed its existence with the subsurface thermal field collected with airborne expendable bathythermograph (AXBT). Churchill and Cornillon[8] traced the subtropical water mass crossing the stream both in NOAA images and hydrographic data, and estimated its influence on the Slope Water north of the stream.

The other major current system, the Kuroshio, has mesoscale meanders and eddies as the Gulf Stream does, and also its particular character of bimodality between the nearly straight path and the large meander path. The trigger for the large meander path has been investigated for several decades. One of the candidates is a small amplitude meander, which develops south of Shikoku and eventually grows to a large meander. Kimura and Sugimoto[9] observed small amplitude meanders, which developed to a large meander but also disappeared without a large meander. A NOAA image has shown potential to describe evolution of small amplitude meanders, while the transition from the straight path to the large meander path may be controlled by the other conditions such as volume transport of the Kuroshio. Hence, we need additional data other than remotely sensed SST; e.g., satellite altimeter data, current meter data, etc.

The Agulhas current also has the thermal structures controlling its dynamics. The current flows southward along the east coast of Africa, turns westward at the southern tip of Africa, and merges into the eastward Antarctic Circumpolar Current. Large (diameter of ~100 km) warm-core rings are produced around the merging point. Lutjeharms and van Ballegooyen[10] used NOAA and Meteosat infrared images to measure sizes and a production rate of the warm-core rings. The westward drifting rings were tracked a distance of ~1000 km away from the production region into the Indian Ocean and are considered to play a significant role in the global water mass circulation.

We now discuss the usage of remotely sensed SST data in current systems weaker than the major current systems. The California Current is an eastern boundary current, which is not associated with as strong SST contrasts as the major currents. However, once northerly winds blow along the coast, the colder subsurface water is upwelled near the coast and spreads to an axis of the California Current. A NOAA image with minimum cloud cover is displayed in Figure 1.2, displaying such thermal features clearly. Rienecker and Mooers[11] showed that cold water streaks were stretched seaward along seaward meanders of the current, where the mesoscale variability was observed also using hydrographic data. Thus, SST features in the weaker currents are sometimes considered to be tracers, which are driven by subsurface dynamic systems.

The other weaker currents are also detected in NOAA images; e.g., the subtropical front by Van Woert,[12] Iceland-Faeroe front by Niiler et al.,[13] the Norwegian Coastal Current by Johannessen et al.,[14] and the coastal current in the Black Sea by Oguz.[15] Instead of temperature, salinity is a dynamically important parameter in high-latitude currents and coastal currents. Hence, SST may not reflect important dynamics in such currents. However, as long as water masses are distinguished with well-defined temperature-salinity properties, SST can be used to detect meandering current positions. The other possibility is that SST features are used as tracers to be advected by subsurface flow field.

1.3 Verification of Mesoscale Variability Detected by Remotely Sensed SST

Since SST features well represent dynamic structures of the major current systems, remotely sensed SST data give reliable description of their mesoscale meanders and

Figure 1.2 The NOAA infrared image collected off the west coast of North America on October 8, 1980. A brighter band along the coast indicates lower temperature originated by coastal upwelling due to northerly winds. Several cold-water tongues extending seaward suggest mesoscale meanders in the California Current. (Courtesy of W. J. Emery.)

detached rings. In particular, the remote sensing data have advantages over *in situ* data in observing qualitative characters such as meander growth and ring separation. Wavelengths and phase speeds of the meanders can also be well determined using the remote sensing data. However, more careful verification is required for mesoscale variability in coastal and high-latitude currents, with dynamics that are mainly controlled by salinity distribution.

Johannessen et al.[14] conducted field experiments combined with remote sensing observations for the Norwegian Coastal Current. A series of NOAA infrared images demonstrated growth of meanders in a 2-week period: a cold (and fresh) water band, identified as the Coastal Current, had the meandering boundary with the warm offshore water, and the meanders grew in time. Near the end of the period, cold water tongues stretched further offshore from the seaward meanders and made cyclonic loops. Hydrographic data revealed that cyclonic eddies developed between consecutive seaward meanders were responsible for forming the tongues. Thus, SST contours were originally arraigned with the Coastal Current inferring dynamic topography, but SST later played a role of a tracer advected by the cyclonic eddies.

An eddy separated from a thermal front is clearly identified just after its formation, although its surface features are modified by atmospheric effects as time proceeds, making the surface appearance more obscure. Pingree and Le Cann[16] tracked an anticyclonic eddy formed of the slope water along the European coast by drifters. The eddy contained warmer water than the surroundings and lived nearly for a year. During fall and winter, the eddy was gradually covered by cold surface water so that it became harder to be identified in NOAA images. As revealed by drifter trajectories and hydrographic data, the rotation transport was ~10 Sv with significant flow extending to a depth of 1000 m. Thus, remotely sensed SST data were useful for detecting the eddy, but ambiguity remained after its surface features were modified by the atmosphere. Determining the transport and vertical profile definitely required a comparison with the drifter and hydrographic data.

1.4 Surface Advective Velocity

As stated above, remotely sensed SST data are capable of describing qualitative characters of mesoscale variability such as meander growth and ring separation, whereas they directly supply little quantitative information. To utilize the SST data more extensively, a technique of deriving advective velocity field has been developed. The first attempt was made by Vastano and Borders,[17] who manually tracked thermal features between two consecutive images and drew velocity vectors from displacement.

Emery et al.[18] developed a more objective method: the maximum cross-correlation was taken between two horizontal gradient maps deduced from SST images so that the best fit was obtained by translating one image with respect to the other. Instead of SST itself, its gradient was used, because thermal features were better characterized by the gradient. The usage of the gradient is equivalent to high-pass filter, which removes SST field with large horizontal scales. Kamachi[19] and Tokmakian et al.[20] extended Emery's method by allowing rotation. Rotation could be important when translation is minor such as near an eddy center, although Emery et al.[21] found that rotation induced minor differences on overall flow field.

Kelly[22] developed an inverse method in which velocity field was determined by satisfying a horizontal heat advection equation. The additional constraint of minimum horizontal divergence was required in order to reduce noises. The basic concept of this method is similar to that of the maximum cross-correlation method of Emery et al.:[18] both methods find the best fit by translating one image with respect to the other. Kelly and Strub[23] compared the two methods, suggesting that the results were consistent with each other. The satellite-derived flow field was further verified against that based on *in situ* data (acoustic Doppler current profiler [ADCP]), near-surface drifters, and altimeter data.

1.5 Numerical and Laboratory Model Simulation

Satellite-derived SST features can be used for initializing and verifying a numerical model. Axis positions of a meandering current are first determined from SST data and taken as an initial condition for a numerical model. Results after numerical integration of the model are compared with another SST field collected later. Numerical simulation of mesoscale meanders and eddies was presented in a series of Ikeda's articles: the Gulf Stream in Ikeda and Apel,[24] the California Current in Ikeda et al.,[25] and the Norwegian Coastal Current in Ikeda et al.[26]

Ikeda et al.[26] modeled the Norwegian Coastal Current by a two-layer quasi-geostrophic model to simulate meander growth and eddy formation deduced from a 2-week time series of NOAA infrared images. As shown in Figure 1.3, the simulation successfully reproduced timescales of meander, growth phase speeds of the meanders, and formation of eddies. The current was found to be baroclinically unstable with transfer of potential energy to kinetic energy contained in the meanders. In addition to the meander growth at the early stage, the model simulated well positions of the cyclonic eddies that had been generated through the current instability.

Robinson et al.[5] conducted combined field-remote sensing-modeling experiments for mesoscale variability in the Gulf Stream, aiming at an operational forecast. An initial state of the numerical model was generated from satellite SST data plus prescribed vertical profiles of meanders and rings. Since the Gulf Stream area is one of the most highly observed regions, we have a great amount of knowledge on the vertical profiles. The simulation was successful for a few weeks, promising its operational use.

Pearce and Griffiths[27] generated a laboratory model of a coastal current. In a rotating system, lighter water was injected on a wall of the model. An along-wall flow was induced toward the direction of long-wave propagation, which is right facing offshore in the Northern Hemisphere. A perturbation grew in the flow as

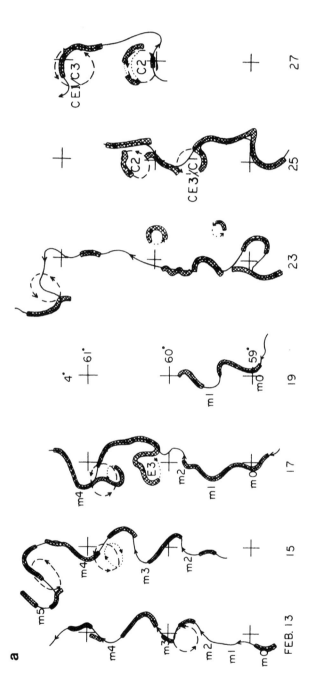

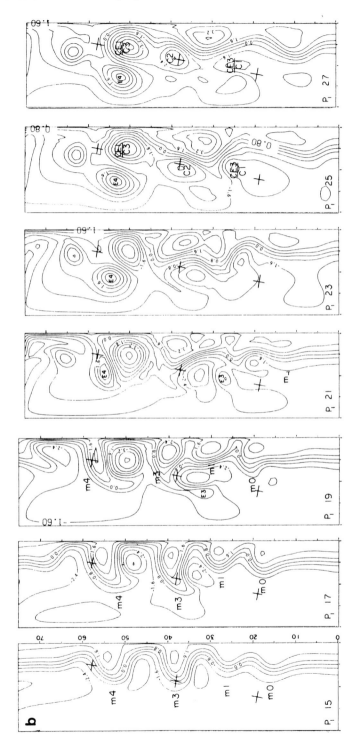

Figure 1.3 (a) Progression of the Norwegian Coastal Current (NCC) axis (solid line), separated anticyclonic eddies (dotted line on the offshore side of the NCC), separated cyclonic eddies (dotted line on the onshore side of the NCC), and cyclonic vortices (broken line) based on NOAA satellite images for February 13–27, 1986. (b) Stream functions from the numerical simulation starting on February 13. (From Ikeda et al., *J. Phys. Oceanogr.*, 19, 20, 1989. With permission of the American Meteorological Society.)

time progressed and produced eddies from large amplitude meanders. Comparison with satellite imagery suggested that the laboratory model was able to describe essential physical mechanisms of meander growth in the coastal current.

1.6 Recommendations

Satellite-derived SST data are useful for detecting axis positions of ocean currents and locations of rings generated from the currents as well as giving qualitative description of meander growth and ring separation. Numerical simulation and forecast require this information for initializing models and also verifying them. Quantitative characters such as current velocities can be extracted using the maximum cross-correlation method. Since cloud cover interferes with SST observations, combination with altimeter data that is corrected regularly is essential for operational use. Note also that the altimeter data directly give quantitative information on the sea surface dynamic height. To utilize the SST data further, assimilation of the data into a numerical model may be useful. This issue will be discussed fully in the following chapter.

References

1. Legeckis, R. V., Application of synchronous meteorological satellite data to the study of time dependent sea surface temperature changes along the boundary of the Gulf Stream, *Geophys. Res. Lett.*, 2, 435, 1975.
2. Bernstein, R. L., Breaker, L., and Whritner, R., California Current eddy formation: ship, air and satellite results, *Science*, 195, 353, 1977.
3. Legeckis, R., A survey of worldwide sea surface temperature fronts detected by environmental satellite, *J. Geophys. Res.*, 83, 4501, 1978.
4. Apel, J. R., Satellite sensing of ocean surface dynamics, *Annu. Rev. Earth Planet. Sci.*, 8, 304, 1980.
5. Robinson, A. R., Spall, M. A., and Pinardi, N., Gulf Stream simulations and the dynamics of ring and meander processes, *J. Phys. Oceanogr.*, 18, 1811, 1988.
6. Richardson, P. L., Gulf Stream ring trajectories, *J. Phys. Oceanogr.*, 10, 90, 1980.
7. Doblar, R. A. and Cheney, R. E., Observed formation of a Gulf Stream cold core ring, *J. Phys. Oceanogr.*, 7, 944, 1977.
8. Churchill, J. H. and Cornillon, P. C., Water discharged from the Gulf Stream north of Cape Hatteras, *J. Geophys. Res.*, 96, 22227, 1991.
9. Kimura, S. and Sugimoto, T., Short-period fluctuations in meander of the Kuroshio's path off Cape Shiono Misaki, *J. Geophys. Res.*, 98, 2407, 1993.
10. Lutjeharms, J. R. E. and van Ballegooyen, R. C., The retroflection of the Agulhas Current, *J. Phys. Oceanogr.*, 18, 1570, 1988.
11. Rienecker, M. M. and Mooers, C. N. K., Mesoscale eddies, jets and fronts off Point Arena, California, July 1986, *J. Geophys. Res.*, 94, 12555, 1989.
12. Van Woert, M., The subtropical front: satellite observations during FRONTS 80, *J. Geophys. Res.*, 82, 9523, 1982.
13. Niiler, P. P., Piacsek, S., Neuberg, L., and Warn-Varnas, A., Sea surface temperature variability of the Iceland-Fare front, *J. Geophys. Res.*, 97, 17777, 1992.
14. Johannessen, J. A., Svendsen, E. S., Sandven, S., Johannessen, O. M., and Lygre, K., Synoptic studies of the three dimensional structure of mesoscale eddies in the Norwegian Coastal Current during winter, *J. Phys. Oceanogr.*, 19, 3, 1989.

15. Oguz, T., La Violette, P. E., and Unluata, U., The upper layer circulation of the Black Sea: its variability as inferred from hydrographic and satellite observations, *J. Geophys. Res.*, 97, 12569, 1992.

16. Pingree, R. D. and Le Cann, B., Anticyclonic eddy X91 in the southern Bay of Biscay, May 1991 to February 1992, *J. Geophys. Res.*, 97, 14353, 1992.

17. Vastano, A. C. and Borders, S. E., Sea surface motion over an anticyclonic eddy on the Oyashio Front, *Remote Sensing Environ.*, 16, 87, 1984.

18. Emery, W. J., Thomas, A. C., Collins, M. J., Crawford, W. R., and Mackas, D. L., An objective method for computing advective surface velocities from sequential infrared satellite images, *J. Geophys. Res.*, 91, 12865, 1986.

19. Kamachi, M., Advective surface velocities derived from sequential images for rotational flow field: limitations and applications of maximum cross correlation method with rotational registration, *J. Geophys. Res.*, 94, 18227, 1989.

20. Tokmakian, R. T., Strub, P. T., and McClean-Padman, J., Evaluation of the maximum cross-correlation method of estimating sea surface velocities from sequential satellite images, *J. Atmos. Oceanogr. Technol.*, 7, 852, 1990.

21. Emery, W. J., Fowler, C., and Clayson, C. A., Satellite image derived Gulf Stream currents compared with numerical model results, *J. Atmos. Oceanogr. Technol.*, 9, 286, 1992.

22. Kelly, K. A., An inverse model for near-surface velocity from infrared images, *J. Phys. Oceanogr.*, 19, 1845, 1989.

23. Kelly, K. A. and Strub, P. T., Comparison of velocity estimates from advanced very high resolution radiometer, *J. Geophys. Res.*, 97, 9653, 1992.

24. Ikeda, M. and Apel, J. R., Mesoscale eddies detached from spatially growing meanders in an eastward-flowing oceanic jet using a two-layer quasi- geostrophic model, *J. Phys. Oceanogr.*, 11, 1638, 1981.

25. Ikeda, M., Mysak, L. A., and Emery, W. J., Observation and modelling of satellite-sensed meanders and eddies off Vancouver Island, *J. Phys. Oceanogr.*, 14, 3, 1984.

26. Ikeda, M., Johannessen, J. A., Lygre, K., and Sandven, S., A process study of mesoscale meanders and eddies in the Norwegian Coastal Current, *J. Phys. Oceanogr.*, 19, 20, 1989.

27. Pearce, A. F. and Griffiths, R. W., The mesoscale structure of the Leewin Current: a comparison of laboratory models and satellite imagery, *J. Geophys. Res.*, 96, 16739, 1991.

2

Mesoscale Variability Revealed with Sea Surface Topography Measured by Altimeters

Motoyoshi Ikeda

In this chapter, the usage of satellite altimeters is reviewed for observing mesoscale variability of the sea surface height, which reflects ocean circulation near the surface. After the historical summary in Section 2.1, observations from the regions of the major current systems are described in Section 2.2 with a focus on verification against *in situ* and drifter data. Although information on the mean sea surface height is usually lost in altimeter data, a method to estimate the mean state is suggested in Section 2.3 in a particular case of a strong current with its axis position varying greatly in time. In Section 2.4, statistical analysis results of mesoscale variability are presented for a large area as a unique contribution of altimeter data. Application to weak currents is discussed in Section 2.5, and corrections for the atmospheric and tidal effects are mentioned in Section 2.6. Data assimilation,

15

which is one of the rapidly developing areas in physical oceanography, is reviewed in Section 2.7. Recommendations are given in Section 2.8.

2.1 Historical Summary

GEOS-3 was the first satellite with an altimeter, which measured the sea surface height for 1975–1978. As presented by Fu et al.,[1] the altimeter explored a new field of global measurements of surface dynamic topography. Following GEOS-3, Seasat was the first dedicated oceanographic satellite with an altimeter, supplying data more accurate than that of GEOS-3. Although the satellite provided only 25-d data with the 3-d repeat cycles in 1978, Cheney et al.[2] showed high variability over the regions in which mesoscale activities had been found energetic using conventional methods. Large root mean square (rms) variabilities of 10–15 cm were observed in the five major current systems: the Gulf Stream, Kuroshio, Agulhas, Antarctic Circumpolar, and Falkland/Brazil confluence.

Using the more recent data collected with Geodetic Satellite (Geosat), which was operated for nearly 3 years with the 17-d repeat cycles, Zlotnicki et al.[3] resolved variability of the sea surface height with periods of a year and shorter. The results from Geosat confirmed the early description with Seasat: there are five major current systems with energetic mesoscale activities. However, Geosat exhibited larger amplitudes of the rms height variability by a factor of three, since periods of the significant part of the mesoscale variability exceed the duration of Seasat; Le Traon et al.[4] and Zlotnicki et al.[3] found seasonal variability of the height as well as seasonal dependence of the height variability, which was significantly higher in fall and winter in the northeastern parts of the Atlantic and Pacific.

There are two new satellites with altimeters. The European Remote Sensing Satellite-1 (ERS-1) launched in 1991 keeps collecting the sea surface height data along with the other various kinds of data. TOPEX/Poseidon launched in 1992 is the first altimeter-dedicated satellite with much higher accuracy. The field of satellite altimeter observation is attaining a rapid development with these new data.

The methods of processing the altimeter data are documented in Appendix D. The sensors have been calibrated carefully, and corrections for atmospheric effects are given. Corrections for solid tides and ocean tides are made based on modeling results. After these corrections, the mesoscale variability in the major current systems is clearly detectable. The temporal variability is determined, while the mean state is normally lost during the processing to remove geoid effects.

2.2 Mesoscale Variability In and Near Major Current Systems and Its Verification

Since the mesoscale variability in the major current systems is clearly identified in the altimeter data, the Gulf Stream and the Kuroshio have been used often for verifying the altimeter data. Willebrand et al.[5] compared altimeter-derived flow

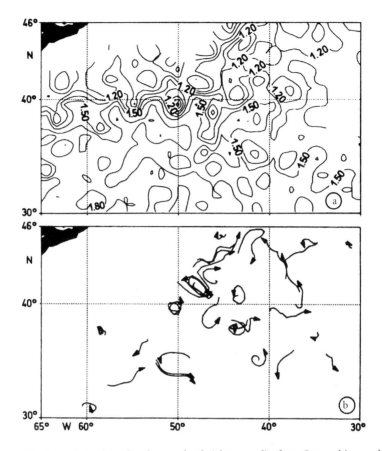

Figure 2.1 (a) Objectively analyzed sea surface height anomalies from Geosat altimeter data for the 17-d period May 24–June 9, 1987 superimposed upon the climatological mean dynamic topography with contour intervals of 15 cm; and (b) trajectories of drifting buoys for the same period. (From Willebrand et al., *J. Geophys. Res.*, 95, 3007, 1990. With permission.)

field with drifter trajectories in the Gulf Stream region. To obtain absolute velocity field from the altimeter data, they superimposed the temporally variable sea surface height on the mean state generated from historical hydrographic data. Since the mean state is subtracted from the altimeter data during the geoid correction, this superposition is necessary. As shown in Figure 2.1, general agreement was achieved between them.

Horton et al.[6] took positions of the Kuroshio axis in the Geosat altimetric sea surface height and confirmed them against thermal structures measured with airborne expendable bathythermograph (AXBT). Although the axis positions were detected with good agreement, sea surface height structures were not always consistent between the two. They attributed the disagreement to the processing of removing the mean state from the altimetric height as well as to a missing absolute velocity due to the assumption of level of no motion in the AXBT data.

For the Gulf of Mexico, Johnson et al.[7] used the Geosat data collected in the Geodetic mission (May 1985–October 1986). Since there were no repeated tracks, sea surface height variability was analyzed over crossover points between the ascending and descending tracks. In a map of mesoscale eddies and the Loop Current, they were able to track the eddies propagating westward at 3–4 cm/sec. Thus, the data from the Geodetic mission are also useful for a mesoscale study.

As stated above, the altimeter data have been verified at least qualitatively for describing mesoscale variability in the major current systems. A more quantitative comparison could give us more confidence on the altimeter data. Joyce et al.[8] compared velocities between the Geosat altimeter data and the data collected with Acoustic Doppler Current Profiler (ADCP) in the Gulf Stream. From the altimeter data, absolute velocities were obtained by estimating a synthetic mean velocity using the method discussed in Section 2.3. Correlation between the two kinds of velocity data was high (~0.7) once the data were smoothed with a 75-km filter.

The altimeter data also provide variability in the major current transport. Fu et al.[1] used the data from the GEOS-3 mission at the crossovers and suggested that the peak-to-peak amplitude of the seasonal cycle of the cross-stream sea surface height difference was ~15 cm, with a maximum in April and a minimum in December. This signal is in good agreement with historical hydrographic observations within their error bars. Using the Geosat data from the Gulf Stream and the Kuroshio, Zlotnicki[9] obtained a maximum sea surface difference in fall. He attributed the difference between these two studies to the distances of the measurements from the current systems.

2.3 Absolute Surface Transport

Information on a mean current velocity is normally lost during the removal of the mean sea surface height in the processing of altimeter data. However, the mean state can be calculated from a series of absolute velocity structures reconstructed under a particular condition. Kelly and Gille[10] and Tai[11] independently developed methods to derive a synthetic mean height by assuming that a current has a Gaussian velocity structure and fluctuates in the crosscurrent direction greater than the current width. The Gulf Stream around 70° W satisfies this condition: it has an e-folding width of 35 km in the crosscurrent direction and shifts its position by 100 km to both sides.

Kelly et al.[12] analyzed the altimeter data used in Joyce et al.[8] further and calculated the mean sea surface height, which was expected to be consistent with the ADCP-derived height minus the time-dependent altimeter-derived height. They actually found good agreement between the two mean states. It is noted here that effects of a centrifugal force ought to be considered in a calculation of the sea surface height from the ADCP data because the Gulf Stream is fast and has a curvature so that a geostrophic balance may not be valid.

Kelly[13] extended the area (73°–46° W) and found that the synthetic mean current was narrower than that derived from historical hydrographic data. The difference is attributable to interannual variability, which is not necessarily captured within the Geosat observation period of 3 years. An additional contributor could be actual current structures more complicated than a single Gaussian function assumed in the mean current calculations. Current bifurcation, which occurs east of 50° W, tends to widen a mean current, because a minor branch, which is omitted in the mean calculations, adds an eastward velocity to the mean flow at a large distance from a main branch. Kelly's method was applied to the Kuroshio area by Qiu et al.[14] and Qiu,[15] giving similar results.

2.4 Statistical and Dynamic Analyses

A statistical analysis is a useful tool for extracting information on spatial and time scales of mesoscale variability from a large data set. The altimeter data are unique in this information on the global view on mesoscale variability. Le Traon et al.[4] and Le Traon[16] analyzed the Geosat altimeter data collected over the North Atlantic. As shown by the correlation coefficients with a 17-d separation in Figure 2.2, the decorrelation timescales of the sea surface height are 20 ~ 50 d in most of the region, or the periods are 80 ~ 200 d. The timescales are smaller in the Gulf Stream and the North Atlantic Current, which contain highly energetic mesoscale variability. The decorrelation spatial scales are 100 ~ 200 km, generally larger south of 30° N than north of that.

Le Traon[16] decomposed the sea surface height data into spatial (along-track) and temporal Fourier components. He converted an along-track propagation speed to a zonal propagation speed, which was 5 ~ 10 cm/sec westward in the North Atlantic except for the Gulf Stream region, where the propagation was eastward. White et al.[17] used the Geosat data from the eastern Pacific and obtained propagation velocities ~5 cm/sec westward, which agreed with the speed of the first-mode baroclinic Rossby wave. This propagation speed has been found commonly by the other authors outside of the major current systems: Bisagni,[18] Matthews et al.,[19] and Tokmakian and Challenor.[20]

Tai and White[21] added a dynamic analysis in the major current systems: computation of Reynolds stress, which indicates nonlinear effects on deceleration or acceleration of a mean flow. The time-average Reynolds stress was found to be positive in the Kuroshio extension region so that the nonlinear effects tended to reduce the Kuroshio. Morrow et al.[22] suggested that the Reynolds stress could play a minor role in the momentum balance of the Antarctic Circumpolar Current. In the bifurcation region of the Gulf Stream around the Newfoundland ridge (50° W), Ikeda[23] showed that the time-dependent Reynolds stress was correlated with deceleration or acceleration only in the second half of the Geosat observations, in which mesoscale variability experienced slower evolution than in the first half.

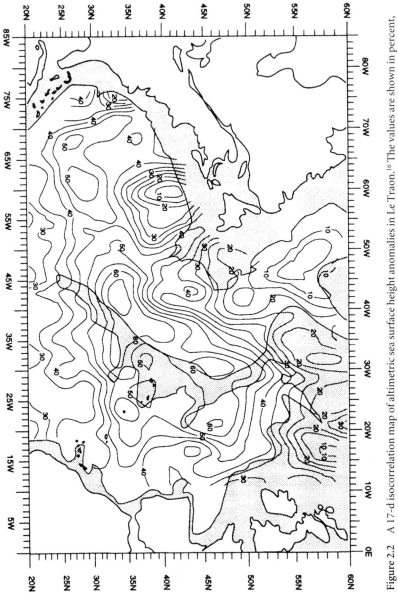

Figure 2.2   A 17-d isocorrelation map of altimetric sea surface height anomalies in Le Traon.[16] The values are shown in percent, and depths shallower than 3000 m are shaded.

2.5 Weak Currents

It has been shown that the altimeter data are verified with the major current systems qualitatively and quantitatively. The data gave also the westward propagation in the less energetic areas consistent with the Rossby wave theory. Although it is harder, verification for weak currents has been attempted as cited below. Stammer et al.[24] related subsurface eddies (Meddy) generated by the Mediterranean outflow with the Geosat altimeter data. The dynamic topography at 500 m relative to 3000 m was found correlated with the altimetric sea surface height.

Zlotnicki et al.[25] conducted careful analyses to obtain correlation of 0.6 ~ 0.9 between the Geosat data and current meter data off the east coast of Africa. The discrepancies were ~1.5 cm/sec and half the typical current speeds. Here the altimeter data were averaged over 140 km, and the current meter data were smoothed over 1 month. The correlation would be low without the averaging and smoothing, indicating that high frequency (1-month period) variability is cut off in the altimeter data, and variability with a small horizontal scale is not well represented in the current meters.

2.6 Correction for Atmospheric and Tidal Effects

Among various corrections described in Appendix D, water vapor effects are potentially significant for mesoscale variability, whereas the other atmospheric effects are insignificant because of their horizontal scales larger than 1000 km. Jourdan et al.[26] showed that contribution of the wet troposphere was more than 10% of dynamic topography, whereas an objective analysis of the dynamic topography reduced the effects because of its shorter timescales (2 d). Minster et al.[27] indicated that simultaneous measurements of water vapor contents could improve the along-track rms mesoscale variability by 3 cm. This improvement is now attempted with TOPEX/Poseidon.

The tidal effects are negligible for energetic mesoscale variability, once the removal of orbital errors also removes major tidal effects with horizontal scales of ~1000 km. However, as suggested by Jacobs et al.,[28] we need to remove the tidal effects carefully for variability with relatively a large (~500-km diameter) horizontal scale and low frequency (~ annual cycle). It is particularly difficult to separate the annual cycle from the M2 tidal effects in the Geosat data, because the M2 tides are aliased at 317 d.

2.7 Data Assimilation

Data assimilation into a numerical model is a useful tool to utilize the altimeter data further. The model screens the data and extracts only the components that are explained with the physics retained in the model. Assimilation methods are divided into two categories: a sequential method and a variational method (Ghil and

Malanotte-Rizzoli).[29] The sequential method is first described and followed by the variational method with two potential problems: one is low cross-track resolution, and the other is a lack of subsurface information.

White et al.[30,31] examined the effects of the low horizontal resolution by introducing the terminology Nyquist wavelength, which is twice the cross-track interval, e.g., the interval is in the range of 110 ~ 140 km for the 17-d repeat cycles of the Geosat data. In assimilation for baroclinic planetary Rossby waves in the California Current with the space and time decorrelation scales of ~400 km and 40–135 d, White et al.[30] showed that the space-time sampling of the Geosat altimeter was appropriate. In assimilation of simulated altimeter data into the same model of the Gulf Stream, White et al.[31] suggested that mesoscale eddies were not well reconstructed in the nonlinear portion near the stream, because the cross- track interval was large relative to the eddy size.

Since surface features are sometimes sensitive to the subsurface flow field (e.g., planetary and topographic Rossby waves), transfer of the surface data into the subsurface is required for reconstructing the surface features. Although the altimeter data supply no direct information on the subsurface field, time sequence of surface data could provide subsurface information, e.g., a Rossby wave speed depends on its vertical profile. Webb and Moore[32] examined the feasibility using altimeter data to reconstruct the subsurface field associated with linear planetary Rossby waves over a flat bottom. They showed that continuous (in time and space) data for 100 d made the separation of the barotropic and first baroclinic waves possible, and hence the subsurface field was reconstructed up to two vertical modes.

In sequential updating assimilation, two methods have been used to reconstruct the subsurface field. One method uses empirical vertical modes observed in the ocean, and the other relies on the mode adjustment that takes place in a model. De Mey and Robinson[33] conducted assimilation for the POLYMODE area only with simulated altimeter data (constructed from the hydrographic and float data) along with a vertical projection based on the empirical modes. The model solution that was constrained by the altimeter data at the sea surface rapidly converged to the data in ocean interior as well.

As a numerical model evolves by itself after surface data input, the surface information is transferred downward by vertical mode adjustment. Holland and Malanotte-Rizzoli[34] studied a simple relaxation (nudging) technique of assimilating altimeter data into a multiple layer quasi-geostrophic model. Examining influence of various data resolutions, they showed that their method worked well with a nearly perfect space-time resolution of the sea surface height, in which data were supplied at every model grid point.

An additional condition can be used at update. Haines[35] used the condition that in a multiple layer quasi-geostrophic model the potential vorticities in the lower layers were unchanged at the time of data injection, where the data were given to the top layer only. He showed that the subsurface field converged to simulated altimeter data in ~200 d around the Gulf Stream. Ikeda[36] selected individual processes important in mesoscale variability and examined the sequential method. A barotropic eddy extending over two tracks was reconstructed in terms of its near-surface structures, while the subsurface field reached only 25% of the simulated data after five repeat cycles (85 d) of assimilation.

A brief description of the variational method may be useful here. An adjoint equation system is derived from the variational principle and optimal control theory. The problem is how to find the minimum in the cost function, which measures differences between an assimilation solution and data, along with constraints formulated by an assumed model by varying control variables. The less well-known components of the model are usually chosen as control variables, e.g., the space of control variables is often composed of initial and boundary conditions. The minimum point is achieved by descending along the directions of gradients of the cost function. An efficient way to calculate the gradients is to solve the adjoint equations, which are obtained by setting partial derivatives of the Lagrange function to zero.

We can expect more efficient vertical transfer in the variational method, because assimilation is not limited by the timescale of mode adjustment now. However, this problem has not been tackled yet using real data. Moore[37] used the variational method, in which a feature model with prescribed vertical profiles was required to generate the first guess of control variables for initializing subsurface structures. Thus, the method is not applicable to a region where no feature model is provided. Schröter et al.[38] applied the variational method only to a reduced gravity model without lower layer motions.

2.8 Recommendations

The altimeter data have been verified for mesoscale variability in the major current systems: time-dependent components exhibit well positions of the meandering currents and separated rings. The mean field, which is necessary to infer absolute velocities, can be retrieved only when a single current shifts back and forth more greatly than its width. It is worthwhile to extend the method to a highly perturbed case and bifurcated currents.

The dynamic analysis has been limited so far to a Reynolds stress calculation. Further analysis such as conversion of potential energy to kinetic energy requires estimates of subsurface field, which could be reconstructed through data assimilation. The subsurface field has to be estimated with errors smaller than the field itself. Since the ordinary sequential updating method needs a long time for vertical mode adjustment, the method may not be appropriate for estimating the subsurface field. The variational method should be examined and established for this purpose.

Verification of altimeter data for subsurface field is a challenging task. Although the data give no direct information on the subsurface field, a time sequence of data combined with data assimilation could infer the subsurface field, which may be compared with hydrographic and current meter data.

Following Geosat, we have ERS-1 and TOPEX/Poseidon, which are currently collecting altimetric sea surface height data. Both satellites work well for the purpose of altimeter measurements, e.g., Proceedings of the First ERS-1 Symposium.[39] We expect mesoscale variability to be better resolved by the ERS-1, which has 35-d repeat cycles and hence track intervals shorter than that of Geosat. TOPEX/Poseidon supplies very accurate altimetric sea surface height, with 10-d repeat cycles but larger track intervals, and is expected to reveal basin-scale circulation. Combination of these two data sets may give us a great opportunity to achieve global measurements of mesoscale variability with high resolutions both in time and space.

References

1. Fu, L.-L., Vazquez, J., and Parke, M. E., Seasonal variability of the Gulf Stream from satellite altimetry, *J. Geophys. Res.*, 92, 749, 1987.
2. Cheney, R. E., Marsh, J. G., and Beckley, B. D., Global mesoscale variability from collinear tracks of Seasat altimeter data, *J. Geophys. Res.*, 88, 4343, 1983.
3. Zlotnicki, V., Fu, L.-L., and Patzert, W., Seasonal variability in global sea level observed with Geosat altimetry, *J. Geophys. Res.*, 94, 17959, 1989.
4. Le Traon, P. Y., Rouquet, M. C., and Bossier, C., Spatial scales of mesoscale variability in the North Atlantic as deduced from GEOSAT data, *J. Geophys. Res.*, 95, 20267, 1990.
5. Willebrand, J., Käse, R. H., Hinrichsen, H. H., and Krauss, W., Verification of Geosat sea surface topography in the Gulf Stream extension with surface drifting buoys and hydrographic measurements, *J. Geophys. Res.*, 95, 3007, 1990.
6. Horton, C. W., Porter, D. L., deWitt, P. W., and Rankin, W. E., Airborne expendable bathythermograph survey of the Kuroshio extension and comparison with simultaneous altimeter measurements, *J. Geophys. Res.*, 97, 7447, 1992.
7. Johnson, D. R., Thompson, J. D., and Hawkins, J. D., Circulation in the Gulf of Mexico from Geosat altimetry during 1985–1986, *J. Geophys. Res.*, 97, 2201, 1992.
8. Joyce, T. M., Kelly, K. A., Schubert, D. M., and Caruso, M. J., Shipboard and altimetric studies of rapid Gulf Stream variability between Cape Cod and Bermuda, *Deep-Sea Res.*, 37, 897, 1990.
9. Zlotnicki, V., Sea level differences across the Gulf Stream and Kuroshio Extension, *J. Phys. Oceanogr.*, 21, 599, 1991.
10. Kelly, K. A. and Gille, S. T., Gulf Stream surface transport and statistics at 69 W from the Geosat altimeter, *J. Geophys. Res.*, 95, 3149, 1990.
11. Tai, C.-K., Estimating the surface transport of meandering jet streams from satellite altimetry: surface transport estimates for the Gulf Stream and Kuroshio extension, *J. Phys. Oceanogr.*, 20, 860, 1990.
12. Kelly, K. A., Joyce, T. M., Schubert, D. M., and Caruso, M. J., The mean sea surface height and Geoid along the Geosat subtrack from Bermuda to Cape Cod, *J. Geophys. Res.*, 96, 12699, 1991.
13. Kelly, K. A., The meandering Gulf Stream as seen by the Geosat altimeter: surface transport, position and velocity variance from 73 to 46 W, *J. Geophys. Res.*, 96, 16721, 1991.
14. Qiu, B., Kelly, K. A., and Joyce, T. M., Mean flow and variability in the Kuroshio Extension from Geosat altimetry data, *J. Geophys. Res.*, 96, 18491, 1991.
15. Qiu, B., Recirculation and seasonal change of the Kuroshio from altimetry observations, *J. Geophys. Res.*, 97, 17801, 1992.
16. Le Traon, P. Y., Time scales of mesoscale variability and their relationship with space scales in the North Atlantic. *J. Marine Res.*, 49, 467, 1991.

17. White, W. B., Tai, C.-K., and DiMento, J., Annual Rossby wave characteristics in the California current region from the Geosat exact repeat mission, *J. Phys. Oceanogr.*, 20, 1297, 1990.

18. Bisagni, J. J., Ocean surface topography measured by the Geosat radar altimeter during the Frontal Air-Sea Interaction Experiment, *J. Geophys. Res.*, 96, 22087, 1991.

19. Matthews, P. E., Johnson, M. A., and O'Brien, J. J., Observation of mesoscale ocean features in the Northeast Pacific using Geosat radar altimetry data, *J. Geophys. Res.*, 97, 17829, 1992.

20. Tokmakian, R. T. and Challenor, P. G., Observations in the Canary Basin and the Azores frontal region using Geosat data, *J. Geophys. Res.*, 98, 4761, 1993.

21. Tai, C.-K. and White, W. B., Eddy variability in the Kuroshio extension as revealed by Geosat altimetry: energy propagation away from the jet, Reynolds stress and seasonal cycle, *J. Phys. Oceanogr.*, 20, 1761, 1990.

22. Morrow, R., Church, J., Coleman, R., Chelton, D., and White, N., Eddy momentum flux and its contribution to the southern ocean momentum balance, *Nature (London)*, 357, 482, 1992.

23. Ikeda, M., Mesoscale variabilities and Gulf Stream bifurcation in the Newfoundland basin observed by the Geosat altimeter data, *Atmos.-Ocean*, 31, 567, 1993.

24. Stammer, D., Hinrichsen, H.-H., and Käse, R. H., Can meddies be detected by satellite altimetry?, *J. Geophys. Res.*, 96, 8118, 1991.

25. Zlotnicki, V., Siedler, G., and Klein, B., Can the weak surface currents of the Cape Verde frontal zone be measured with altimetry?, *J. Geophys. Res.*, 98, 2485, 1993.

26. Jourdan, D., Boissier, C., Braun, A., and Minster, J.-F., Influence of the wet tropospheric correction in mesoscale dynamic topography as derived from satellite altimetry, *J. Geophys. Res.*, 95, 17993, 1990.

27. Minster, J.-F., Jourdan, D., Normant, E., Brossier, C., and Gennero, M.-C., An improved special sensor microwave imager water vapor correction for Geosat altimeter data, *J. Geophys. Res.*, 97, 17859, 1992.

28. Jacobs, G. A., Born, G. H., Parke, M. E., and Allen, P. C., The global structures of the annual and semiannual sea surface height variability from Geosat altimeter data, *J. Geophys. Res.*, 97, 17813, 1992.

29. Ghil, M. and Malanotte-Rizzoli, P., Data assimilation in meteorology and oceanography, *Adv. Geophys.*, 11, 141, 1991.

30. White, W. B., Tai, C.-K., and Holland, W. R., Continuous assimilation of Geosat altimetric sea level observations into a numerical synoptic ocean model of the California Current, *J. Geophys. Res.*, 95, 3127, 1990.

31. White, W. B., Tai, C.-K., and Holland, W. R., Continuous assimilation of simulated Geosat altimetric sea level into an eddy-resolving numerical ocean model. I. Sea level differences, *J. Geophys. Res.*, 95, 3219, 1990.

32. Webb, D. J. and Moore, A., Assimilation of altimeter data into ocean models, *J. Phys. Oceanogr.*, 16, 1901, 1986.

33. De Mey, P. and Robinson, A. R., Assimilation of altimeter eddy field in a limited-area quasi-geostrophic model, *J. Phys. Oceanogr.*, 17, 2280, 1987.

34. Holland, W. R. and Malanotte-Rizzoli, P., Assimilation of altimeter data into an ocean model: space verses time resolution studies, *J. Phys. Oceanogr.*, 19, 1507, 1989.

35. Haines, K., A direct method for assimilating sea surface height data into ocean models with adjustments to the deep circulation, *J. Phys. Oceanogr.*, 21, 843, 1991.

36. Ikeda, M., Sequential updating assimilation of simulated Geosat altimeter data from mesoscale features into a two-layer quasi-geostrophic model, *J. Oceanogr.*, 49, 697, 1993.

37. Moore, A. M., Data assimilation in a quasigeostrophic open ocean model of the Gulf Stream region using adjoint method, *J. Phys. Oceanogr.*, 21, 398, 1991.

38. Schröter, J., Seiler, U., and Wenzel, M., Variational assimilation of Geosat data into an eddy-resolving model of the Gulf Stream extension area, *J. Phys. Oceanogr.*, 23, 925, 1993.

39. Proc. 1st ERS-1 Symp. Space at the Service of Our Environment, ESA SP-359.

3

Synthetic Aperture Radar on ERS-1

Johnny André Johannessen, Robert A. Shuchman, and Ola M. Johannessen

3.1 Introduction

The first European Space Agency (ESA) remote sensing satellite (ERS-1) was successfully launched on July 17, 1991 carrying a suite of microwave and infrared instruments. Since launch, a total of approximately 400,000 synthetic aperture radar (SAR) images, each 100×100 km with a center incidence angle of 23°, have been collected over the ocean and sea ice surfaces. Unlike the infrared radiometers and visible imaging, the SAR is unaffected by cloud cover and visible light conditions; and unlike the radar altimeter, the SAR obtains a two-dimensional image of the surface. In this chapter we will provide an assessment of the capabilities and uniqueness of the vertical polarization, ERS-1 C-band SAR in imaging of upper ocean circulation features.

For the ocean surface, high resolution (16×20 m) SAR images are principally formed by resonant Bragg scattering whereby the transmitted 0.056-m radar waves, insensitive to clouds, fog, and visible light conditions, are reflected back to the antenna by short gravity waves of approximately the radar wavelength.[1] These surface waves are formed primarily in response to the surface wind stress. Spatial variations of these short waves are frequently observed in SAR images as pointed out in several articles. These variations are usually induced by larger scale features and processes such as long gravity waves, variable wind velocities, and stratification in the marine atmospheric boundary layer; and variable currents associated with upper ocean circulation features including fronts, eddies, upwelling and internal waves, tidal circulation, and bottom topography.[2-10]

In spite of the remarkable imaging capabilities reported from Seasat[2-4,8,9] and SIR-A/B,[11-13] the lack of spaceborne SAR data, between Seasat (launched in July 1978 and failed in October 1978), (Space Shuttle flight A (SIR-A) in November 1981 and flight B (SIR-B) in October 1984), Almaz-1 (launched in March 1991 and ended in October 1992) and ERS-1 (launched in July 1991) has to some extent limited the development of SAR application areas in oceanography. Most of the systematic studies of radar backscatter from the ocean have meanwhile been obtained through airborne SAR imaging campaigns, in particular, providing essential understanding of SAR imaging of wind waves and swell.[5,7] However, since the launch of Almaz-1 and ERS-1, spaceborne SAR data have been regularly available; and a large number of validation studies have been conducted that are promising for improved quantitative understanding of the SAR imaging mechanisms and subsequent application in oceanography including air-sea interaction.[14]

In Section 3.2 we will briefly provide an overview of some of the experiments reported in Reference 14. A review of typical ERS-1 SAR image expressions of mesoscale variability including frontal structures and eddies is discussed in Section 3.3. Examples on applications of SAR images in ocean circulation dynamics study are presented in Section 3.4, followed by a summary in Section 3.5.

3.2 ERS-1 SAR Validation Studies

More than 5 years prior to the launch of ERS-1 it was recognized that the collection and preparation of suitable *in situ* data sets coincident with the satellite observations (in particular, in meteorology and oceanography) were necessary in order to validate the satellite sensor data. In view of this, ESA released an Announcement of Opportunity whereby investigators were invited to propose field campaigns in support of the ERS-1 validation studies. Approximately 32 ERS-1 SAR validation studies for ocean and coastal regions were proposed, of which about 11 focused on imaging capabilities of mesoscale circulation features (Table 3.1). A brief presentation of some of these validation experiments, in particular, the Norwegian Continental Shelf Experiment-NORCSEX '91, is given in this chapter. For further detailed description of these 11 studies, the readers are referred to the proceedings from the First and Second ERS-1 symposiums.[14,15]

During the commissioning phase from August 15, to December 15, 1991, ERS-1 orbited the earth in a 3-d exact repeat cycle. In this period a large number of SAR

Table 3.1 Overview of ERS-1 SAR Validation Experiment for Mesoscale Ocean Variability Studies

Principal Investigator	Study Region	Objectives
Nilsson et al.	East Australia	Mapping the East Australian Current with SAR
Gower	West coast of Canada	Surface feature mapping with SAR
Johannessen, J. A. et al.	Norwegian Coast	Coastal ocean studies with ERS-1 SAR during NORCSEX
Scoon and Robinson	English Channel	SAR imaging of dynamic features in the English Channel
Keyte et al.	Iceland-Faeroe frontal region	Comparison of ERS-1 SAR and NOAA AVHRR images of the Iceland-Faeroe front
Shemdin et al.	Gulf of Alaska	SAR ocean imaging in the Gulf of Alaska
Tilley and Beal	Gulf Stream	ERS-1 and ALMAZ SAR ocean wave imaging over the Gulf Stream and Grand Banks
Alpers and La Violette	Gibraltar and Messina Straits	Study of internal waves generated in the Strait of Gibraltar and Messina by using SAR
Font et al.	Western Mediterranean	Comparison of ERS-1 SAR images to in situ oceanographic data
Liu and Peng	Gulf of Alaska	Waves and mesoscale features in the marginal ice zone
Johannessen, O.M. et al.	Greenland and Barents Seas	ERS-1 SAR ice and ocean signature validation during SIZEX

Note: Detailed reports on these studies including highlights are found in Reference 9.

images were regularly obtained from the ascending orbit across the western coast of Norway as part of ESA calibration-validation campaign and NORCSEX '91.[16,17] The NORCSEX '91 study[17] focused on the ERS-1 C-band SAR imaging capabilities, in particular, investigating: (1) how frontal boundaries associated with wind, current, and internal waves lead to characteristically different backscatter modulation that in turn can be distinguished and classified in SAR imagery; and (2) under what environmental conditions this classification is possible. The fast SAR processing and distribution system from Tromsø Satellite Station and Nansen Environmental and Remote Sensing Center (NERSC) provided analyzed images onboard R/V Håkon Mosby within 3 h of SAR data acquisition. The ship was then steered into regions were the SAR images expressed interesting features. In turn, near coincident *in situ* data were obtained providing the temporal and spatial variations in the upper ocean currents, the atmospheric boundary layer stratification, and the near surface wind vector necessary to interpret the different SAR image expressions.

Several of the validation studies reported in Table 1, on the other hand, were conducted without access to near-real time spaceborne SAR data. In ocean regions with prominent and permanent frontal boundaries, i.e., the Gulf Stream and the East Australian Current, this lack of near-real time SAR data did not pose any apparent problems in the design of the ship survey.[18,19] Near coincident ERS-1 SAR, Almaz-1 SAR, and NOAA advanced very high-resolution radiometer (AVHRR) images were obtained and supported by *in situ* ship measurements from which it

Table 3.2 Mesoscale Ocean Current Variability: Relationship between Types of Ocean Current and Ice Edge Features Imaged by the SAR and the Dominating Environmental Conditions that Lead to the Image Expressions

	SAR Image Expressions				
	Natural Film	Wave-Current Interaction		Wind Stress	Sea Ice
Type		Shear/Convergence	Convergence		
Eddies	X	X		X	X
Fronts					
nonthermal	X	X	X		
thermal	X	X		X	X
Internal waves	X		X		X

Note: that for sea-ice applications the ice acts as a tracer (i.e., spiral of ice reflecting eddies and regular spaced ice bands in connection with internal waves).

is possible to examine the usefulness of ERS-1 C-band SAR in mapping ocean current boundaries. Additional observational evidence of the SAR-imaging capabilities of the mesoscale circulation patterns along the ice edge was provided by the two studies that focused on sea ice imaging capabilities of SAR in the marginal ice zone of the Gulf of Alaska and the Greenland and Barents Seas.[20,21]

Moreover, in the Iceland-Faeroe frontal region systematic collection of SAR data spanning its lifetime will provide important information on how the radar signature from this particular frontal region varies with season and environmental conditions. The oceanographic and meteorological conditions associated with each of these images will be derived from a combination of historical data, available ship observations, and sea surface temperature pattern derived from AVHRR images. Systematic examinations of SAR images are also being conducted off the coast of Norway,[22] with particular aim at characterizing the ERS-1 SAR detection capabilities of oil spill and natural slicks.

A brief overview of the major findings from these validation studies, expected to provide new and updated understanding of the ERS-1 SAR-imaging capabilities of mesoscale ocean variability, is provided in the next section.

3.3 SAR Image Expressions

Expressions of mesoscale variability in radar images can be characterized by three types, i.e., eddies including jets and vortex pairs, meandering fronts and internal waves (Table 3.2). The dominating surface environmental conditions that cause these expressions are the presence of natural film (surfactants), ice floes and grease ice mirroring the upper ocean current features, short wave-current interaction along shear and convergence dominated frontal boundaries, and rapid shifts in the surface wind stress connected with thermal fronts. In the remainder of this chapter we will concentrate on image expressions of eddies and frontal boundaries in ice-free waters, and potential application in ocean circulation dynamic studies.

Eddies

The eddy depicted in the first SAR image (Figure 3.1a) received at Tromsø Satellite Station on August 21, 1991 is located in the southern region of Frohavet off the central west coast of Norway.[23] The 100 × 100-m median filtered SAR image manifests the existence of an eddy by the dark low backscattering lines (200–500 m wide) outlining a cyclonic spiral with an approximate diameter of 5 km.

These spiral lines are interpreted as expressions of small-scale turbulence aligned in the direction of the larger scale eddy orbital motion. The turbulence, in turn, leads to convective motion in the water that can bring organic material present in the upper layer to the surface where it can remain as a microlayer of natural surface film.[24] As the concentration of this surface film (surfactant molecules) increases, it can reach sufficient surface tension to inhibit growth of capillary and short gravity waves. In addition, the film edge may reflect the short waves that propagate at oblique angles to the edge, thus limiting the advance of short wave roughness through the slick covered region. In turn the lack of small-scale surface roughness prevents the radar echo from the surface to retain sufficient strength, leading to manifestation of low backscatter, dark features, or surface slicks. Further reduction in the surface roughness can result if the turbulent upwelling alters the surface current so that it affects the wave growth-relaxation rate.[25,26] The SAR image displays no variations of the dark spiraling lines due to radar look angle dependence, and no narrow bright lines are depicted. This suggests that modulations resulting from wave-current interactions along the eddy frontal boundaries such as reported by Johannessen et al.[14] can be ignored.

The spiraling lines indicate convergence toward the eddy center. Similar converging spiral structures have been observed in aircraft photography and shore-based radar images of ice edge eddies.[27,28] Indications of the convergence suggest that this cyclonic eddy may be important for the distribution and concentration of chlorophyll a, algae, and pollutants such as oil spill. Nilsson et al.[19] are also reporting on SAR expressions of a fine-scale, cyclonic meander ending in a hammerhead that has no counterpart in the AVHRR thermal image. They interpret the presence of the dark, narrow lines that express the feature to be connected with convergence of organic matter at the surface. Several eddies of similar size, shape, and structure are also identified in the collection of ERS-1 SAR images (ERS-1 SAR image calendar 1992, ESA/ESRIN, Italy) from the Bay of Naples, Italy; West Frisian Islands, The Netherlands; and Halland, Sweden and Øresund, Denmark. Such features have also been observed through sun glitter in high resolution SPOT and SIR-A/B photographs under cloud-free conditions.[11-13]

In the 6-h period centered at the time of SAR data acquisition, weather observations report 5 m/sec southwesterly winds along the coast and western part of Frohavet where the eddy was detected, dropping to 2–3 m/sec southwesterly winds at midnight. In the same period the winds remain less than 2–3 m/sec along the eastern part of Frohavet. Apart from these standard weather observations along the

a

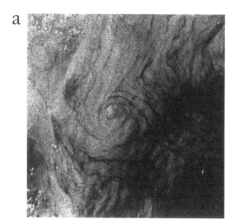

b

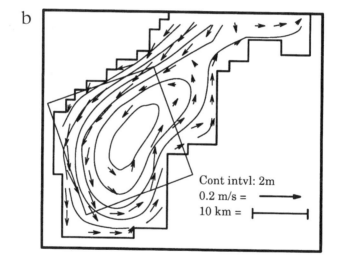

Figure 3.1   (a) SAR image received at Tromsø Satellite Station (Norway) of the cyclonic eddy in Frohavet obtained at 2110 UTC on August 21, 1991. The image is a blowup of a standard scene (100 × 100 km) with a size of 20 × 20 km. The pixel size in this full resolution image is 16 m in azimuth and 20 m in range. The bright point return in far range is caused by a ship, whereas larger bright returns in near range and along image top are coastal islands. Data reception and SAR processing were conducted at Tromsø Satellite Station. Image was processed at Nansen Environmental and Remote Sensing Center. Arrows mark azimuth and range (look) directions. (From Johannessen, J.A. et al., *Int. J. Remote Sensing*, 14, 2203, 1993. With permission. Copyright ESA.) (b) Upper layer circulation pattern obtained from RG model.[17] Arrows mark flow direction and strength. Contour lines indicate departure from initial upper layer thickness of 50 m at 5-m intervals. The model domain is shown by the steplike boundary outlining the shallow water isobath (<50 m) of Frohavet. The rotated square box represents the 20 × 20 km coverage of the SAR image.

coast, no surface data are available to offer sufficient analysis of the physical and biochemical conditions in the eddy, in particular, along the dark spiraling lines.

The threshold wind speed value for the C-band waves is estimated by Donelan and Pierson[29] to be about 3.25 m/sec at 10-m height above the surface. Hence, the fully developed 0.07–0.08-m waves necessary to provide resonant Bragg backscatter are formed in the eddy region, and the presence of surface film at sufficient concentration can in turn form areas of slicks that lead to the backscatter contrast of about 6–10 dB. In addition, the backscatter front and region of low-radar return in the eastern sector are caused by the wind dropping below threshold. The expressions of dark spiral lines are expected to disappear at higher winds (7m/sec) since wind-induced mixing in the upper layer will redistribute the surface slicks and prevent such damping.[30]

Since we have no access to a sequence of SAR images in this case, the temporal characteristics of this eddy, i.e., whether it is forming, fully developed, or decaying, cannot be determined. In turn, we are not entirely sure that the observed 5-km eddy diameter represents its fully developed size. However, these synoptic manifestations of surface convergence in the eddy, rotational direction, and horizontal dimension are clearly important for studies of mesoscale coastal circulation and water quality. Moreover, they offer an excellent opportunity for comparison and validation of model simulations of surface current pattern. This is further discussed in Section 3.4.

Alternatively, mesoscale and submesoscale eddies in spaceborne SAR images are also expressed as series of curved boundaries characterized by concentric curvilinear lines of bright (and dark) radar cross section. This is reported by Liu and Peng[20] during a near-real time eddy watch study in the Shelikof Strait in Alaska. Font et al.,[31] moreover, found similar expressions of mushroom-like dipole eddies with a diameter of about 10 km in SAR images from the channel between the Balearic Islands and Catalan coast of Spain. Although no complete explanations for the radar expression are provided, they imply roughness change through short wave-current interaction along boundaries of convergence and divergence along the eddy margins similar to Johannessen et al.[10] Scoon and Robinson[32] also consider this mechanism to cause the dark and bright radar cross section along the frontal margin of counterrotating eddies in the central English Channel.

Large-scale (<100 km) eddies such as the warm water, Gulf Stream, rings may also be expressed through changes in the atmospheric boundary layer stratification and hence the wind stress induced by the strong (>8°C) thermal gradients at winds up to 7 m/sec.[33,34] On the other hand, radar cross-section expression of cold rings due to corresponding wind stress change are less possible due to weaker surface temperature gradients.[35] None of the studies reported in Reference 14 includes discussions of this type of imaging mechanism.

Frontal Boundaries

The temporal and spatial variations of the near-surface current across the Norwegian Coastal Current (NCC) were obtained by two current meter moorings and the

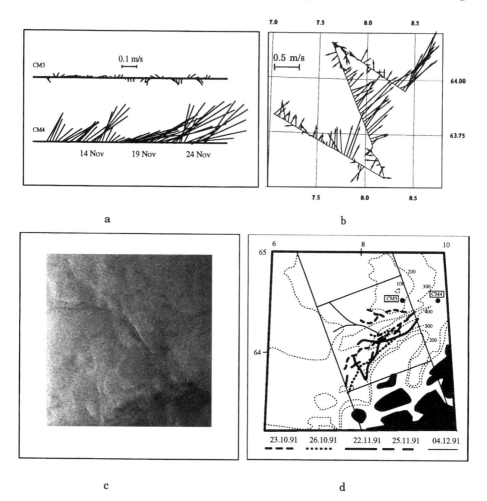

Figure 3.2 (a) Temporal variations in the current at 25-m depth below surface at CM3 and CM4; (b) spatial (and temporal) variations in the current at 15-m depth across the Norwegian Coastal Current; (c) ERS-1 SAR image (100 × 100 km) expression obtained on December 4 showing both bright and dark fronts; (d) extraction of SAR image boundaries overlaid topography. The 100 × 100 km region covered by the SAR image is also shown.

ship mounted Acoustic Doppler Current Profiler (ADCP) (Figure 3.2a and b) during NORCSEX '91. The maximum current reaches 1.25 m/sec in the NCC while the horizontal current shear across the front reaches up to a maximum of about 0.5 × 10^{-3} s^{-1} or roughly five times the Coriolis parameter. The convergence along the front is at least an order of magnitude less. The hydrographic structure in the upper 200 m was measured with a towed undulating Seasoar, and the sea surface temperature was recorded with a surface thermistor. These data show that the density (current) front is maintained by the salinity change, while the temperature remained almost constant at 8°C with a change across the NCC front of less than 0.5°C.

From the sequence of 17 ascending SAR scenes regularly collected during NORCSEX '91, evidence of frontal features assumed to express the shearing frontal boundary are clearly identified in 5 images, approximately 30% of the total collected. During these five ascending SAR overpasses, the wind speed remained between 4 and 10 m/sec, while the boundary layer stratification remained unstable ($T_{air} - T_{sea} = -3°C$) until November 20 when it gradually became neutral as a result of increase in the air temperature. An example of the SAR image expression of the frontal boundary is shown in Figure 3.2c, obtained on December 4. The front is seen to meander across the image, appearing bright (+3 dB) in the center area and dark (−2 dB) in near range in comparison to the mean radar cross section. In Figure 3.2d all the frontal boundaries detected in the five images are shown. The frontal orientation is mostly aligned in the mean northeastward direction of the current, with decreasing spacing between the boundaries toward the channel between Haltenbanken and the coast due (as expected) to topographic steering of the coastal current.[36] There is also a tendency to find a concentration of the fronts along the coastal shelf break; and in a few places, single fronts seem to split into several meandering fronts. If these meanders become unstable, they may evolve into mesoscale eddies with radar backscatter characteristics similar to those discussed in the previous section.[20,31]

We interpret these expressions to be the manifestation of features within the current boundary zone mentioned above. The negligible sea surface temperature change and the relatively constant stratification in the atmospheric boundary layer rule out any thermal front impact on the radar image expression.[33] We therefore conclude that these frontal boundaries (Figure 3.2c and d) are expressions of surface roughness change induced by short wave-current interaction along the current front between the coastal water and the offshore water of Atlantic origin. Due to the relatively large time gap between the SAR-detected frontal features, it is not possible to study the temporal characteristics of the frontal boundary.

Evidence of similar frontal structures are reported by Tilley and Beal[18] in the Gulf Stream. They argue that the linear features did not always coincide with the changes in sea surface temperature. They therefore assume that velocity gradients or current shears might be the source of contrast in the ERS-1 SAR images. Distinct frontal features connected to wave-current interaction along a current shear zone are also observed in the SAR images in the western Mediterranean by Font et al.[31] without the existence of a sea surface temperature boundary. In many of the SAR images discussed by Nilsson et al.,[19] the current flow pattern is expressed by pairs of sharp ridges related to wave-current shear interaction.

Changes from bright to dark radar backscatter return along the fronts (Figure 2c) are seen in practically all the radar images of this shear-convergence type. (Note that the converging, nonthermal front identified in Table 3.2 refers to tidal fronts.) It can reflect sensitivities to both changes in the dominating environmental parameters and the radar look direction. In particular, the wind direction and the peak propagation direction of the short gravity waves relative to a meandering current boundary can cause variations in the amplitude of the Bragg resonant waves. This, in turn, can produce perturbations as seen in the radar return. On the other hand, the perturbed radar backscatter can also be dominated by the relative importance of current shear to current convergence or upwelling along the front. If this is the case, the radar images can provide information about the importance of vertical processes and eventual biological activities along the current front.

Nilsson et al.[19] also assume that the SAR expressions of sharp, bright, and dark ridges are related to the current shear along the boundary of the East Australian Current . This is also in agreement with the NORCSEX '91 results. Moreover, ERS-1 SAR images obtained along the coast of Norway since NORCSEX '91 repeatedly show meandering boundaries of the NCC, i.e., appearing both bright and dark . From these images it is also evident that accumulation of natural film sometimes takes place along frontal boundaries.[37] However, this is outnumbered by surface slick expressing mesoscale and submesoscale eddies, in particular, during spring and summer. Gower[38] reports similar findings along the west coast of Canada. This is perhaps an indirect indication of the spring and summer bloom in the upper layer. Such productivity is usually reported to appear in patches. After the launch of Seastar in spring of 1994, near synoptic Sea-Viewing Wide Field-of-View Sensor (SeaWIFS) and SAR images can be examined for these particular expressions revealing whether any correlation exists between surface roughness and ocean color.

In contrast to the narrow bright and dark SAR frontal expressions discussed above, ocean thermal fronts can be expressed as a steplike change in radar cross section due to rapid shift in wind stress caused by the horizontal temperature gradient.[33] Similar steplike changes in radar cross section are also documented to be the manifestation of mesoscale wind field variations.[10,17] From these characteristic differences in radar expression it might still be possible to classify the ocean frontal observations according to thermal or nonthermal boundaries, in particular, if wind speed and direction were known in the imaged region.

Tilley and Beal[18] furthermore estimate the current field based on the SAR images using only information about the spatial evolution of the wave field across the Gulf Stream. Under the assumptions that (1) the current boundaries are straight and parallel, (2) the current profile is only a function of crosscurrent position, and (3) the incident wave system, in the absence of a current would be spatially homogeneous, the unknown peak current was predicted to be about 2 m/sec. Similar techniques are used by Barnett et al.[39] and Liu et al.[40] to locate and estimate orbital speed in mesoscale eddies. This offers promise that the SAR might be a useful tool to quantify unknown surface currents.

3.4 Integrated Application of SAR Data

It is evident that the detection capabilities of SAR are promising in monitoring the temporal and spatial variability of the upper ocean circulation. This is particularly useful in regions where the use of spaceborne infrared radiometers is hampered by extensive cloud cover. We will demonstrate this by giving a few examples on how SAR data might be integrated in a system dedicated for application in mesoscale ocean variability monitoring.

SAR Image Simulation

The short wave-current interaction mechanism has been analytically explored by Lyzenga and Bennet[41] and Lyzenga[42] using a SAR image simulation model. The model (ERIM Ocean Model [EOM]) solves the wave action spectral transport equation given that the magnitudes of the current shear and vector wind are quantified. In the case with a current that varies only in one direction (x), the interaction of the short surface waves with the current field can be written (to the first order):[42]

$$\left(C_{gx} + U\right)\partial N/\partial X - \left(K_x \, dU/dX + K_y \, dV/dX\right)\partial N/\partial K_x = -\beta N_o f \qquad (1)$$

where C_{gx} is the x component of the wave group velocity; U and V are the x and y components of the surface current; K_x and K_y are the x and y components of the wave number; N is the wave action spectral density (defined as the wave energy density divided by the wave frequency); N_o is the equilibrium spectral density; β is the initial wave growth rate due to wind input;[26] and $f = N/N_o - 1$ represents the fractional spectral perturbation or deviation from the equilibrium state.

Several experiments were conducted to simulate spectral perturbations, or changes in the radar cross section, across ocean current fronts. Only when the strength of the ocean current shear approaches 5×10^{-3} s^{-1} and a wind speed of 5 m/sec is prescribed parallel to the front, do the largest fractional spectral perturbation reach about 1. Moreover, this occurs at ocean surface wavelengths of the order of 1 m. Both at longer wavelengths where the group velocity becomes much larger than the current speed, and at shorter wavelengths where the relaxation rate (typically on the order of 0.1–1.0 s^{-1}) plays a predominant role, this spectral perturbation decreases. However, when the tilt effect from this 1-m wave is included in the C-band Bragg model,[42] the corresponding maximum change across the current shear reaches 1 dB. In addition, both bright (positive) and dark (negative) perturbations are reproduced in the model for different look directions relative to the frontal orientation. In comparison we observe bright/dark backscatter variations ranging from +/–1 to 3 dB across the frontal boundaries with a maximum shear of only 0.5 × 10^{-3} s^{-1} at about 15-m depth. There is no suggestion in the hydrographic structure of the upper 15 m that the current shear at the surface could be as large as the one used in the model.

Additional simulation studies are necessary to obtain better consistency between the EOM model and SAR observations. The EOM model might, for example, be improved through systematic simulation of ERS-1 SAR images with fairly well-quantified coincident *in situ* surface conditions. Moreover, systematic comparison with other simulation models such as suggested by Holliday et al.[43] and Thompson et al.[44] would be valuable as well. It is nevertheless evident that the combination of SAR observations and simulations can become a powerful tool to obtain better quantitative understanding of the imaging mechanisms. In turn, this could lead to better estimates of geophysical parameters and processes that dominate the mesoscale ocean variability.

Current Model-SAR Comparison

Validation of mesoscale ocean circulation model results is often poor due to lack of observational data. Combination of SAR observations (and IR images when available) with results from fine resolution coastal ocean circulation models (i.e., quasi-geostrophic, primitive equation) can become an attractive tool in such validation-by-comparison efforts. In turn, wider applications of both the satellite SAR images and the model results can be expected. This is clearly demonstrated by Johannessen et al.[23] comparing the simultaneous occurrence of the eddy obtained in the ERS-1 C-band SAR image (discussed in Section 3.3) and a primitive equation, one-layer, reduced gravity model (Figure 1a and b).

In that article, they initialize the upper layer thickness in the model to be 50 m, and the density difference between the upper and lower layer as 1 kg m^{-3}. This gives a typical eddy scale (or internal Rossby radius of deformation) of about 5 km. The model grid size of 1.8 km, therefore, resolves this length scale. The model is forced by observed winds, which favor propagation of a downwelling Kelvin wave northeastward along the coast prior to the SAR observation. At a speed of approximately 1 m/sec, determined by the upper layer thickness, the wave reaches Frohavet in time to produce positive vorticity as it travels around the semiclosed bay. The model dynamics produce a cyclonic eddy, in good agreement with the SAR observed eddy, in response to the horizontal current shear (barotropic) instability (Figure 1b). The cyclonic orbital motion in the eddy is about 0.20 m/sec. The convergence toward the eddy center, qualitatively indicated in the SAR image, cannot be accurately quantified from the model output in this case due to insufficient spatial resolution in the model.

The cyclonic eddy remains present in the simulated surface current for about 70 h. Since no sequence of SAR images of the eddy is available, the model estimate of the temporal characteristic cannot be validated. Such a possibility would significantly strengthen the results. However, eventual disappearance of the eddy expression in the SAR image is not completely a documentation that no eddy is present in the upper layer, since the surface roughness characteristics of this type of eddy expressed in the SAR image are sensitive to the wind field usually disappearing at winds above 7 m/sec.

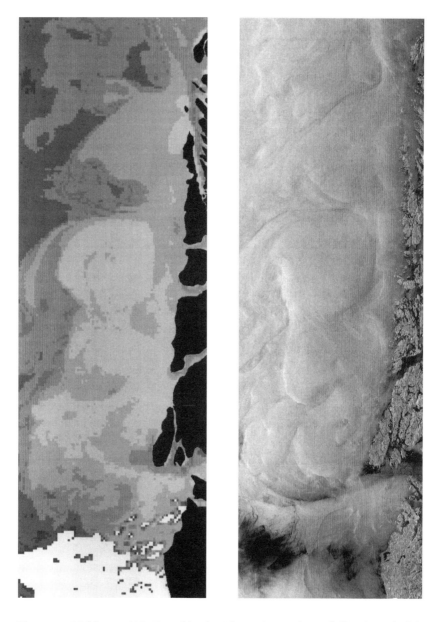

Figure 3.3 Multisensor SAR-IR combination. The two images, i.e., IR (left) and SAR (right) are obtained 6 h apart on October 3, 1992. In the IR image the sea surface temperatures decrease from 14°C (light gray) to 12°C (dark gray), black is land masking, and white represents clouds. The imaged region is about 300 km long and 100 km wide.

Multisensor IR-SAR Combination

An excellent documentation that frontal features associated with mesoscale circulation patterns is expressed in SAR images is shown in Figure 3.3. This image was obtained off the west coast of Norway. The NOAA AVHRR image (left) and the ERS-1 SAR image (right) were obtained 6 h apart on October 3, 1992 under northerly winds of about 5 m/sec and air temperature from 12 to 14°C. There is a remarkable relationship between the sea surface temperature field and the sea surface roughness field. The IR image shows the mesoscale meandering pattern of the Norwegian Coastal Current off the west coast of Norway, with typical scales of about 50 km. Smaller mesoscale scale features are also observed. The temperature decrease from 14°C (white) in the coastal water to 12°C (gray) in the Atlantic water offshore. The maximum temperature front is about 1.5°C across a distance of 2 km.

The ERS-1 SAR image expresses features with configuration and orientation in good agreement with those seen in the IR image, both at the 50-km scale and at smaller scales of about 10 km. The image of three scenes (i.e., 300 km long) gives a much better appreciation of the oceanographic features along the coast than is typically available in one scene. The SAR frontal boundaries are recognized with both dark and bright radar cross section of varying cross frontal width, in particular, the bright fronts.

As in the NORCSEX '91 case, the temperature fronts are weak along the current boundaries, i.e., the current changes are maintained through the salinity gradient. Hence, we interpret the SAR image expressions to be a manifestation of wave-current interaction along the current (thermal) fronts. It is evident that this SAR image interpretation is strongly supported by the near coincident sea surface temperature field mapped in the cloud-free satellite IR image. The combined analyses, therefore, support the multisensor approach to improved mesoscale ocean monitoring, and will in turn provide better initialization of thermodynamic models.

3.5 Summary

Preliminary results from the ERS-1 SAR image validation studies and review of previous results from Seasat and SIR-A/B SAR studies reveal that under moderate winds between about 3 and 10 m/sec the SAR provides information on the mesoscale circulation pattern manifested by the expressions of current boundaries including the presence of meanders and eddies. The corresponding imaging mechanisms seem to alternate between:

1. Damping of short gravity waves by the presence of surface slicks presumably accumulated in the direction of the current by small-scale turbulent motion
2. Short gravity wave-current interaction along the shear-convergence dominated current boundary
3. Changes in wind stress induced by strong gradients in the sea surface temperature
4. Long wave-current refraction

5. Presence of sea ice mirroring the upper ocean circulation pattern

From the number of SAR images examined from the coast of Norway, the presence of natural film is frequently seen expressing spiraling eddies during the summer months, while meandering fronts throughout the year are mostly expressed by short wave-current interaction. Eddies expressed by natural film can lead to strong damping in radar cross section of up to 6–10 dB, while short wave-current interaction can lead to departure in the short wave energy from equilibrium and subsequent change of about +/–3 dB in radar cross section. Ocean circulation features expressed through rapid shifts in wind stress due to ocean thermal fronts are rare in the seas surrounding Norway. On the other hand, this imaging mechanism is possibly more active in waters with strong sea surface temperature gradients such as in the Gulf Stream.

The imaging mechanisms and their associated radar expressions (i.e., narrow, bright, and dark curved structures) are sensitive to both the near surface wind speed, wind direction relative to the frontal orientation, atmospheric boundary layer stratification, and radar wavelength. Damping by natural films appears possible up to about 7 m/sec. Furthermore, above 10–12 m/sec the spatial distribution of the short gravity waves is predominantly reflecting the wind stress field, the wind wave, and eventual swell field. Hence, for increasing winds, eddies expressed by presence of natural film will vanish before eddies expressed by short wave-current interaction. In contrast, no features may be expressed at winds below about 3 m/sec, since resonant Bragg waves of about 0.07–0.08-m wavelength are not developed.

An operational marine monitoring and forecasting system requires improved knowledge of (1) the local and regional surface and subsurface current conditions, atmospheric forcing, and fresh water supply; (2) the chain of events from meteorological via oceanographic to the chemical-biological conditions that leads to the bloom of a toxic algae; (3) the drift pattern and dispersion of any oil spill or toxic algae; and (4) the relationship between the physical environment and the biological systems as well as the driving mechanisms for the ecological variability. Regular access to multisensor, high-resolution satellite data, integrated with *in situ* measurements and numerical models, is an important and necessary element in such a system including data assimilation.[45] With the launch of SeaWIFS, ERS-2, and Radarsat in 1995, one can in combination with AVHRR satellite data provide significantly improved temporal and spatial coverage. In particular, the wide swath (500-km) coverage obtained with Radarsat provides improved capabilities to regularly monitor temporal and spatial variabilities on 3–10 d and 10–100 km. In the meantime we expect to develop better SAR image classification algorithms and gain improved quantitative understanding of SAR imaging mechanisms through an integrated use of SAR simulation models and an increasing number of dedicated ERS-1 SAR image validation studies. Satellite SAR, independent of clouds, fog, and light conditions, may therefore play a key role in the success of such an integrated system.

Acknowledgments

Core support for this study was provided by the Norwegian Space Center (NSC) under the contract Norwegian ERS-1 Application Project, Geophysical Institute, University of Bergen; Statoil; European Space Agency (ESA) under a SAR Feasibility Study contract; the Oceanographer of the Navy, U.S., and NATO Collaborative Research Grants Program-CRG 910551.

References

1. Wright, J.W., Detection of ocean waves by microwave radar: the modulation of short gravity-capillary waves, *Boundary Layer Meteorol.*, 13, 87, 1978.
2. Beal, R.C., DeLeonibus, P.S., and Katz, I., Eds., *Spaceborne Synthetic Aperture Radar for Oceanography*, John Hopkins University Press, Baltimore, MD, 1981, 215 pp.
3. Vesecky, J.F. and Stewart, R.H., The observations of ocean surface phenomena using imagery from the Seasat synthetic aperture radar: an Assessment, *J. Geophys. Res.*, 87, 3397, 1982.
4. Kirwan, A.D., Ahrens, T.J., and Born, G.H., Eds., Seasat Special Issue II, *J. Geophys. Res.*, 88, 1531, 1983.
5. Hasselmann, K., Raney, R.K., Plant, W., Alpers, W., Shuchman, R.A., Lyzenga, D., Rufenach, C.L., and Tucker, M.F., Theory of synthetic aperture radar ocean imaging: a MARSEN view, *J. Geophys. Res.*, 90, 4659, 1985.
6. Alpers, W.R., Theory of radar imaging of internal waves, *Nature (London)*, 314, 245, 1985.
7. Hughes, B.A. and Gasparovic, R.F., Introduction, JOWIP and SARSEX Special Issue, *J. Geophys. Res.*, 93, 1988.
8. Fu, L.L. and Stewart, R.H., Some examples of detection of oceanic mesoscale eddies by the Seasat synthetic aperture radar, *J. Geophys. Res.*, 88, 1844, 1983.
9. Fu, L.L. and Holt, B., Seasat Views Oceans and Ice with Synthetic Aperture Radar, Jet Propulsion Laboratory Publication, Jet Propulsion Laboratory, Pasadena, CA, 1982, 200 pp.
10. Johannessen, J.A., Shuchman, R.A., Johannessen, O.M., Davidson, K.L., and Lyzenga, D.R., Synthetic aperture radar imaging of upper ocean circulation features and wind fronts, *J. Geophys. Res.*, 96, 10411, 1991.
11. Ford, J.P., Cimino, J.B., and Elachi, C., Space Shuttle Columbia Views the World with Imaging Radar: The SIR-A Experiment, California Institute of Technology, *JPL 82-95*, 1983, 179 pp.
12. Ford, J.P., Cimino, J.B., Holt, B., and Ruzek, M., Shuttle Imaging Radar Views the Earth from Challenger: The SIR-B Experiment, California Institute of Technology, *JPL 86-10*, 1986, 135 pp.
13. Beal, R.C., Monaldo, F.M., Tilley, D.G., Irvine, D.E., Walsh, E.J., Jackson, F.C., Hancock, D.W., III, Hines, D.E., Swift, R.N., Gonzales, F.I., Lyzenga, D.R., and Zambresky, L.F., A comparison of SIR-B directional ocean wave spectra with aircraft scanning radar spectra, *Science*, 232, 1531, 1986.
14. *Proc. 1st ERS-1 Symp., Space at the Service of Our Environment*, ESA Publication Division, Noordwijk, Vol. 1 and 2, ESA SP-359, 1993.
15. *Proc. 2nd ERS-1 Symp., Space at the Service of Our Environment*, ESA Publication Division, Noordwijk, ESA SP-361, 1994.
16. Attema, E., The ERS-1 geophysical validation programme for wind and wave data products, *Proc. ERS-1 Geophys. Validation*, Wooding, M., Eds., European Space Agency, ESA WPP-36, 1992, 1.

17. Johannessen, J.A., Shuchman, R.A., Davidson, K., Frette, Ø., Digranes, G., and Johannessen, O.M., Coastal ocean studies with ERS-1 SAR during NORCSEX'91, *in Proc. 1st ERS-1 Symp., Space at the Service of Our Environment,* ESA Publication Division, Noordwijk, ESA SP-359, 1993, 113.

18. Tilley, D.G. and Beal, R.C., ERS-1 and Almaz ocean wave imaging over the Gulf Stream and Grand Banks, *in Proc. 1st ERS-1 Symp., Space at the Service of Our Environment,* ESA Publication Division, Noordwijk, ESA SP-359, Vol. 2, 1993, 729.

19. Nilsson, C.S., Tildesley, P., and Petersen, J., Mapping of the East Australian current with the ERS-1 SAR and shipborne studies, *In Proc. 1st ERS-1 Symp., Space at the Service of Our Environment* ESA Publication Division, Noordwijk, ESA SP-359, Vol. 1, 1993, 95.

20. Liu, A.K. and Peng, Ch.Y., Waves and mesoscale features in the marginal ice zone, *in Proc. 1st ERS-1 Symp., Space at the Service of Our Environment,* ESA Publication Division, Noordwijk, ESA SP-359, Vol. 1, 1993, 343.

21. Johannessen, O.M., Sandven, S., Campbell, W.J. and Shuchman, R.A., Ice studies in the Barents Sea by ERS-1 SAR during SIZEX'92, *In Proc. 1st ERS-1 Symp., Space at the Service of Our Environment* ESA publication division, Noordwijk, ESA SP-359, Vol.1, 1993, 277.

22. Wahl, T., Eldhuset, K., and Skøelv, Å., Ship traffic monitoring using ERS-1 SAR, *In Proc. 1st ERS-1 Symp., Space at the Service of Our Environment* ESA Publication Division, Noordwijk, ESA SP-359, Vol. 2, 1993, 823.

23. Johannessen, J.A., Røed, L.P. and Wahl, T, Eddies detected in ERS-1 SAR images and simulated in reduced gravity model, *Int. J. Remote Sensing,* 14, 2203, 1993.

24. Vesecky, J.F., Kasischke, E.S., and Shuchman, R.A., Marine Microlayer Effects on the Observations of Internal Waves by SAR, *Environmental Research Institute of Michigan research report to DARPA,* 1988, 65.

25. Caponi, E.A., Crawford, D.R., Yuen, H.C., and Saffman, P.G., Modulation of radar backscatter from the ocean by a variable surface current, *J. Geophys. Res.,* 93, 12249, 1988.

26. Plant, W.J., A relationship between wind stress and wave slope, *J. Geophys. Res.,* 87, 1961, 1982.

27. Johannessen, J.A., Johannessen, O.M., Svendsen, E., Shuchman, R.A., Manley, T., Campbell, W.J., Josberger, E.G., Sandven, S., Gascards, J.C., Olaussen, T.I., Davidson, K.L., and Van Leer, J., Mesoscale eddies in the Fram Strait marginal ice zone during the 1983 and 1984 Marginal Ice Zone Experiment, *J. Geophys. Res.,* 92, 6754, 1987.

28. Wakatsuchi, M. and Ohshima, K.I., Observations of ice-ocean Eddy street off the Hokkaido Coast in the Sea of Okhotsk through radar images, *J. Phys. Oceanogr.,* 20, 585, 1990.

29. Donelan, M.A. and Pierson, W.J., Radar scattering and equilibrium ranges in wind-generated waves with application to scatterometry, *J. Geophys. Res.,* 92, 4971, 1987.

30. Scott, J.C., Surface films in oceanography, *ONRL Workshop Report,* C-11,-86, 1986, 19.

31. Font, J., Martinez, A., Gorriz, E.G., et al, Comparison of ERS-1 SAR images of the Western Mediterranean to *in situ* oceanographic data: PRIM-1 Cruise, *in Proc. 1st ERS-1 Symp., Space at the Service of Our Environment,* ESA Publication Division, Noordwijk, ESA SP-359, Vol. 2, 1993, 883.

32. Scoon, A. and Robinson, I.S., SAR imaging of dynamical features in the English Channel, *in Proc. 1st ERS-1 Symp., Space at the Service of Our Environment,* ESA Publication Division, Noordwijk, ESA SP-359, Vol. 1, 1993, 119.

33. Askari, F., Geernaert, G.L., Keller, W.C., and Raman, S., Radar imaging of thermal fronts, *Int. J. Remote Sensing,* 14, 275, 1993.

34. Lich, D.E., Mattie, M.G., and Mancini, L.J., Tracking of Warm Water Ring, in *Spaceborne Synthetic Aperture Radar for Oceanography,* Beal, R.C., DeLeonibus, P.S., and Katz, I., Eds., John Hopkins University Press, Baltimore, MD, 1981, 215 pp.

35. Cheney, R.E., A search for cold water rings, in *Spaceborne Synthetic Aperture Radar for Oceanography*, Beal, R.C., DeLeonibus, P.S., and Katz, I., Eds., John Hopkins University Press, Baltimore, MD, 1981, 215 pp.

36. Haugan, P.M., Evensen, G., Johannessen, J.A., Johannessen, O.M., and Pettersson, L.H., Modeled and observed mesoscale circulation during NORCSEX'88, *J. Geophys. Res.*, 96, 1991.

37. Hovland, H.A., Johannessen, J.A., and Digranes, G., Norwegian Surface Slick Report, *Nansen Environmental and Remote Sensing Center Special Rep.*, Bergen, Norway, November 1993.

38. Gower, J.F.R., Wind and surface features in SAR images: the Canadian Program, *in Proc. 1st ERS-1 Symp., Space at the Service of Our Environment*, ESA Publication Division, Noordwijk, ESA SP-359, Vol. 1, 1993, 101.

39. Barnett, T.P., Kelly, F., and Holt, B., Estimation of the two-dimensional ocean current shear field with a synthetic aperture radar, *J. Geophys. Res.*, 94, 16087, 1989.

40. Liu, A.K., Jackson, F.C., and Walsh, E.J., A case study of wave-current interaction near an oceanic front, *J. Geophys. Res.*, 94, 16189, 1989.

41. Lyzenga D.R. and Bennett, J.R., Full-spectrum modeling of synthetic aperture radar internal wave signatures, *J. Geophys. Res.*, 93, 1988.

42. Lyzenga, D.R., Interaction of short surface and electromagnetic waves with ocean fronts, *J. Geophys. Res.*, 96, 10765, 1991.

43. Holliday, D., St-Cyr, G., and Woods, N.E., A radar imaging model for small to moderate incidence angles, *Int. J. Remote Sensing*, 7, 1809, 1986.

44. Thompson, D.R., Gotwols, B.L., and Sterner, R.E., A comparison of measured surface wave spectral modulations with predictions from a wave-current interaction model, *J. Geophys. Res.*, 93, 12339, 1988.

45. Johannessen, J.A., Røed, L.P., Johannessen, O.M., Evensen, G., Hackett, B., Pettersson, L.H., Haugan, P.M., Sandven, S., and Shuchman, R.A., Monitoring and modeling of the marine coastal environment, *Photogramm. Eng. Remote Sensing*, 59, 351, 1993.

4

Coastal Tides and Shelf Circulation by Altimeter

Guoqi Han

4.1 Introduction

Advances in understanding the shallow-water tide and shelf circulation have been made in the last century through observational, theoretical, and numerical studies. However, even after more than a century of efforts, oceanic tides over shallow waters are not well known yet, e.g., those on the offshore part of the northwest European continental shelf are not always known with confidence to the accuracy of 10% in amplitude and 10° in phase.[1] More accurate models of leading oceanic tidal constituents are in turn required to extract low-frequency (subinertial) shelf circulation. It has been recognized in the last decade that such models rely not only on the knowledge of tidal astronomical potential and governing hydrodynamic equations, but also on the direct observations of tidal fields, as in the case of Schwiderski's[2] semiempirical model. To estimate shelf circulation, local measurements and observations have to be made in most instances,[3] which are extremely expensive and therefore considerably limited. Satellite altimetry produces unique global measurements of instantaneous sea surface heights over a period of several years, with regular temporal and spatial distribution, and thus has the great potential to help scientists understand the shallow-water tide and shelf circulation.

In principle, any accurate altimetric measurements of the height of a satellite above the sea surface, when coupled with a precise determination of a satellite orbit and the geoid, would allow mapping of the sea surface topography, and provide displacement of the sea surface from the geoid due to surface circulation including the time-averaged mean. Even if the geoid is not determined precisely, repeated observations can provide temporal variability of the sea surface height. With precision of a few centimeters, altimetric measurements will improve tidal charts when coupled with tide-gauge data and/or hydrodynamic models. The measurements permit estimations of a variable geostrophic surface current and changes in its mass transport when combined with conventional data of internal density structure.[4]

Altimetric data have been widely used to detect mesoscale variability[5-12] and oceanic tides[13-20] in the deep ocean. However, application to the continental shelf and slope is more challenging. Three major problems prevent the application. First, the dynamic processes on the continental shelves are strongly influenced by the coastal boundary and shallow (~100 m) bottom topography and further complicated by buoyancy flux due to river runoff and solar heating, as described by Csanady,[21] LeBlond,[22] and Mork.[23] Hence, dominant dynamic variabilities are at periods of several days to weeks and at cross-shelf scales of tens of kilometers.[24] High frequencies and small horizontal scales of variability are simply problematic for the altimeter sampling. Second, an altimeter does not work well near the coast due to tracking problems when crossing the land-sea boundary.[25] Finally, the various environmental corrections such as the wet tropospheric effect are more uncertain in coastal areas than in the deep ocean. Therefore, it tends to be much more difficult to interpret shallow-water than deep-ocean dynamic processes from altimetric measurements.

So far, there have been only a few applications of altimeter height data to the tides over the continental shelves,[1,26,27] and few attempts on the low-frequency variations over the shelf and the shelf break except for Han et al.[28] These studies demonstrated particular difficulties in applying altimeter height data to shallow-water tide and circulation, but presented techniques to overcome the difficulties. Thus, they suggested potential of altimeter height data in application to shallow-water processes. In this chapter, two applications in shallow waters are discussed: (1) tidal variability in Section 4.2 and (2) low-frequency shelf circulation variability in Section 1.3. The various environmental corrections and orbital error adjustment are not discussed, because they have been fully explained by Cheney et al.[25] and in Appendix D.

4.2 Tidal Variability

A number of methods for extracting tidal information from altimetric height data have been developed in the last decade. They can be classified into: (1) conventional methods, in which the tidal analysis is independently performed on the altimetric height data at each spatial location; and (2) inverse methods, in which spatial correlation is included in the tidal solution as defined by the *a priori* spatial covariances in the tidal signal.

The conventional methods are purely temporal analyses, essentially analogous to the analysis of common tide-gauge data. The methods are suitable for either the

Table 4.1 Aliased Periods of Ocean Tides Due to Repeat Sampling at the Geosat 17-d, ERS-1 35-d, and TOPEX/Poseidon 10-d Missions

	Constituent	Geosat	ERS-1	TOPEX/Poseidon
Aliased period (d)	M_2	317	95	62
	N_2	52	97	50
	S_2	169	∞	59
	K_2	88	182	87
	K_1	175	365	173
	O_1	113	75	46
	Q_1	74	133	69
	P_1	4,470	366	89

shallow waters or the deep oceans. Over the continental shelf, where the horizontal scales of the tides are small, these methods are not constrained by large satellite ground-track intervals, which are consequences of short (a few days) sampling periods. Harmonic analysis and response analysis fall into the conventional method category.

Harmonic analysis is subject to the usual limitations of data duration and resolvability of leading tidal constituents. Since altimeter sampling periods are usually much longer than tidal periods, tidal frequencies are aliased into lower frequencies by altimeter sub-Nyquist sampling. Given a satellite sampling period T, the aliased periods T' for tidal constituents are calculated by:[29]

$$T' = \left[f - T^{-1} \quad \text{integer} \left(fT + 0.5 \right) \right]^{-1} \tag{1}$$

where f is the original tidal frequency, and the integer function returns the largest integer less than or equal to its argument. Table 4.1 lists the aliased periods of the eight leading tidal constituents by Geosat, TOPEX/Poseidon, and ERS-1. As an example, look at constituent M_2 by Geosat sampling. First, 317 d or roughly 19 repeat cycles are required for a complete cycle of aliased M_2 at a particular location. With this duration all the leading semidiurnal constituents are resolvable from each other. Second, it seems hard to separate a certain semidiurnal constituent S_2 from K_1, a diurnal one. Fortunately, there is additional phase information that enables clear separation of these two constituents at any location if data from both ascending and descending passes are available.[30] Third, as the aliased period of M_2 is close to a year, any purely temporal analysis of M_2 will be contaminated by the annual cycle. Likewise both K_1 and S_2 will be contaminated by the semiannual cycle.

Minster et al.[27] analyzed the Geosat altimeter data to identify variations of the sea level off the mouth of the Amazon River. The harmonic analysis was used to map the M_2, N_2, and O_1 cotidal charts. The strong variations of the sea level of up to 6 m were observed, and were found to be dominated by tidal signals and strongly correlated with bathymetry. Overall, the altimetric cotidal charts for M_2 locally agreed well with results of a high-resolution finite element tidal model of the North Atlantic Ocean developed for the TOPEX/Poseidon mission.

As discussed above, it is difficult to resolve all tidal constituents using harmonic analysis because of the limited duration of altimeter data. This difficulty can be overcome by response analysis, in which once the response weights of each tidal species (e.g., semidiurnal or diurnal tides) have been determined, a solution for any tidal constituents in this species could be obtained, even though some constituents are not resolved with the satellite sampling period.[16] Although the temporal resolution is less important for an orthotide response analysis, accuracy would be improved if two leading constituents in a specific tidal species are synodically separated within the observation duration.[30]

Using response analysis along with the first 11 months of the Geosat altimetry records, Cartwright and Ray[26] mapped the tides of the area near the Patagonian Shelf. Then, they computed the tidal power fluxes into the shelf areas from the tidal maps through Laplace's tidal equations. The realistic and plausible results were obtained for the flow of tidal energy from the ocean to the shelf in that region.

Woodworth and Thomas developed a modified form of response analysis.[1] They analyzed Geosat altimeter data over the period from November 1986 to August 1988 and derived the parameters of the semidiurnal ocean tides over the northwest European continental shelf. In order to reflect the consequence of shallow waters and nonlinear processes, response analysis was performed by means of a linear regression of the data against the equilibrium tide (with a zero lag and a 2-d lag), producing estimated admittance curves with four fitted parameters. Harmonic parameters for M_2 (Figure 4.1) and S_2 were found to be compatible with those of previous empirical tidal maps and were estimated to be accurate to approximately 6 cm. This accuracy was comparable to, or better than, that provided by numerical models of the area. They concluded that considering the coastal waters overall, the leading tidal constituents can be derived from Geosat and subsequent altimetry to the accuracy that is not previously possible from conventional tide-gauge measurements or from numerical modeling.

Inverse methods are joint analyses in time and space. A set of *a priori* covariances, which are explicitly estimated from all the available information concerning the tides and the various errors embedded in the altimetric measurements, quantifies the spatial and spectral properties of the tides and the errors. The methods simultaneously extract several tidal constituents by an inversion of the data and the corresponding errors associated with these constituents by computing the *a posterior* covariances.[14,31] The methods have been successfully applied to recover the tides in deep oceans[13,14,20] and in Asian semienclosed seas.[32] Nevertheless, in cases when an altimeter has a few day sampling periods, the spatial correlation scales of tides in shallow waters are smaller than the satellite ground-track intervals; and therefore the methods may not be applied in shallow waters.

Recently, Mazzega and Berge[32] have mapped the M_2, S_2, N_2, K_2, K_1, O_1, P_1, and Q_1 cotidal charts in the Sea of Okhotsk, the Sea of Japan, the East China Sea along with the Bo Hai and Yellow Seas, and the Indonesian Seas by inverting combined sets of tide-gauge data and a reduced set of TOPEX/Poseidon altimeter data.[32] The inverted results were compared with the tide-gauge harmonic constants that were not used in the inversion. It was concluded that the results from the altimeter data had accuracy comparable to Schwiderski's[2] global tidal model, and Cartwright and Ray's[16] results extracted from Geosat altimetry.

4.3 Low-Frequency Variability

In addition to recovery of tidal signals from altimetry, another important application of altimetry in physical oceanography is to detect the low-frequency (subinertial) sea surface height variability such as annual cycles. Here the tides are required to be removed from the raw altimetric height data.

In deep oceans, the tidal effects are usually removed by the oceanic tides computed from global tidal models; for example, for Geosat data, Schwiderski's model is used to estimate amplitudes and phases for four semidiurnal, four diurnal, and three long period constituents on a $1° \times 1°$ grid with an accuracy of 10–20 cm.[28] The tidal errors have the horizontal scales on the order of 1000 km, and hence can be further reduced by removing the linear or quadratic trend in the orbit error correction. In addition, the low-frequency mesoscale variability, which is of interest, is energetic in deep oceans and has the horizontal scales on the order of 100 km. Therefore, tidal corrections are not crucial for interpretation of the nontidal oceanic mesoscale phenomena in deep oceans.

In shallow waters, however, tides are often stronger than low-frequency signals and have horizontal scales of ~100 km due to the local coastline and bottom topography. Furthermore, global tidal models may not be adequate for shallow seas; for example, Schwiderski's model is based on a scheme of hydrodynamic interpolation between sites at which tidal measurements have been made. The M_2 amplitudes and phases from this model are significantly different from both tide-gauge and altimeter data in the Amazon mouth,[27] and over the Scotian Shelf and the Grand Banks.[28] If Schwiderski's model were used for the oceanic tide correction to altimeter data on the shelves, the corrected sea surface variability would be severely contaminated by model errors.

As pointed out in Section 4.2, semidiurnal and diurnal constituents are aliased into fluctuations with longer periods by altimeter sub-Nyquist sampling. If some tidal constituents happen to be aliased into variations of a period of half a year or a year, their errors may contaminate the semiannual or annual cycle of the sea surface changes. Hence tidal correction errors can be a crucial factor to mislead an interpretation of low-frequency dynamic features.

As a new application of Geosat altimeter data to the shallow seas, Han et al.[28] recently studied the annual variations of the sea surface slopes and the tidal slopes along the ascending ground tracks over the Scotian Shelf and the Grand Banks. An effective way of removing the tidal effects is to use a more accurate local tidal model, instead of Schwiderski's model. Han et al.[28] applied a local tidal model called the BIO model to make ocean tide correction. The BIO model, which estimated the tidal amplitudes and phases for three semidiurnal and two diurnal constituents (M_2, N_2, S_2, K_1, and O_1) on a $4' \times 4'$ grid, solved the fully nonlinear equations of

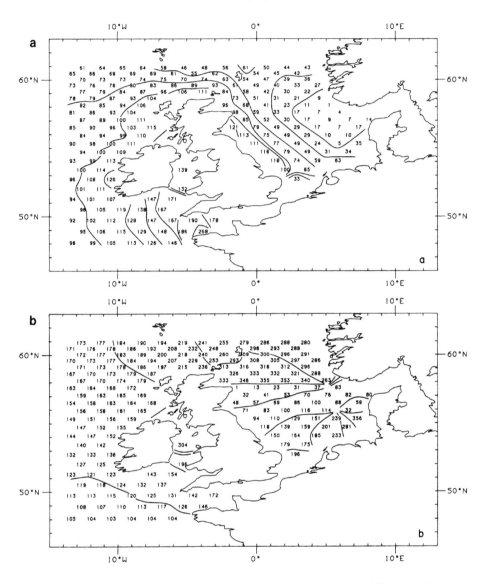

Figure 4.1   Geosat-derived M_2 distribution of (a) amplitude in centimeters and (b) Greenwich phase lag in degrees, and (c) distribution of V in centimeters for the Geosat M_2 tide compared with the improved Flather model over the northwest European continental shelf. (From Woodworth, P.L. and Thomas, J.P., *J. Geophys. Res.*, 95, 3561, 1990. With permission.)

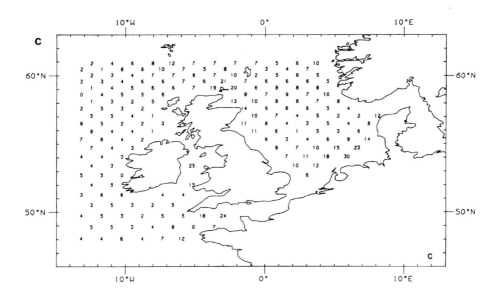

Figure 4.1 (continued)

motion.[33] Elevation conditions for open boundaries were derived primarily from observations. The BIO model was in good agreement with the tide-gauge data. The error of tidal elevation was estimated to be ~3 cm by comparing the model results with observed tidal elevations. The BIO model also agreed well with altimeter data, but significantly differs from Schwiderski's model, as shown in Table 4.2.

Another key component to an effective separation of the low-frequency variability of interest from the aliased tidal constituents is harmonic analysis, because any local tidal model is still subject to certain errors, even though it is better than Schwiderski's model over the continental shelf. With a few years duration of the Geosat altimeter data, some aliased tidal constituents could be hard to separate from the annual or semiannual signal. The solution may be subject to large statistical uncertainties. In order to estimate the uncertainty of the solution accurately, they considered the effects of serially correlated errors. The autocorrelation of errors was estimated using the resulting residuals after running ordinary least squares, and tested with the Durbin-Watson statistic.[34]

Since the Scotian Shelf and the Grand Banks lie near the Gulf Stream characterized by high sea surface height variability, the height data are sensitive to the bias-tilt correction for orbit errors along ground tracks with an arc length of about 2500 km. Since the dynamic variability of interest had the horizontal dimensions on the order of 100 km, much smaller than the arc length over which the orbit errors are corrected, the sea surface slopes over the shelf and the shelf break were more robust

Table 4.2   Estimated Amplitudes and Phases of the Aliased M_2 Component (317-d period) and Their 90% Confidence Intervals for Slopes over the 2° Latitudinal Range

	Track No.	A	B	C
Amplitude	16	34.2 ± 3.59	19. ± 11.2	24. ± 4.18
(10^{-7})	19	12.0 ± 2.70	4.1 ± 1.06	11. ± 3.94
	22	10.2 ± 2.21	4.4 ± 2.15	8.2 ± 3.19
	25	9.3 ± 3.66	3.9 ± 2.06	10. ± 3.31
	28	7.1 ± 4.42	2.2 ± 1.58	5.8 ± 4.52
	31	4.6 ± 4.42	1.9 ± 1.24	5.8 ± 4.58
	17	4.9 ± 4.52	3.1 ± 1.67	5.1 ± 3.84
	20	4.6 ± 2.55	1.6 ± 1.51	7.7 ± 6.23
	23	6.5 ± 1.82	1.8 ± 0.85	8.0 ± 3.83
	26	7.0 ± 1.38	2.2 ± 0.57	7.4 ± 3.10
	29	5.6 ± 1.38	2.6 ± 0.72	3.6 ± 5.54
	15	5.3 ± 1.91	2.9 ± 2.08	6.2 ± 2.39
Phase	16	0.9 ± 0.11	2.2 ± 0.65	0.9 ± 0.17
(rad)	19	0.4 ± 0.23	0.7 ± 0.26	0.4 ± 0.36
	22	5.7 ± 0.22	1.3 ± 0.51	5.6 ± 0.40
	25	4.7 ± 0.41	4.5 ± 0.56	4.8 ± 0.32
	28	3.3 ± 0.72	2.9 ± 0.82	3.5 ± 0.89
	31	1.7 ± 1.27	0.6 ± 0.73	1.5 ± 0.91
	17	6.1 ± 1.18	6.1 ± 0.56	0.4 ± 0.86
	20	5.3 ± 0.58	5.0 ± 1.26	5.6 ± 0.94
	23	4.7 ± 0.28	5.6 ± 0.50	4.7 ± 0.50
	26	4.0 ± 0.20	4.5 ± 0.27	4.4 ± 0.42
	29	2.9 ± 0.25	2.9 ± 0.28	2.6 ± ∞
	15	2.0 ± 0.37	2.0 ± 0.80	1.1 ± 0.40

Note: The tracks are shown in Figure 1. A: BIO model; B: Schwiderski's model; C: Raw altimeter data. Reference for phase is Geosat repeat cycle 32. From Han, G.Q. et al., *Atmos.-Ocean,* 31, 591, 1993. With permission.

than the height itself. Therefore, the slopes were estimated by linear regression, and then the time series of the slopes were systematically analyzed.

The annual amplitudes and phases, as well as their 90% confidence intervals of the sea surface slopes over the shelf and the shelf break are shown in Figure 4.2. The annual variation of the sea surface slopes over the western Scotian Shelf was found to be consistent with the geostrophic surface currents calculated from historical hydrography on the inner Halifax line.

4.4 Summary

The ability of satellite altimetry to provide tidal and low-frequency variabilities in shallow waters has been demonstrated from Geosat height data, and data from the ongoing TOPEX/Poseidon mission. Another mission, ERS-1, is also flying. Besides, several more missions are scheduled in the next decade. The precision of observations reaches 3 ~ 5 cm rms for TOPEX/Poseidon,[35] much better than 5 ~ 11cm rms for Geosat,[36] and may be improved in future missions.

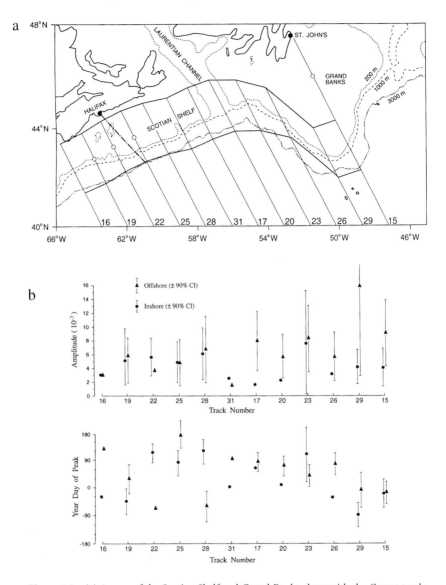

Figure 4.2   (a) A map of the Scotian Shelf and Grand Banks along with the Geosat tracks (thin solid lines with two-digit labels for the numbers of the Geosat ascending ground tracks) and the ranges for slope calculations limited by the two thick solid lines. A chained straight line off Halifax indicates the position of the Halifax Hydrographic Section. (b) The amplitude and the year day of maximum annual slope estimated from the time series of the BIO-corrected sea surface slopes over the shelf (inshore) and the shelf break (offshore) of the Scotian Shelf and Grand Banks, along with their 90% confidence intervals (CI) if the annual amplitudes are significantly different from zero. (From Han, G.Q. et al., *Atmos.-Ocean*, 31, 591, 1993. With permission.)

The temporal and spatial resolutions of sampling are important,[37] particularly in shallow waters. A satellite orbit should repeat often so that aliasing of high-frequency variability can be avoided. Meanwhile, a high cross-track resolution is desirable. The two requirements are incompatible, and a reasonable compromise should be reached. The cross-track resolutions of Seasat and ERS-1 with 3-d repeat cycles are adequate for a study of deep ocean tides, but they are not suitable for a study of tides in shallow waters. The Geosat 17-d exact repeat mission, the TOPEX/Poseidon 10-d repeat sampling, and the ERS-1 35-d mission provide a reasonable spatial resolution for shallow-water tides.

Another constraint to be considered is to avoid tidal aliases that contaminate important low-frequency signals; in particular, the predominant M_2 tidal alias should be shorter than half a year. Geosat aliased the M_2 tidal component into 317 d per cycle, close to the annual period. TOPEX/Poseidon aliases the M_2 tide at a period of 2 months. ERS-1 35-d repeat mission aliases the M_2 tidal component at a period of 3 months. Therefore, both the M_2 tide and the annual cycle may be better extracted from TOPEX/Poseidon and ERS-1 data than Geosat data.

It can be concluded, therefore, that satellite altimetry promises to become a powerful tool for better understanding ocean tides in shallow waters; this also will help the study of low-frequency circulation and other dynamic processes in these regions, with important application to many branchs of geophysics and with practical benefits to marine engineering by the end of this century.

References

1. Woodworth, P.L. and Thomas, J.P., Determination of the major semidiurnal tides of the northwest European continental shelf from Geosat altimetry, *J. Geophys. Res.*, 95, 3061, 1990.
2. Schwiderski, E.W., On charting global ocean tides, *Rev. Geophys. Space Phys.*, 18, 243, 1980.
3. Robinson, I.S., *Satellite Oceanography: An Introduction for Oceanographers and Remote-Sensing Scientists*, Ellis Horwood, West Sussex, England, 1985, chap. 1.
4. Maul, G.A., *Introduction to Satellite Oceanography*, Martinius Nijhoff, Dordrecht, The Netherlands, 1985, chap. 14.
5. Matthews, P.E., Johnson, M.A., and O'Brien, J.J., Observation of mesoscale ocean features in the Northeast Pacific using Geosat radar altimetry data, *J. Geophys. Res.*, 97, 17829, 1992.
6. Kelly, K.A., The meandering Gulf Stream as seen by the Geosat altimeter: surface transport, position, and velocity variance from 73 to 46°W, *J. Geophys. Res.*, 96, 16721, 1991.
7. Kelly, K.A. and Gille, S.T., Gulf Stream surface transport and statistics at 69°W from the Geosat altimeter, *J. Geophys. Res.*, 95, 3149, 1990.
8. Qiu, B., Recirculation and seasonal change of the Kuroshio from altimetry observations, *J. Geophys. Res.*, 97, 17801, 1992.
9. Qiu, B., Kelly, K.A., and Joyce, T.M., Mean flow and variability in the Kuroshio extension from Geosat altimeter data, *J. Geophys. Res.*, 96, 18491, 1991.
10. Le Traon, P.Y. and Rouquet, M.C., Spatial scales of mesoscale variability in the North Atlantic as deduced from Geosat data, *J. Geophys. Res.*, 95, 20267, 1990.
11. Stammer, D. and Boning, C.W., Mesoscale variability in the Atlantic Ocean from Geosat altimetry and WOCE high-resolution numerical modeling, *J. Phys. Oceanogr.*, 22, 732, 1992.
12. White, W.B., Tai, C.K., and Dimento, J., Annual Rossby wave characteristics in the California Current Region from the Geosat Exact Repeat Mission, *J. Phys. Oceanogr.*, 1990.
13. Jourdin, F., Francis, O., Vincent, P., and Mazzega, P., Some results of heterogeneous data inversions for oceanic tides, *J. Geophys. Res.*, 96, 20267, 1991.

14. Mazzega, P. and Jodin, F., Inverting Seasat altimetry for tides in the northeast Atlantic, in *Advances in Tidal Hydrodynamics*, Parker, B., Ed., John Wiley & Sons, New York, 1991, 569.

15. Schwiderski, E.W., High-precision modeling of mean sea level, ocean tides, and dynamic ocean variations with Geosat altimeter signals, in *Advances in Tidal Hydrodynamics*, Parker, B., Ed., John Wiley & Sons, New York, 1991, 593.

16. Cartwright, D.E. and Ray, R.D., Oceanic tides from Geosat altimetry, *J. Geophys. Res.*, 95, 3069, 1990.

17. Sanchez, B.V. and Cartwright, D.E., Tidal estimation in the Pacific with application to Seasat altimetry, *Mar. Geodesy*, 12, 81, 1988.

18. Woodworth, P.L. and Cartwright, D.E., Extraction of the M_2 ocean tide from Seasat altimeter data, *J. Geophys. J.R. Astr. Soc.*, 84, 227, 1986.

19. Mazzega, P., M_2 model of the global ocean tide derived from Seasat altimetry, *Mar. Geodesy*, 9, 335, 1985.

20. Mazzega, P., The M_2 oceanic tide recovered from Seasat altimetry in the Indian Ocean, *Nature (London)*, 302, 514, 1983.

21. Csanady, G.T., *Circulation in the Coastal Ocean*, D. Reidel, Dordrecht, 1982, chap. 1.

22. LeBlond, P.H., Tides and their interaction with other oceanographic phenomena in shallow water (review), in *Advances in Tidal Hydrodynamics*, Parker B., Ed., John Wiley & Sons, New York, 1991, 357.

23. Mork, M., On the dynamics of coastal currents, in *Coastal Oceanography*, Gade, H.G., Edwards, A., and Svendsen, H., Eds., Plenum Press, New York, 1983, 1.

24. Smith, P.C. and Schwing, F.B., Mean circulation and variability on the eastern Canadian continental shelf, *Cont. Shelf Res.*, 11, 977, 1991.

25. Cheney, R.E., Douglas, B.C., Agreen, R.W., Miller, L., Porter, D.L., and Doyle, N.S., Geosat altimeter geophysical data record—user handbook, *NOAA Technical Memorandum NOS-NGS-46*, Rockville, MD.

26. Cartwright, D.E. and Ray, R.D., New estimates of oceanic tidal energy dissipation from satellite altimetry, *Geophys. Res. Lett.*, 16, 73, 1989.

27. Minster, J.F., Genco, M.L., and Brossier, C., Variations of the sea level in the Amazon estuary, *Cont. Shelf Res.*, 15, 1287, 1995.

28. Han, G.Q., Ikeda, M., and Smith, P., Annual variation of sea surface slopes on the Scotian Shelf and Grand Banks from Geosat altimetry, *Atmos.-Ocean*, 31, 591, 1993.

29. Jacobs, G.Q., Born, G.H., Parke, M.E., and Allen, P.C., The global structure of the annual and semiannual sea surface height variability from Geosat altimeter data, *J. Geophys. Res.*, 97, 17813, 1992.

30. Cartwright, D.E., Detection of tides from artificial satellites (review), in *Advances in Tidal Hydrodynamics*, Parker, B., Ed., John Wiley & Sons, New York, 1991, 547.

31. Mazzega, P., The solar tides and the sun-synchronism of satellite altimetry, *Geophys. Res. Lett.*, 16, 507, 1989.

32. Mazzega, P. and Berge, M., Ocean tides in the Asian semi-enclosed seas from TOPEX/Poseidon, *J. Geophys. Res.*, 99, 24867, 1994.

33. de Margerie, S. and Lank, K.D., Tidal circulation of the Scotian Shelf and Grand Bank, Contract report to the Department of Fisheries and Oceans of Canada, Contract No. 08SC. FD901-5-X515, ASA Consulting, Dartmouth, 1986.

34. Sen, A. and Srivastava, M., *Regression Analysis—Theory, Methods, and Applications*, Springer-Verlag, New York, 1990.

35. TOPEX/Poseidon Joint Verification Team, TOPEX/Poseidon joint verification plan, NASA Jet Propulsion Laboratory, California, 1992.

36. Sailor, R.V. and LeSchack, A.R., Preliminary determination of the Geosat radar altimeter noise spectrum, *Johns Hopkins APL Tech. Dig.*, 8, 182, 1987.

37. Parke, M.E., Stewart, R.H., and Farless, D.L., On the choice of orbits for an altimetric satellite to study ocean circulation and tides, *J. Geophys. Res.*, 92, 11,693, 1987.

5

Linear and Nonlinear Internal Waves in Coastal and Marginal Seas

John R. Apel

5.1. Introduction

It has been known for over two decades that certain types of internal waves have surface signatures that are recognizable in images of the sea surface. The sensors that are known to reveal internal waves include cameras; scanners operating in the visible and very near infrared; and more recently, imaging radars, both real and synthetic aperture. Even thermal infrared and microwave radiometers have detected (but not necessarily imaged) internal wave surface signatures.

This chapter discusses the physical and theoretical basis for the imaging. It is necessarily succinct, but the reader may find more extensive developments of the material presented here in the references.

Figures 5.1 and 5.2 are examples of packets of continental shelf internal waves captured by the Multispectral Scanner (MSS) on the U.S. spacecraft Landsat, and the synthetic aperture radar (SAR) on the European Space Agency ERS-1, respectively. Both images are of wave packets in the New York Bight, taken just north of

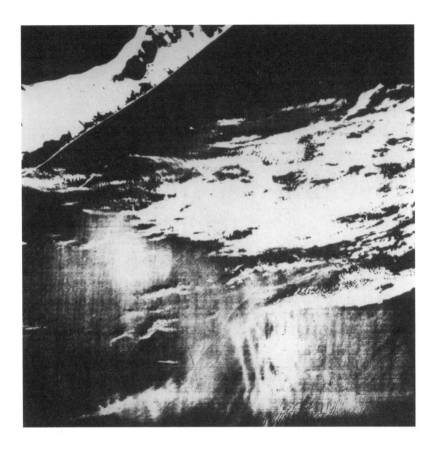

Figure 5.1 Landsat MSS image of eastern Long Island area for July 23, 1973, showing packets of tidally generated internal waves in lower central portion of image. Wavelengths are about 500 m and interpacket separations are about 15 km (From Apel, J. R., Byrne, H. M., Proni, J. R., and Charnell, R. L., *J. Geophys. Res.*, 89, 10,529, 1984. With permission.) Data courtesy of NASA and U.S. Geological Survey.

the Hudson Canyon. The waves have certain ubiquitous characteristics that they display at almost all locations where they have been observed, including (1) their appearance in groups or packets separated by a few tens of kilometers, and usually propagating toward shore; and (2) a strong tendency to vary uniformly in wavelength and crest length, with the longest wavelengths and crest lengths appearing at the front of the packet and the shortest at the rear. Figure 5.3 is a schematic representation of these features, plus some other characteristics that only become apparent with *in situ* measurements (Apel, 1980).

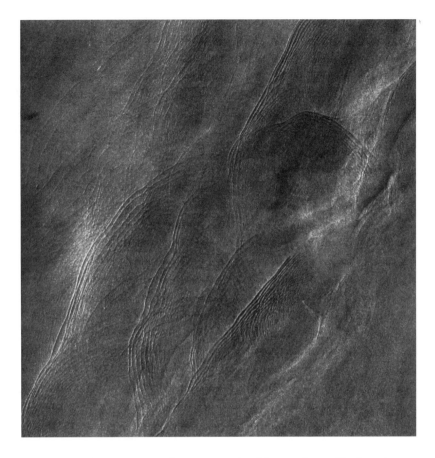

Figure 5.2   Synthetic aperture radar image at C-band from ERS-1 satellite from the same region as Figure 1 (N 39:37:20, W 72:35:40), made on July 18, 1992 at 15:34:35 UT, illustrating quasi-periodic internal wave signatures as viewed in 5.5-cm radar illumination. Image dimensions about 90×90 km^2. Data courtesy of European Space Agency.

Several questions arise as a result of such observations. For example, what mechanisms allow the internal waves, with amplitudes that are very small at the surface even if large at depth, to appear to such a wide range of sensors? Why are they so coherent as to be recognizable over a large span of space and timescales? What relation do they have to deep ocean internal waves, with spectra as described by Garrett and Munk (1975) implying that very little coherence exists, and that energy levels are approximately the same across the globe? Some answers to these questions have been gleaned over two decades of research in many areas of the world, although many other questions, including mechanisms of generation and dissipation, remain to be addressed.

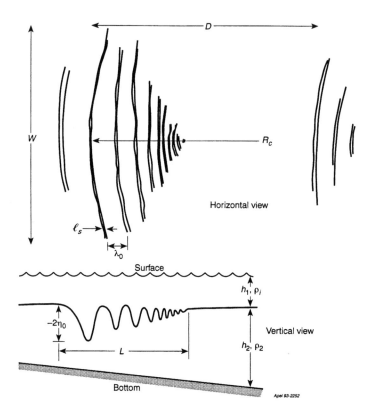

Figure 5.3 Schematic of internal soliton packets on the continental shelf, showing charac-teristic variations in the amplitude, wavelength, and crest length of internal waves, along with definitions of length scales. See Table 5.1.

5.2 Internal Waves Near the Continental Shelf

Observations

Internal waves on the continental shelf have been observed visually for many years prior to their detection by remote sensors. In the early 1970s, their characteristic signatures were identified first in aircraft photography and soon after in Landsat (then ERTS) multispectral scanner imagery (e.g., Figure 5.1). Subsequent studies with Landsat showed that the packets were repeatable to a considerable degree and were most likely tidally generated (Apel et al., 1975a). Later, field programs in the area demonstrated that the waves were highly nonlinear and appeared at approxi-mately 12.5- and 25-h intervals during the summer months, thus establishing their tidal origin (Apel et al., 1975b). However, in the autumn, wind mixing reduced the stratification in shelf waters to the point that internal wave activity could no longer be supported, so that no signatures were seen in the images.

The first published reports of internal waves observed with imaging radar appeared at about the same time. The Jet Propulsion Laboratory (JPL) L-band SAR on the National Aeronautics and Space Administration (NASA) CV-990 aircraft detected features off the Malaspina Glacier in Alaska, where the intrusion of glacial meltwater into the ocean formed a front that developed the coherent oscillations characteristic of solitons (Elachi and Apel, 1976). These features displayed the same kind of light/dark/normal contrast modulations along the propagation direction as did the MSS imagery.

With the flight of Seasat in 1976, hundreds of observations of internal waves were reported, the initial ones coming from the Gulf of California. There, in highly stratified subtropical waters, tidal flows in the straits near the island of Tiburon generated copious quantities of wave packets that were visible with high contrast in the SAR imagery. SAR-equipped spacecraft such as the U.S. Shuttle Imaging Radar A and B (SIR-A and -B), the Soviet Almaz, and the European Space Agency ERS-1 have all yielded excellent images of internal waves at many locations around the world in both deep and shallow water. Indeed, the SAR is the premier sensor for such observations, being capable of all-weather, day/night observations with controlled geometry and calibrated illumination.

Generation

Tidal action appears to be the dominant mechanism for generation of the coherent oceanic internal waves that are observable with remote sensors; riverine or glacial intrusions provide secondary generating mechanisms close to shore. Also, the boundaries of intense current systems such as the Gulf Stream appear to be sources of coherent wave packets that propagate at large angles to the current direction. In waters having shallow pycnoclines, even ships can excite internal wave wakes, which have some of the characteristics of internal shock waves (Apel and Gjessing, 1989). Since both tides and bottom topography are global features, one expects tidally excited internal waves to occur wherever stratified waters exist. The requirement for generation is most likely that the current velocity, u, should exceed the local phase velocity of the internal wave, c_p, that is, the internal Froude number, Fr, should exceed unity, where Fr is, for a two-layer model,

$$\mathrm{Fr} = \frac{u^2/\Delta l_z}{g\Delta\rho/\rho} \tag{1}$$

Here ρ is the density, $\Delta\rho$ its change across the interface, g the acceleration of gravity, and Δl_z the vertical scale of the density gradient, $\Delta\rho/\Delta l_z$. Thus the generation of an internal wave by current flow across a bathymetric feature is a kind of shock process in density-supported oscillations. Since internal wave phase speeds are typically on the order of 0.25–1.0 m/sec, only modest tidal currents are required for their formation. Such flows occur over underwater sills; near continental shelf breaks, islands, straits, and shallow banks; and even on midocean ridges. Indeed, the generation process is likely to be as ubiquitous as the combination of bathymetry, stratification, and current flow. If this is true, as it appears to be, then the entire rim and island population of an ocean basin can be sources of internal waves.

At least two mechanisms have been advanced to explain the generation process at the shelf break. The first, formulated by Rattray (1960), has the barotropic tide scattering into baroclinic modes at the shelf break when the vertical angle of the tidal propagation vector coincides with the angle of the shelf slope. The second is a kind of lee wave mechanism, wherein tidal flow directed offshore results in an oscillation of the pycnocline down-current from the shelf break that remains fixed as long as the tide is steady (Maxworthy, 1979). However, as the tidal ellipse is swept out, the reversal of the current releases the lee wave from its down-current phase-locked position, which then propagates onshore, where it evolves independently of further tidal action except for advective effects. Although there is some experimental support for both of these processes at the shelf edge, at sills or similar geometries the second mechanism is clearly dominant (Halpern, 1971; Apel et al., 1985).

Direct observations of the generation of internal solitary waves at the Nova Scotian shelf break have been made by Sandstrom and Elliott (1984). The wave packets were generated during the phase of the tide when the current was directed offshore. The evolution of the packet occurred with surprising rapidity; fully developed solitary wave characteristics appear within 2–3 h of its birth. This observation is in contradistinction to soliton packets growing from disturbances at a sill in the Sulu Sea in the Philippines, which propagated on the order of 100 km before they had become fully modulated solitons (Apel et al., 1985).

At the edge of the continental shelf, other processes appear to be active in generating internal waves. Upwelling, especially near regions of strong bathymetric relief, appears to generate internal waves. Figure 5.4 shows one instance of such a disturbance, as observed by the SAR on the Soviet spacecraft Almaz-*1* (Etkin and Smirnov, 1991). On the basis of preliminary calculations, the present writer believes that the upwelling serves to excite internal waves in canyons and similar regions of high relief.

Propagation

Remote sensors have provided overviews of coherent internal waves to a degree that has occurred with few other oceanic processes. From the studies to date, it appears that waves generated at the continental shelf break tend to propagate mainly toward shore. They exist in groups or packets, each containing several oscillations that vary in their wavelengths, λ, and crest lengths, W (cf. Figure 5.3). Table 5.1 gives typical scales for the quantities shown on the figure for midlatitude continental shelf waves.

It should be emphasized, however, that coherent internal waves have been observed with widely differing characteristics from the ones listed here. For example, in the island archipelago of the Far East, solitons with peak-to-trough

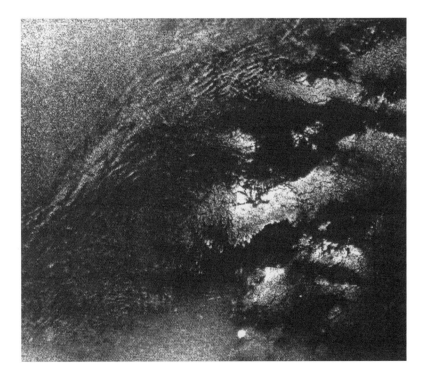

Figure 5.4   Image of upwelling and internal wave generation occurring at the edge of the continental shelf west-southwest of Ireland, taken by the 10-cm SAR on the Russian satellite Almaz-1 in July 1991. Dimensions approximately 40 x 40 km², resolution about 25 m. (Data courtesy of of V.S. Etkin and A.L. Smirnov.)

Table 5.1   Typical Scales for Continental Shelf Internal Waves

L (km)	$2\eta_0$ (m)	h_1 (m)	h_2 (m)	l_s (m)	λ_0 (m)	W (km)	D (km)	R_c (km)	$\Delta\rho/\rho$
1–5	0–15	20–50	100	100	50–500	0–30	15	25–∞	0.001

Note: See Figure 5.3 for definitions of mathematical quantities.

amplitudes of 100 m and phase speeds of 2.5 m/sec have been reported, and a crest length of over 90 km has been observed with the Landsat MSS in the Sulu Sea in the Philippines (Apel et al., 1985).

Typical dynamic quantities for the waves are as follows:

Dynamic Quantities

Brunt-Väisälä frequency, $N/2\pi$	10 cycles/h
Radian frequency, ω	0.001–0.005 rad/sec
Phase speed, c_p	0.20–1.0 m/sec
Group speed, c_g	0.10–0.50 m/sec
Current speed, u	0.10–1.0 m/sec
Packet lifetime, τ_{life}	24–48 h
Interpacket period, τ_{gen}	12.5–25 h

Attenuation

Internal waves on the continental shelf soon encounter shoaling water that affects their propagation speeds and degree of nonlinearity. Near the region where the pycnocline depth is roughly half the total depth, they appear to undergo various transformations. These processes have only been poorly studied, but there are indications that soliton-like internal waves may fragment into smaller, shorter daughter solitons, or may disperse by breaking up or developing various types of vorticity. In this region increased amounts of suspended sediments often exist, as have been observed with acoustic echo-sounders (Apel et al., 1975b; Proni and Apel, 1975). Since the bottom currents associated with the waves are large enough to resuspend sediments (e.g., 25–30 cm/sec), breaking on the sloping bottom may be the process that ultimately destroys the waves. Their signatures have generally disappeared from the imagery taken in this domain. This process apparently injects large amounts of nutrients into the food chain in the region (Sandstrom and Elliott, 1984), and it is likely that other areas also similarly benefit from the effects of internal waves.

In deep water where they are free to propagate, there is direct evidence of the possibility of internal wave lifetimes of several days. In the Sulu Sea, a 60-m high internal soliton was followed by ship for over 2 d as it propagated across 300 km of the sea, during which time it decayed to a 20-m amplitude (Apel et al., 1985). Solitary waves do not disperse as do linear waves, but in the laboratory appear instead to remain as solitons to arbitrarily small amplitudes (Kao, 1991). This would indicate that it could take several days for a freely propagating solitary wave packet to decay into the background internal wave noise. Liu et al. (1985) report an effective horizontal eddy viscosity for Sulu Sea solitons of $A_h \approx 10$ m^2/sec and demonstrate that as they spread out with an increasing radius of curvature, R_c, they decay as $1/r$ (and $1/t$), rather slow rates.

Little is known about the direct processes leading to an eddy viscosity of the size invoked by Liu et al., but local, incipient shear-flow instability could be a cause; other dissipative mechanisms include weak scattering by background internal turbulence, radiation to the overhead surface wave field (the process by which their surface signatures are made visible), and bottom frictional drag. The relative strengths of these mechanisms are not known.

5.3 Internal Wave Observations in Deep Water

Because of the tidal/bathymetric influences on internal wave generation and the apparent need for phase coherence for signature recognition to occur, remote sensing of internal waves has been focused on the continental shelves, island arcs, straits, and similar geographic areas. However, theoretical studies suggest that the shelf break generation process also launches an offshore-traveling disturbance that propagates down to the bottom, reflects/refracts upward, and reappears near the surface several hundred kilometers at sea (New, 1989; Pingree and New, 1989). Now deep-ocean internal waves are characterized by a more or less global energy spectrum, random phases, and root mean square (rms) amplitudes of roughly 5 m (Garrett and Munk, 1975). If the process of propagation offshore is also active wherever the generation of an onshore component occurs, then the observed deep-ocean internal wave field could originate in part from the bathymetric features around its margins. A measurement at a given point in the deep sea would show the sum of many waves arriving from a variety of points and directions, and the resultant signal would have random amplitudes, phases, and propagation directions, as is observed. Such incoherent fields may not be detectable in an image because they presumably have no pattern recognizable to the eye. Nevertheless, coherent internal wave signatures have been observed in Seasat imagery near the Mid-Atlantic Ridge in 1000-m deep water (Barber, 1993). Russian investigators report internal-wavelike surface signatures of an 800-m high seamount in 5500 m of water off Kamchatka, as observed by the SAR on the Soviet spacecraft Kosmos 1870 (Etkin et al., 1991).

It seems likely, therefore, that even in deep water, currents flowing past topographic features that protrude into the permanent thermocline can generate groups of internal waves. The process may be direct, perhaps via some variant of shear-flow instability, or more indirect, such as the mechanisms thought to be active in the seamount signature (Etkin et al., 1991). However, not enough high-resolution imagery has been taken of the open ocean to establish a statistically significant sample of the occurrence of coherent internal waves in the deep ocean.

5.4 Soliton Dynamics

Although not all internal waves are solitons, the ones observable by remote sensors appear to possess solitary-like characteristics much of the time. I will therefore discuss only this class of wave, realizing that the soliton theory reduces to that for linear, small-amplitude waves as the wave amplitude, η, goes to zero.

The simplest equation demonstrating solitary-wave characteristics was derived by Korteweg and deVries (1895) to explain coherent, shallow-water pulses observed in a Scottish canal by Russell (1844). Since then, many other nonlinear equations have been discovered and in certain cases solutions found that exhibit nonlinear, cohesive nature; see Whitham (1974) for a good introductory treatment. Here we shall confine our attention to internal waves describable by the weakly nonlinear Korteweg-deVries equation. We will start with the classic two-layer model and generalize slightly to continuously distributed density profiles. More comprehensive treatments are available in the literature (Ostrovsky and Stepanyants, 1989; Apel and Gonzales, 1983).

Referring to the two-layer model of Figure 5.3: for each layer $i = 1, 2$, let the density be ρ_i, the mean speed be u_i, and the layer thickness be h_i. The vertical displacement of an isopycnal surface is $\eta(x,z,t)$, the phase speed of long-length infinitesimal-amplitude waves is c_0, and the wave number is k (in rad/m). As is known from shallow-water theory for this case (see Apel, 1987, for example), the linear phase speed is

$$c_0 = \left(g \frac{\Delta\rho}{\rho} \frac{h_1 h_2}{h_1 + h_2} \right)^{1/2} = \left(g' h' \right)^{1/2} \tag{2}$$

where g is the acceleration of gravity and g' and h' are the effective gravity and depth, as given by the center term of Equation 2. The wave speed depends on the harmonic mean of the layer depths; in the deepwater limit, that mean becomes asymptotically the same as the upper layer depth, h_1. Thus there is a limiting speed in the problem, c_0, which is a necessary condition for the existence of a solitary-like wave.

Now, however, if the wave amplitude becomes appreciable, its generally downward character (Figure 5.3) effectively deepens the upper layer, leading to an increase in speed given approximately by:

$$c \approx \left[g'(h' + \eta) \right]^{1/2} \geq c_0 \tag{3}$$

The wave then tends toward a discontinuous shocklike front, which is resisted by linear dispersion of the Fourier components of the waveform. A balance between the nonlinear cohesive forces and the linear dispersion results in a wave of permanent shape, that is, a soliton.

The soliton does not require the discontinuous layering of the present simple model for its existence. Any continuously stratified layer of fluid, where the density behaves as $\rho = \rho(z)$, can support internal waves, linear or solitary. In such a fluid, buoyancy oscillations can exist for frequencies less than the Brunt-Väisälä frequency, N, where

$$N^2 = -\frac{g}{\rho} \frac{d\rho}{dz} \tag{4}$$

is established by the vertical density derivative, and for ω greater than the inertial frequency, $f = 2\Omega\sin\Lambda$; here Ω is the rotational rate of earth and Λ is the latitude.

Both linear and solitary internal waves are governed by the internal gravity wave equation for the vertical structure function, $\Phi_n(z)$ (Phillips, 1977; Apel, 1987):

$$\frac{d^2\Phi_n}{dz^2} + k^2\left(\frac{N^2(z)}{\omega_n^2} - 1\right)\Phi_n = 0 \qquad (5)$$

where the horizontally traveling wave is assumed to be of the form:

$$\eta(x, z, t) = \sum_n \eta_{0n}\Phi_{0n}(z)A(x - c_n t) \qquad (6)$$

and where n is the modal index for the vertical structure function: $n = 1, 2, \dots$. Here η_{0n} are amplitudes giving the vertical scales of the wave, $A(x - c_n t)$ is the normalized amplitude function, and c_n is the phase speed for mode n.

For the case of solitary internal waves, the evolution equation for $A(x - c_n t)$ is given by the Korteweg-deVries equation:

$$\frac{\partial A}{\partial t} + c_{on}\left(\frac{\partial A}{\partial x} + \gamma\frac{\partial^3 A}{\partial x^3} + \alpha A\frac{\partial A}{\partial x}\right) = 0 \qquad (7)$$

In Equation 7, the dispersion coefficient, γ multiplies the dispersive term, A_{xxx}, while the factor α is the nonlinear coefficient (Apel and Gonzalez, 1983). There are two well-known solutions to the internal-wave Korteweg-deVries equation: the single-pulse soliton and the onoidal (or Jacobi elliptic) function. However, I shall treat only the former, since it demonstrates the most important characteristics of the solitary wave. The single-pulse solution for the amplitude function is given by:

$$A(x, t) = -\mathrm{sech}^2\left(\frac{x - c_n t}{2L}\right) \qquad (8)$$

which has the form of an isolated pulse moving to the right with a nonlinear speed, c_n, approximately given by (for small wave numbers):

$$c_n \simeq c_0\left(1 - \gamma k^2 + 2\alpha\eta_{0n}/3\right) \qquad (9)$$

Here c_0 is the infinitesimal-amplitude, long-wavelength speed which, for a two-layer model, is given by Equation 2, and L measures the width of the pulse. In general, however, c_n must be obtained from solutions to the eigenvalue equation, Equation 5, subject to the boundary conditions at the surface, $z = 0$, and the bottom, $z = -H$; these solutions quantize the eigenfunctions and eigenvalues. Usually the so-called rigid-lid conditions are imposed that have $\Phi_n(0) = \Phi_n(-H) = 0$. The downward-displaced soliton pulse is illustrated in Figure 5.5, and demonstrates the bell-shaped characteristic of the Korteweg-deVries solution.

Figures 5.6–9 show solutions for solitary internal waves and associated quantities in coastal waters. Plots of density anomaly, $\sigma_t = \rho(z) - 1000$ (in kg/m³), and of buoyancy frequency, Equation 4, are shown in Figure 5.6. The first two eigenfunctions of Equation 5 are shown in Figure 5.7 for a wave number, $k/2\pi = 2.6$ cycles/km; the associated dispersion relation, $\omega_n = \omega(k)$ is shown in Figure 5.8, and the phase and group speeds, ω_n/k, and $\partial\omega_n/\partial k$, are shown in Figure 5.9. These quantities were obtained by numerical solution of Equation 5, given the profile for $N(z)$ in Figure 5.6.

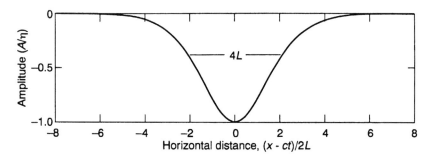

Figure 5.5 Soliton shape and parameters. The width parameter, $4L$, measures the full width of $\mathrm{sech}^2(x)$.

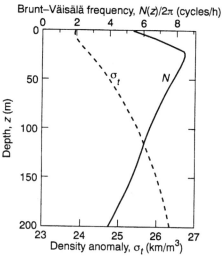

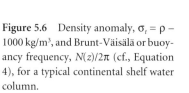

Figure 5.6 Density anomaly, $\sigma_t = \rho - 1000$ kg/m³, and Brunt-Väisälä or buoyancy frequency, $N(z)/2\pi$ (cf., Equation 4), for a typical continental shelf water column.

The current velocity is $\mathbf{u} = (u,0,w)$. From the linear kinematic boundary condition, $\partial\eta/\partial t = w$; from the continuity equation in two dimensions, $\partial u/\partial x = -\partial w/\partial z$, one obtains for the horizontal velocity for mode n

$$u_n = -\eta_{0n}c_n \frac{d\Phi_n}{dz}\,\text{sech}^2\left(\frac{x-c_nt}{2L}\right) \tag{10}$$

and for the vertical velocity

$$w_n = \frac{\eta_{0n}c_n}{L}\,\Phi_n(z)\,\text{sech}^2\left(\frac{x-c_nt}{2L}\right)\tanh\left(\frac{x-c_nt}{2L}\right) \tag{11}$$

For use in imaging theory ahead, one component of the strain rate is required, $\partial u/\partial x$:

$$\frac{\partial u_n}{\partial x} = -\frac{\eta_{0n}c_n}{L}\frac{d\Phi_n}{dz}\,\text{sech}^2\left(\frac{x-c_nt}{2L}\right)\tanh\left(\frac{x-c_nt}{2L}\right) \tag{12}$$

This quantity is to be evaluated at the surface, $z = 0$, for use in imaging calculations.

The strain rate of Equation 12 describes the alternating compressive and tensile effects of the internal currents that render them visible on the ocean surface. In the region of phase of converging surface currents, those surface waves with group speeds that are near the phase speeds of the internal waves are swept together and amplified, while that part of the phase having diverging surface currents reduces the overlying surface wave spectral content. This results in alternating regions of enhanced and diminished surface wave spectral density, regions that are rougher

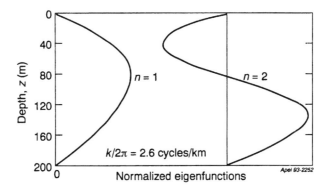

Figure 5.7   Eigenfunctions $\Phi_n(z)$, of the first two modal solutions, $n = 1, 2$, of Equation 5 at a wave number $k/2\pi = 2.6$ cycles/km, for the density structure of Figure 5.6.

and smoother than the average. Electromagnetic radiation incident on the surface is thus scattered differentially in the rough and smooth portions, and an image constructed from such scattered radiation — say a photograph made in visible light or a radar image — will be a map of the roughness variations. The map thus mirrors the underlying internal wave *strain rates*, not the amplitudes or currents. However, these quantities are related by Equations 6–16.

The mathematical description of the straining processes is grounded in the equation for conservation of the *wave action* spectrum (Phillips, 1977; Apel, 1987). Wave action density, N, is wave energy density per unit surface area divided by the Doppler frequency of the wave, $\omega_d = \omega - \mathbf{k} \cdot \mathbf{u}$, where $\mathbf{u}$ is the horizontal current velocity at the surface. Thus we have

$$N(\mathbf{k}, \mathbf{x}, t) = \frac{\Psi(\mathbf{k}, \mathbf{x}, t)}{\omega - \mathbf{k} \cdot \mathbf{u}} \tag{13}$$

where $\Psi(\mathbf{k}, \mathbf{x}, t)$ is the wave number spectrum, which varies locally in space and time due to the advective and straining effects of internal wave currents.

The conservation equation (or more properly, the *nonconservation equation*) for action spectral density, $N(\mathbf{k}, \mathbf{x}, t)$, is derived from the energy density equation for surface waves (Phillips, 1977). It states that the action spectrum changes along characteristics in $(\mathbf{k}, \mathbf{x})$ space according to:

$$\frac{dN(\mathbf{k}, \mathbf{x}, t)}{dt} = \frac{\partial N}{\partial t} + \nabla_k N \cdot \frac{d\mathbf{k}}{dt} + \nabla_k N \cdot \frac{d\mathbf{x}}{dt} = -\frac{N(N - N_{eq})}{\tau} \tag{14}$$

where $N_{eq}(\mathbf{k})$ is the equilibrium action density, and τ is the *relaxation time*, a measure of how long a perturbed wave spectrum takes to relax back to its equilibrium state. This equation may be integrated along characteristics in $(\mathbf{x}, \mathbf{k})$ space, defined by:

$$\frac{d\mathbf{x}}{dt} = \mathbf{c}_g + \mathbf{u}$$

$$\frac{d\mathbf{k}}{dt} = -\mathbf{k} \cdot \nabla \mathbf{u} \tag{15}$$

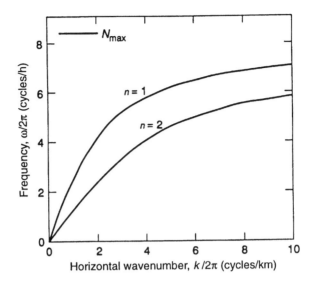

Figure 5.8 Dispersion relation, $f_n = f(k/2\pi)$ [or $\omega_n = \omega(k)$] as calculated from the density profile of Figure 5.6.

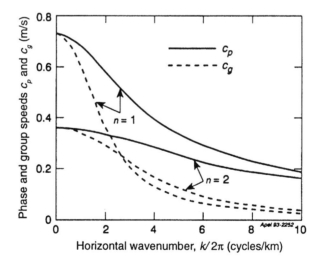

Figure 5.9 Phase and group speeds, ω_n/k and $\partial\omega_n/\partial k$, for the conditions above.

Thompson et al. (1988) give the solution for N in terms of a perturbation, P, as:

$$P = N_{eq} \int_{-\infty}^{t} \sum_{i,j=1}^{2} k_i' \frac{\partial u_j}{\partial x_i'} \frac{\partial\left(1/N_{eq}\right)}{\partial k_i'} \exp\left(-\int_{-t'}^{t} \frac{dt''}{\tau(k'')}\right) dt' \qquad (16)$$

where the perturbed action density spectrum is

$$N = \frac{N_{eq}}{1 + P} \qquad (17)$$

Equation 16 is a one-dimensional temporal integral along the paths defined by Equations 15a and b, and gives the distortion of the spectrum as the surface waves encounter the converging and diverging phases of the internal wave field caused by the strain rates of Equation 12. See Hughes (1978) or Thompson et al. (1988) for a more complete treatment of the effects described here.

5.5 Radar Cross Section

The modifications to the surface wave spectrum from internal waves can be quite significant. An internal wave having a maximum strain rate of perhaps 2×10^{-3} s^{-1} can cause an increase in spectral level of order 200%. This increased roughness is readily seen in remote images of the sea surface even under rather windy conditions. Because the roughness has a definite relationship to the underlying internal wave position and phase, it can be used to map the internal wave field. It is even possible to estimate the internal wave amplitudes with somewhat coarse accuracy by using certain of the kinematic features of the images, along with the historical subsurface density profile of the water column (Apel and Gonzalez, 1983).

From the knowledge of the equilibrium height spectrum and its perturbations, one may calculate the radar backscatter from the roughness variations. Radar scattering theory for rough surface scatter is quite complicated, and only a cursory description can be given here (see Beckmann and Spizzichino, 1963; Valenzuela, 1978; or Apel, 1987, 1994 for expositions on this subject in the oceanic context). In the theoretical treatment, the ocean appears via two independent descriptors: (1) the surface wave probability distribution for either heights, η, or slopes, $\partial\eta/\partial x$ and $\partial\eta/\partial y$; and (2) the surface wave autocovariance function, $R_\eta(x')$, where x' is the lag distance between two measurements of the height. The probability distribution function (PDF) measures the statistics of the vertical displacements, and the autocovariance function measures the correlations and distributions of hills and valleys. At zero lag distance, the autocovariance yields the height variance, σ_η^2.

Holliday (1991) gives a formulation for the radar scattering cross section per unit area of the ocean surface, σ^0, which, with slight modifications, is as follows:

$$\sigma_{v,h}^0(\mathbf{q},\theta) = \frac{q^4}{\pi q_z^2}\left|R_{v,h}(\theta,\varepsilon_r)\right|^2 \int_{-\infty}^{\infty}\int L(\mathbf{x}')\exp(-i2\mathbf{q}_h\cdot\mathbf{x}')$$

$$\times \exp\left\{-4q_z^2\sigma_n^2\left[1-\rho(\mathbf{x}')\right]\right\}d\mathbf{x}'^2 \tag{18}$$

The quantities of relevance to Equation 22 are σ^0 = radar cross section per unit area; q, Ω = EM wave vector and frequency; q_h, q_z = horizontal, vertical components; $R_{v,h}(\theta,\varepsilon_r)$ = reflection coefficients for vertical (v); and horizontal (h) polarizations; $\varepsilon_r(\Omega)$ = relative dielectric function for seawater; $L(\mathbf{x}')$ = illumination function on surface; θ, ϕ = incidence and azimuth angles; $\mathbf{x}'$ = sea level coordinates; η = ocean surface wave height; $R_\eta(\mathbf{x}')$ = autocovariance of heights; σ_η^2 = variance of height spectrum = $\langle\eta^2\rangle = R_\eta(0)$; $\rho(\mathbf{x}')$ = autocorrelation of heights = $R_\eta(\mathbf{x}')/\sigma_\eta^2$; k, ω = ocean wave vector and frequency; and $\Psi(\mathbf{k})$ = height spectrum of surface wave vectors = $\mathscr{F}_2[R_\eta(\mathbf{x}')]/(2\pi)^2$.

Often evaluation of the very complicated integrals of Equation 18 is not required in order to estimate the *signal contrast* of an internal wave image. Instead, for small variations of the current strain rate, it may be assumed that the relative modulation of the spectrum, $\delta\sigma^0/\sigma^0$ is mirrored in the relative modulation of the cross section, viz:

$$\frac{\delta\sigma^0}{\sigma^0} \simeq \frac{\delta N}{N_{eq}} = \frac{\delta\Psi}{\Psi_{eq}} \tag{19}$$

This means that the solution to Equation 14 can be used to estimate the image contrast, provided that the fractional modulation is small. Much of the research work involving the analysis of internal wave properties has used this formula; it has also been used to calculate the contrast in images of shallow sand banks in the sea, where the current strain rates are derived from simple potential flow calculations (Alpers and Hennings, 1984).

The equilibrium surface wave directional spectrum in such calculations is often taken as a simple power-law/cosine formula such as:

$$\Psi_{eq}(\mathbf{k}) = \Psi_0\exp(-k_p^2/k^2)k^{-4}\cos^4(\theta/2) \tag{20}$$

Here $k_p = g/u^2$ is the surface wave number of the spectral peak; u is the wind speed at 10-m height; ϕ is the azimuth angle relative to the wind; and Ψ_0 is the spectral constant, which is approximately 0.00190 for mks units and k in rad/m (Apel, 1994).

Figures 5.10–12 are images and graphs showing internal waves in the New York Bight via their surface signatures in SAR images made at 1.2 (L-band) and 9.5 (X-band) GHz (Figure 5.10); *in situ* current meter measurements made simultaneously with the SAR image (Figure 5.11); and theoretical cross-sectional modulations computed using the formalism presented in Equations 15–17 and 18–20, as shown in Figure 5.12. The agreement between observation and theory is quite good, as may be seen by a comparison of the solid and dotted lines in Figure 5.12. (Apel et al., 1988).

5.6 Summary

From the brief description presented here, it is clear that for the case of coherent internal waves on the continental shelf, remote sensing can yield quantitative measurements of hydrodynamic properties of subsurface flows by using a combination of imagery, *in situ* measurements, hydrodynamic theory, and radar scatter theory. In this area, remote sensing has been ahead of the in-water observations and has provided a broad view of the phenomenon, as well as guidance to the development of theoretical models to describe the results and to predict additional properties. Here, the agreement is quite satisfactory, although significant anomalies remain in the theoretical calculations at other combinations of frequency, polarization, incidence angle, and air/sea stability conditions (Apel, 1994).

This is not to say that remote sensing has solved the internal wave problem in its entirety — far from it. The preponderance of internal wave energy in the ocean is probably quasi-random and incoherent, and presents little that is recognizable by way of surface patterns. Nevertheless, a suspicion exists that a fraction of what appears as internal chaos in point measurements may actually be quasi-orderly wavelike motions, to be revealed as such by the more Olympian view provided by remote sensing. In this case, to paraphrase an ancient Chinese proverb, "A single image may be worth a thousand thermistor time series."

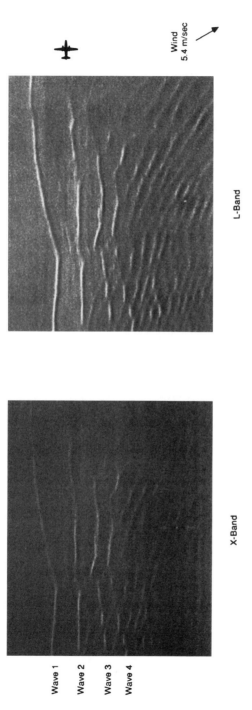

Figure 5.10 Synthetic aperture radar signatures of internal solitary waves on the continental shelf off New York, similar to the ones in the previous figures. Radar frequencies: X-band (10 GHz, left) and L-band (1.2 GHz, right). Data courtesy of ERIM.

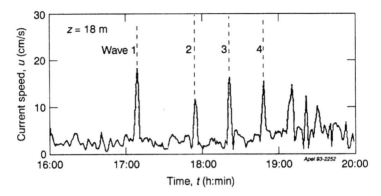

Figure 5.11 Current meter measurements of solitary internal wave of Figure 5.10.

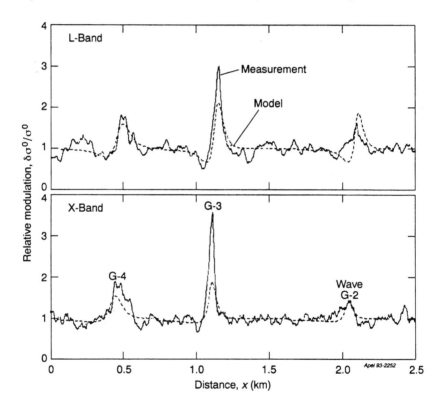

Figure 5.12 Comparison between observed (solid) and calculated (dotted) signatures in SAR images of the solitons of Figure 5.10 for L-band (upper) and X-band (lower) SAR images, using theory of Equations 14–20.

References

1. Alpers, W. R. and Hennings, I., A theory of the imaging mechanisms of underwater bottom topography by real and synthetic aperture radar, *J. Geophys. Res.*, 89, 10,529, 1984.
2. Apel, J. R., Byrne, H. M., Proni, J. R., and Charnell, R. L., Observations of oceanic internal waves from the earth resources technology satellite, *J. Geophys. Res.*, 80, 865, 1975.[a]
3. Apel, J. R., Proni, J. R., Byrne, H. M., and Sellers, R. L., Near-simultaneous observations of intermittent internal waves from ship and spacecraft, *Geophys. Res. Lett.*, 2, 128, 1975.[b]
4. Apel, J. R., Satellite sensing of ocean surface dynamics, *Annu. Rev. Earth Planet. Sci.*, 8, 393, (Annual Reviews, Palo Alto), 1980.
5. Apel, J. R. and Gonzalez, F. I., Nonlinear features of internal waves off Baja California as observed from the SEASAT imaging radar, *J. Geophys. Res.*, 88, C7, 4459, 1983.
6. Apel, J. R., Holbrook, J. R., Tsai, J., and Liu, A. K., The Sulu Sea internal soliton experiment, *J. Phys. Oceanogr.*, 15, 1625, 1985.
7. Apel, J. R., *Principles of Ocean Physics*, Academic Press, London, 1987.
8. Apel, J. R., Gasparovic, R. F., Thompson, D. R., and Gotwols, B. L., Signatures of surface wave/internal wave interactions: experiment and theory, *Dynamics Atmos. Oceans.*, 12, 89, 1988.
9. Apel, J. R. and Gjessing, D. T., Internal wave measurements in a Norwegian fjord using multifrequency radar, *Johns Hopkins APL Tech. Dig.*, 10, 295, 1989.
10. Apel, J. R., An improved model of the ocean surface wave vector spectrum and its effects on radar backscatter, *J. Geophys. Res.*, 99, 16,269, 1994.
11. Barber, B. C., On the dispersion relation for trapped internal waves, *J. Fluid Mech.*, 252, 31, 1993.
12. Beckmann, P. and Spizzichino, A., *The Scattering of Electromagnetic Waves from Rough Surfaces*, Pergamon Press, Oxford, 1963.
13. Elachi, C. and Apel, J. R., Internal wave observations made with an airborne synthetic aperture imaging radar, *Geophys. Res. Lett.*, 3, 647, 1976.
14. Etkin, V. S., et al., Radiohydrophysical Aerospace Research of Ocean, Academy of Science of the USSR, Space Research Institute, Pr. 1749, 1991.
15. Etkin, V. S. and Smirnov, A. L., private communication, 1991.
16. Garrett, C. J. R. and Munk, W. H., Space-time scales of internal waves: a progress report, *J. Geophys. Res.*, 80, 291, 1975.
17. Halpern, D., Semidiurnal tides in Massachusetts Bay, *J. Geophys. Res.*, 76, 6573, 1971.
18. Holliday, D., Backscattering of electromagnetic waves from a perfectly conducting, gently sloping surface of finite amplitude, *IEEE Trans. Ant. Prop.*, 39, 251, 1991.
19. Hughes, B. A., The effect of internal waves on surface wind waves: theoretical analysis, *J. Geophys. Res.*, 83C, 455, 1978.
20. Kao, T. W., private communication, 1991.
21. Korteweg, D. J. and deVries, G., On the change of form of long waves advancing in a rectangular canal, and on a new type of stationary waves, *Phil. Mag.*, 39, 422, 1895.
22. Liu, A. K., Holbrook, J. R., and Apel, J. R., Nonlinear internal wave evolution in the Sulu Sea, *J. Phys. Oceanogr.*, 15, 1613, 1985.
23. Maxworthy, T., A note on the internal solitary waves produced by tidal flow over a three-dimensional ridge, *J. Geophys. Res.*, 84, 338, 1979.
24. New, A. L., Internal tides in the Bay of Biscay, *Deep-Sea Res.*, 36, 735, 1989.
25. Ostrovsky, L. A., Stepanyants, Yu. A., Do internal solitons exist in the ocean, *Rev. Geophys.*, 27, 293, 1989.
26. Phillips, O. M., *The Dynamics of the Upper Ocean*, 2nd ed., Cambridge University Press, 1977.
27. Pingree, R. D. and New, A. L., Downward propagation of internal tidal energy in the Bay of Biscay, *Deep-Sea Res.*, 36, 735, 1989.
28. Proni, J. R. and Apel, J. R., On the use of high-frequency acoustics for the study of internal waves and microstructure, *J. Geophys. Res.*, 80, 1147, 1975.

29. Rattray, M., Jr., On the coastal generation of internal tides, *Tellus*, 12, 54, 1960.
30. Russell, J. S., *Report on Waves*, John Murray, London, 1844.
31. Sandstrom, H. and Elliott, J. A., Internal tide and solitons on the Scotian Shelf: a nutrient pump at work, *J. Geophys. Res.*, 89, 6415, 1984.
32. Thompson, D. R., Gotwols, B. L., and Sterner, R. E., II, A comparison of measured surface wave spectral modulations with predictions from a wave-current interaction model, *J. Geophys. Res.*, 93, 12,339, 1988.
33. Valenzuela, G. R., Theories for the interaction of electromagnetic and oceanic waves — A review, *Boundary Layer Meteorol.*, 1(4) 61, 1978.
34. Whitham, G. B., *Linear and Nonlinear Waves*, John Wiley & Sons, New York, 1974.

6

Basin-Scale Oceanic Circulation from Satellite Altimetry

Pierre-Yves Le Traon

6.1 The Challenge

Understanding of general circulation dynamics has progressed a lot in the past decade through theoretical and numerical studies (e.g., Luyten et al.;[1] Semtner and Chervin;[2] Bryan and Holland;[3] Colin de Verdière[4]). However, because of the space and time variability of the ocean, oceanographers are faced with a huge observational problem. The ocean cannot be viewed as an unchanging system but several decades of *in situ* measurements are needed to observe it globally (e.g., Levitus[5]). Our description of the general oceanic circulation thus remains qualitative and, to say the least, very little is known about its variation in time. Much progress can be expected from the use of satellite altimetry. Satellite altimetry has unique capabilities for producing the necessary global, synoptic view of the oceans. Given the high vertical coherence of oceanic variability, this surface information is a strong constraint for inferring, together with *in situ* measurements, the three-dimensional oceanic circulation (e.g., Wunsch and Gaposhkin[6]).

Determining the general circulation of the ocean and of its variability is one of the major goals of satellite altimetry (e.g., Stewart et al.[7]). This is also the most challenging one. First, to observe the absolute circulation, there is a need for a precise, independent geoid (the geoid is the equipotential surface of the earth gravity field to which a motionless ocean surface would conform). The importance of the geoid stems from the fact that it varies by as much as 100 m, i.e., about 2 orders of magnitude more than the oceanic circulation. There has been steady progress in obtaining more accurate geoid models (e.g., Nerem et al.;[8] Rapp et al.;[9] Lerch et al.[10]) and the absolute circulation estimation is now within reach, at least qualitatively. A dedicated gravimetric mission is, however, needed to significantly enhance our knowledge of the geoid at intermediate scales (wavelengths shorter than 4000 km). Since the geoid evolves at geologic time scales, it can be assumed to be time invariant for oceanography and is not needed to study temporal variations of the sea level due to ocean dynamics. The variation of the basin-scale circulation is thus theoretically much more accessible. However, this signal, contrary to the mesoscale oceanic signal, is weak. It also decorrelates less easily from altimetric measurement errors and, in particular, from the orbit error caused by imperfect knowledge of the spacecraft position. The basin-scale oceanic circulation and its variability as assessed from satellite altimetry thus require careful analysis. The satellite altimeter mission must also be particularly optimized to achieve this ambitious goal.

In the past, four satellites carried altimeters for earth observation: Skylab (1973) GEOS-3 (1975), Seasat (1978), and Geosat (1985–1989). Two are operating in 1993: European Remote Sensing Satellite-1 (ERS-1, launched in 1991) and TOPEX/Poseidon (launched in 1992). Several are scheduled in the next decade. Fu[11] reviewed the first attempts to estimate the basin-scale circulation with GEOS-3 and Seasat. The U.S. Navy Geosat altimeter operated on a near-repeat orbit (17.05-day cycle) for almost 3 years (November 1986–June 1989). Geosat was particularly suitable for mesoscale observations (Cheney et al.;[12] Douglas and Cheney;[13] Le Traon[14]) but less for basin-scale circulation. However, it gave much better results than GEOS-3 and Seasat. The European Space Agency (ESA) satellite ERS-1 (35-day cycle for the oceanographic phase) is an improved alimetric mesoscale mission with microwave radiometer onboard. We do not expect, however, significant progress as far as basin-scale oceanic circulation is concerned. TOPEX/Poseidon (10-day repeat cycle), the joint U.S./France mission, was specifically optimized to observe the basin-scale oceanic circulation. Typical accuracy of the TOPEX/Poseidon orbit is 3 cm (e.g., Nouel et al.[15]) which is almost an order of magnitude better than the best Geosat or ERS-1 orbits (20 cm). This makes possible, for the first time, the observation of the basin-scale circulation and of its variations.

The chapter is organized as follows. A brief overview of analysis methods and altimeter measurement errors is given in Section 6.2. Section 6.3 focuses on the most important error for basin-scale oceanic variability, i. e., the orbit error. The accuracy of existing global geoids is discussed in Section 6.4. The observation of the absolute circulation by an altimeter is addressed in Section 6.5 while Section 6.6 deals with large-scale oceanic variability studies. The main prospects for satellite altimetry in basin-scale circulation studies are given in the conclusion.

6.2 Analysis of Altimetric Data for Large-Scale Oceanography

Oceanic Signal Extraction from Altimetry

The altimetric observation of the sea surface topography S can be described by (see Appendix D):

$$S = N + \zeta + \zeta' + \varepsilon_{orb} + \varepsilon + n \tag{1}$$

N is the geoid, ζ is the permanent dynamic topography, and ζ' is its variable part. ε_{orb} is the orbit error while ε are other altimeter errors (propagation effects in the troposphere and the ionosphere, tides, electromagnetic bias, inverse barometer effect) and n is the altimeter measurement noise (white noise).

Present geoids are not generally accurate enough to estimate globally the absolute dynamic topography $\zeta + \zeta'$, except at very long wavelengths (see Section 6.4). A dedicated gravimetric mission is needed to enhance our knowledge of the geoid. An alternative approach would be to estimate a synthetic geoid by combining *in situ* and/or model data with altimeter data (e.g., Glenn et al.[16]) or *a priori* information on the oceanic signal (Kelly and Gille[17]). The variable part of the dynamic topography ζ' is, however, easily extracted since no prior knowledge of the geoid height is needed. The most commonly used method is the so-called repeat track method (collinear analysis) (see Appendix D). The crossover difference technique (Fu and Chelton;[18] Tai and Fu[19]) does not require a repeating orbit since only differences at crossovers of the satellite tracks are analyzed. This makes the analysis more difficult and the spatial sampling is too coarse for mesoscale studies but allows estimation of the large-scale variability. It also allows estimation of the sea level variability between different altimetric missions (Miller et al.;[20] Jacobs et al.[21]). A third method uses an altimetric mean sea surface as a reference (De Mey and Menard[22]). It is the simplest but requires precise altimetric mean sea surfaces, particularly at short wavelengths, which are not yet available. Although it should be possible to obtain accurate estimations after the TOPEX Poseidon and ERS-1 missions (Blanc and Le Traon[23]), this approach will remain less precise during the next few years.

Measurement Errors

Altimeter measurements of sea surface topography are affected by a large number of errors (see Appendix D). For studies of large-scale oceanic variability, most of these corrections must be taken into account since they are large scale and can significantly contaminate the oceanic signal. Some of these errors can be corrected with dedicated instrumentation. TOPEX/Poseidon is a good example of what is required. It carries a dual-frequency altimeter for precise ionospheric corrections. A three-channel microwave radiometer (TOPEX Microwave Radiometer) is used to correct the range delay effects of moisture and identify the rainfall contamination of radar returns. For these two corrections residual errors are thus less than 2 cm root mean square (rms). For satellites without a radiometer or a dual frequency altimeter, model outputs are needed. Errors will vary depending on models but TOPEX/Poseidon will allow us to estimate the spectral characteristics of model errors. Electromagnetic bias and inverse barometer effect can be partly deduced from the analysis of altimeter data themselves (e.g., Fu and Glazman;[24] VanDam and Wahr;[25] Gaspar et al.;[26] Fu and Pihos[27]). Inverse barometer effect correction depends on the frequency-wave number oceanic response to pressure forcing. Recent numerical and empirical studies suggest that the ocean generally responds as an inverted barometer on timescales of a few days to months except maybe in tropical regions (Ponte;[28] Fu and Pihos[27]). This correction should thus generally be applied if one wants to study the dynamic response of the ocean (Tai[29]). Oceanic tidal errors deserve special attention. Tidal signal is the most important variable signal in altimetric data. This signal is only partially corrected using global ocean tide models (e.g., Schwiderski;[30] Cartwright and Ray[31]). The residual errors are then aliased at certain periods depending on the repeat period of the satellite (Parke et al[32]). This can cause important problems for the interpretation of large-scale altimetric signals (Jacobs et al.;[33] Zlotnicki[34]). More accurate global tide models based on hydrodynamic models and/or TOPEX/Poseidon data will be available soon and should be used in the future (Le Provost[34a]). The repeat period of the satellite should also be chosen to avoid an aliasing near dominant oceanic periods (e.g., annual or semiannual periods). The M_2 tide is thus aliased near 60 days for TOPEX/Poseidon which is much less of a problem than the Geosat 317-days aliasing period or ERS-1 aliasing of the solar diurnal tide to the time-mean signal.

Last, but not least, comes the orbit error. This error is caused by imperfect knowledge of the spacecraft position in the radial direction. It has been actually the largest error on altimetric measurements of sea surface topography. With TOPEX/Poseidon, it is comparable to tidal uncertainty. This error will depend on the quality of the satellite tracking system. For TOPEX/Poseidon, precise orbit determination is achieved via three distinct tracking systems (Doppler Orbitography and Radiopositioning Integrated by Satellite [DORIS], Laser, Global Positioning System [GPS]) that provide almost global coverage of satellite orbits, together with the very high altitude (1336 vs. 800 km for Seasat, Geosat, or ERS-1) that reduces drag on the satellite. The radial orbit error is thus about 3 cm rms. This is to be compared to the 20 cm accuracy of the best orbits available for Geosat and ERS-1. Orbit errors can be reduced by adjusting the altimeter data. This important aspect of altimeter data analysis requires a minimum knowledge of orbit error theory. The next section is thus especially devoted to orbit error. Finally, all tracking networks must be tied into a precisely controlled International Earth Reference System (IERS), and a number of ground-based validation stations must be maintained to verify altimeter measurements and cross-validate data from different missions. The use of data from different altimetric missions furthermore requires that orbits be calculated with the same geopotential model to avoid discontinuities due to different geographically correlated orbit errors (see Section 6.3).

6.3 Orbit Error

Orbit Error Theory

A complete review of orbit error theory is beyond the scope of this chapter. The interested reader is referred to Kaula,[35] Wagner,[36] Engelis,[37] Schrama,[38] or Balmino[39] for a thorough discussion. A minimum amount of knowledge, however, is required to make best use of altimetric data.

Precise orbit determination methods generally consist of a numerical integration of the equations of motion over one arc (about 6 days for the GEM-T2 Geosat orbit, about 10 days for the TOPEX/Poseidon orbits) and an adjustment of the initial state vector and certain parameters of the force models (see, e.g., Tapley[40]). The two main forces acting on a satellite are gravitational forces (conservative forces) and surface forces (non-conservative forces). Gravitational forces are principally due to the earth's gravity field. They also include the gravitational attraction of the moon and the sun (and the closest and largest planets) and the perturbing effects of oceanic and solid tides. The surface forces consist of atmospheric drag and radiation pressures. Atmospheric drag acts in a very complex way due to changes in the atmosphere density caused by the action of the sun and the complex shapes of satellites. Because of the uncertainty in the knowledge of these forces, the equations of motion are only approximate. The initial position and velocity measurements are estimated from a tracking system but also with remaining errors.

The theoretical radial orbit error due to gravitational model uncertainties was first derived by Kaula.[35] He determined its frequencies as linear combinations of three frequency components:

$$F_{\text{lmpq}} = \left(1 - 2p + q\right)\left(\dot{\omega} + \dot{M}\right) - q\dot{\omega} + m\left(\dot{\Omega} - \dot{\theta}\right) \qquad (2)$$

where $\dot{\omega} + \dot{M}$ is the orbital frequency (1 cycle per revolution), $\dot{\omega}$ is the frequency of the rotation of the perigee, and $\dot{\Omega} - \dot{\theta}$ is the frequency of the rotation of the earth with respect to the precessing orbital plane (nodal day). l, m, p and q are integers. The frequencies thus correspond to the one cycle per revolution and its harmonics frequencies modulated in time by daily and m-daily frequencies. It can be shown (e.g., Schrama[38]) that all these gravitationally induced orbit errors (except those associated with the slow precession of the argument of perigee) are the same for each repeat sample of a ground track. This is not surprising since the satellite experiences the same gravitational forces along the track.

Initial state vector errors and errors due to the mismodeling of non-conservative forces (drag, radiation pressures) can also be derived analytically (Engelis;[37] Denker and Rapp;[41] Balmino;[39] Chelton and Schlax[42]). They have a constant part and a one-cycle and two-cycle-per-revolution frequency part with time-varying amplitudes. The above theoretical background yields two main conclusions. First, the orbit error is mainly at long wavelengths with a dominant error at one cycle per revolution (i.e., 40,000 km) modulated in time. Other frequencies at subharmonics and longer periods do also arise, due mainly to geopotential errors. After a repeat-track analysis, however, most of the orbit error due to gravity errors cancels out. The remaining error has a much simpler spectrum dominated by the one cycle per revolution frequency. Note, however, that empirical parameters are generally adjusted in the orbit calculation (e.g., one-cycle-per-revolution signal with daily amplitudes). As a consequence, residuals can also have errors at periods other than one cycle per revolution.

Orbit Error Reduction

The orbit error can be reduced by analyzing the altimeter data. To study the oceanic variability, only the variable part of the orbit error needs to be estimated. This can be done via a global or regional minimization of crossover or repeat-track differences (relative to a mean or to a given cycle). These differences do not contain any geoid signal and are dominated by the orbit error. As explained in the previous section, only the non-gravitational orbit error will be seen on the repeat-track differences. This does not matter, however, for the study of the oceanic variability. Crossover differences will see, in addition, part of the geopotential orbit error but will also miss the so-called geographically correlated orbit error, i.e., the orbit error which depends only on the geographic location (e.g., Tapley and Rosborough[43]). Note that if the mean oceanic signal is to be recovered, the global, simultaneous dynamic adjustment of orbit error, geoid, and dynamic topography described in Section 6.4 is probably the better method (although it is much more complex).

Crossover or repeat-track differences also contain the oceanic signal. The main problem is thus to separate the orbit error from this oceanic signal. One needs first to assume an *a priori* analytical form or an *a priori* spectrum of the orbit error. As explained in the previous section, a one-cycle-per-revolution sinusoid modulated in time is a good analytical expression for the variable part of the orbit error (e.g., Chelton and Schlax;[42] Francis and Bergé[44]). To study the absolute signal, a more complex functional form depending on geoid error should be used (e.g., Engelis;[37] Houry et al.[45]). A more empirical approach, commonly used, is to approximate the orbit error by a first or second degree polynomial over a given arc length. The polynomial should adequately represent the orbit error (e.g., Tai[46]) and also minimize the oceanic signal removal (Le Traon et al.;[47] Tai[48]). There is thus a trade-off between long-arc and/or low-degree polynomial that minimizes the oceanic signal removal and short-arc and/or high-degree polynomial that better represents the orbit error (Le Traon et al.[47]).

Commonly used methods that are based on the repeat-track analysis and a polynomial adjustment, or even a sinusoidal approximation, are not well adapted, however, for basin-scale variability studies. They remove the along-track long-wavelength oceanic signals together with the orbit error. Tai[48] recommends that a second degree polynomial adjustment on arcs longer than 10,000 km be used to avoid removing too much of the gyre to global-scale variability. This adjustment, however, already removes a large part of the global sea level seasonal cycle which is characterized by a large-scale slope between the two hemispheres. Thus, although the repeat-track analysis methods can be used for specific studies, they are not generally recommended for basin-scale variability studies.

To minimize the oceanic signal removal, one should use more complex methods using cross-track information. Global crossover minimization can thus be used to estimate the orbit error without removing too much of the large-scale oceanic signal (e.g., Tai and Fu;[19] Tai;[49] Le Traon et al.[50]). It will actually only remove the oceanic signal which has a signature on crossover differences. If the adjustment is limited to crossovers with a time lag smaller than typically 10 days, most of the large-scale oceanic signal will not be seen on the crossover differences and will be preserved. The main difference with respect to the repeat-track analysis is that the time scale differences between the orbit error and the oceanic signal are explicitly taken into account. More generally using inverse techniques, the orbit error signal could be obtained through a global adjustment taking into account not only the spatial but also the temporal characteristics of the orbit error and of the oceanic signal (Mazzega and Houry;[51] Blanc et al.[52]). This should provide a better estimation of the oceanic variability signal.

These orbit error reduction techniques should be used, anyway, with care with TOPEX/Poseidon data. TOPEX/Poseidon orbits have already an accuracy of about 3 cm; and the orbit error is probably no longer the dominant error compared, for example, to oceanic tide errors. This will make the separation even more difficult.

6.4 Geoid

This section is focused on global geoid models. Good gravimetric geoids may also exist in limited areas, but they are no longer available for basin-wide scales and they will not be discussed here. The importance of the geoid is twofold. The accuracy of the geoid directly translates as accuracy in the absolute dynamic topography (see Equation 1). It also partly determines the accuracy of orbit error since a large part of the orbit error is due to geopotential errors (see Section 6.3). The geoid N is generally described in terms of spherical harmonics which are natural basic functions for signal representation on a sphere:

$$N = \sum N_{nm} Y_{nm}(\theta, \lambda) \qquad (3)$$

The Y_{nm} are conventional spherical harmonics and the N_{nm} are constants. θ and λ are latitude and longitude. This series is given up to a maximum degree which determines the resolution of the model (the wavelength is approximately equal to 40,000 km divided by the degree). Large-scale features of the geoid are relatively well known through satellite geodesy since satellite trajectories are mainly affected by low degrees and low orders (small n and m) of the earth gravity field. They can provide relatively accurate estimates of the geoid up to degree 20 or even 50 (Rapp[53]). GEM-9, the geoid used to estimate the absolute circulation with GEOS-3 data (Mather et al.[54]), was complete up to degree and order 20 (Lerch et al.[55]). The error on GEM-9 was estimated at 1.9 m and the omission error 2.5 m (due to higher degree terms). However, the error was only 30 cm for degree and order below 6. GEM-L2 was about twice as accurate as GEM-9 (Lerch et al.[56]) and was used to estimate the absolute circulation from Seasat and GEOS-3 data (Cheney and Marsh[57]). Further progress was made with the GEMT-1 and GEMT-2 and more recently GRIM4-S2 or GEMT-3S geoid models (Reigberg et al.;[58] Marsh et al.;[59] Lerch et al.[10]). For example, the estimated uncertainty complete to degree and order 4 is only 3 cm for GEMT-1. GEMT-3S is one of the latest geoids using only tracking data. Its estimated uncertainty complete to degree and order 10 and 50 is about 18 and 160 cm, respectively, (Lerch et al.[10]). All these geoids use only satellite tracking data and are altimetry independent. They are still too much in error to determine the absolute circulation with a useful accuracy (e.g., Roemmich and Wunsch[60]). This situation is unlikely to change until a dedicated gravitational mission is flown.

To improve and extend the satellite-only model, one can introduce additional measurements such as gravity anomalies and altimeter data. Satellite altimetry and surface gravity data contain a rather direct and unattenuated measurement of the short-wavelength geoid and are clearly indispensable (for the present time) to get the intermediate and short wavelengths of the geoid. The use of altimeter data allows much more accurate estimation of the geoid. To use altimetric data in the geoid estimation, one need, however, to remove a model of the dynamic topography or to simultaneously solve it in the solution. Early attempts considered first that the dynamic topography was negligible (which, as far as first geoid estimations were concerned, was sensible but clearly unacceptable for oceanography) or used the Levitus[5] dynamic topography, model as a correction. More recently, simultaneous estimations of the geoid, the dynamic topography, and the orbit error have been achieved (Tapley et al.;[61] Nerem et al.;[8] Marsh et al.;[59] Lerch et al.;[10] Bode et al.[62]). The simultaneous improvement of N, ζ, and orbit error is possible (and useful) only because the orbit error depends on the geoid. These estimations generally start from an *a priori* estimation for N (e.g., geoid from satellite tracking data only) and ζ (e.g., the Levitus solution). Altimeter data are then introduced (generally together with surface gravity data). They provide an estimation of $N + \zeta$ with an error dominated by the orbit error. N is adjusted to reduce the orbit error which gives a new estimate for N and ζ through Equation 1. However, the separability is still limited to the long wavelengths because the orbit error depends mainly on the low degree terms of the geoid. ζ is thus estimated up to a certain degree (e.g., 10) which means that, in these estimations, altimeter data provide also directly the geoid for the higher degrees (once the orbit error is corrected). As a consequence these geoids are much more accurate than the geoids using only satellite tracking data. This improvement is, of course, dramatic at higher orders, but the inclusion of altimetric data also has a favorable impact on the low orders because it improves the deconvolution of the higher degree terms (Lerch et al.[10]). GEMT-3, which is a combination of GEMT-3S, altimeter data, and surface gravimetry data, is developed up to degree 50 and has an overall accuracy of 59 cm. For degrees below 20, the accuracy is of 25 cm (Lerch et al.[10]). GEMT-3 is a preliminary version of the Joint Gravity Model-2 (JGM-2) geoid which is developed as a joint effort to provide a good geoid model for the TOPEX/Poseidon mission. The Ohio State University geoid OSU91A (Rapp et al.[9]) model is developed up to degree and order 360. It has an accuracy of 11 and 25 cm for degrees 20 and 50, respectively. These smaller formal errors are due to different weighting philosophies and do not really mean that OSU91A is more accurate than GEMT-3. Note finally that these errors are based on global averages and that the mean value over the ocean is almost twice as small because altimeter data are used in the solution.

When using these geoids for oceanography to derive the absolute signal, one should always remember that these geoids are associated with a dynamic topography model developed up to a certain degree. They are not independent from altimetric data and they certainly have absorbed part of the oceanic signal. In particular, they directly contain the oceanic signal above the cutoff dynamic topography expansion degree. We are thus faced with an unsolved problem. It would be highly preferable for oceanography to work with altimetry independent geoids. This is not possible in a useful fashion, until a dedicated gravimetric mission is flown. We are thus almost constrained to use altimetry-dependent geoids although we know that one should be extremely cautious in interpreting oceanographic results.

6.5 Mean Oceanic Circulation from Altimetry

As explained in the previous section, estimations of the general (mean) oceanic circulation by altimetry have been limited to the longest wavelengths of the oceanic circulation (mainly to degree <10, i.e., wavelength > 4000 km [e.g., Tai;[63] Marsh et al.;[59] Nerem et al.;[8] Lerch et al.[10]]) due to the lack of a precise geoid at smaller scales. The first attempts to estimate the absolute dynamic topography from GEOS-3 and Seasat data showed that the very low degrees of the oceanic circulation (degree 4–6) were detectable (e.g., Cheney and Marsh;[57] Tai;[63] Tai and Wunsch;[64] Engelis[65]). See also Fu[11] for a review. These estimations used the geoid available at the beginning of the 1980s (GEM-9, GEM-L2). Using the more accurate GEMT-1 geoid model, Tai[66] provided reasonable estimates up to degree 8 but the solution still showed large errors. All these estimations used an altimetric mean sea surface (e.g., the Seasat mean sea surface calculated by Rapp[67]). The mean sea surface was adjusted to minimize crossover differences to reduce orbit error. The geographically correlated orbit error (see Section 6.3) cannot be observed from crossovers and was thus directly mapped in the solution. In addition, several arcs were fixed that introduced additional errors. As a result the orbit error in these estimations also contributed significantly. The geoid was subtracted, and the dynamic topography was filtered to get the low order and low degrees. This filtering is complicated by the fact that the dynamic topography is only defined over the ocean. There are several ways to deal with the problem (Tai;[63] Tai and Wunsch;[64] Tai[66]). The most satisfactory one, however, is to define functions that are orthogonal over the oceans only (Hwang[68]).

The next class of dynamic topography solutions from altimetry comes from the simultaneous solution for geoid, orbit error, and dynamic topography discussed in the previous section. Marsh et al.[59] obtained a solution combining GEM-T1, surface gravimetry, and Seasat altimetric data. The corresponding dynamic topography model was developed up to degree 10 and showed good qualitative agreement with existing *in situ* measurements. The estimated accuracy was about 15 cm. Denker and Rapp[41] did a similar calculation but using 1 year of Geosat data instead of Seasat data.

The OSU91A sea surface topography model was based on GEM-T2, surface gravimetry data, and Geosat data (Rapp et al.[9]). It showed a significant improvement compared to previous solutions. The expansion was carried up to degree 10. An expansion up to degree 15 did not give better results, because of the geoid error. The error on the sea surface topography was estimated to less than 10 cm for the 10 degree expansion. Note that for this solution the degree (1, 0) was fixed at the value implied by a harmonic analysis of the Levitus dynamic topography since it was not separable from the a constant (over one repeat cycle) one-cycle-per-revolution orbit error. Nerem et al.[8] performed a similar calculation and found a typical error of 15 cm over the ocean for an expansion up to degree 15. Lerch et al.[10] combined GEOS-3, Seasat and Geosat data, surface gravimetry, and GEMT-3S geoid to yield the GEMT-3 geoid and a sea surface topography model accounting for temporal variations between the three altimetric missions. The sea surface topography was also developed up to degree 15. As already explained, differences between error estimates do not necessarily reflect differences in real accuracy. They may also occur because different weighting were applied to data and because different calibrations of error covariance were used. A careful examination of how errors were obtained is thus necessary in order to use them. They generally proved to be rather optimistic (Martel and Wunsch[69]).

The advantage of solving simultaneously for geoid, orbit error, and dynamic topography will become less clear with TOPEX/Poseidon. TOPEX/Poseidon orbit error is less than 5 cm with the geographically correlated orbit error less than 2 cm (Fu[70]). The dynamic topography can thus be obtained almost directly by subtracting existing geoids from a mean sea surface. This means that geoid errors will almost directly translate into dynamic topography errors. Figure 6.1 shows the mean dynamic topography obtained with 10 months of TOPEX/Poseidon data using the OSU91A geoid model. The solution is based on $4° \times 4°$ averages. There is again a significant improvement compared to previous solutions. The figure clearly shows the subtropical gyres and the western boundary currents such as the Gulf Stream, the Kuroshio, and the Brazil/Falklands confluence region. The Antarctic Circumpolar Current is also plain to see. This dynamic topography agrees very well with the Levitus[5] climatology (Nerem et al.[71]). However, the major limitation still remains the geoid.

These dynamic topography models and their error covariance models can be combined with other *in situ* measurements to estimate the mean oceanic circulation (e.g., Wunsch and Gaposchkin;[6] Roemmich and Wunsch;[60] Martel and Wunsch;[72] Mercier et al.;[73] Le Traon and Mercier[74]). Satellite altimetry can now probably provide useful constraints for the large scales of the absolute circulation (e.g., Martel and Wunsch[69]). These large-scale constraints will be better assessed using global inverse models. In any case, the dynamic topography error estimates must be accurate to let the inverse model extract the relevant information from altimeter data.

6.6 Large-Scale Oceanic Variability

We know little about the variation of the basin-scale oceanic circulation over time. The contribution of satellite altimetry is thus of great importance for climate and global change studies. As explained in Section 6.2, no geoid knowledge is required to observe the large-scale oceanic variability. This signal is thus theoretically much more accessible by satellite altimetry although it is small, on the order of 10 cm, and does not separate well from altimetric measurement errors. The large-scale sea level variability comprises the seasonal and interannual variations of the gyre-scale circulation (e.g., Levitus[75]) and of western boundary currents (e.g., Qiu et al.[76]). There is also the seasonal cycle of the sea level due to thermal expansion associated with temperature fluctuations above the seasonal thermocline. Its amplitude is typically of 10 cm (Gill and Niiler[77]). Global warming is another cause of large-scale sea level variability which will probably translate into a nonevenly distributed sea level rise (e.g., Manabe et al.[78]). Finally, there is, of course, the seasonal changes of the oceanic circulation in the tropical regions and the sea level changes associated with the El Niño effect. These are described in detail by in Chapter 8 and will not be discussed further in this chapter.

Numerous studies have focused on the change of position and intensity of western boundary currents (e.g., Kelly and Gille;[17] Qiu et al.;[76] Zlotnicki[79]). These changes were rather well detected with Geosat altimeter data because they were of relatively small scales. Only a few results are given here to illustrate the contribution of satellite altimetry for these studies. Kelly and Gille[17] found that the annual transport cycle in the Gulf Stream reached a minimum in May and

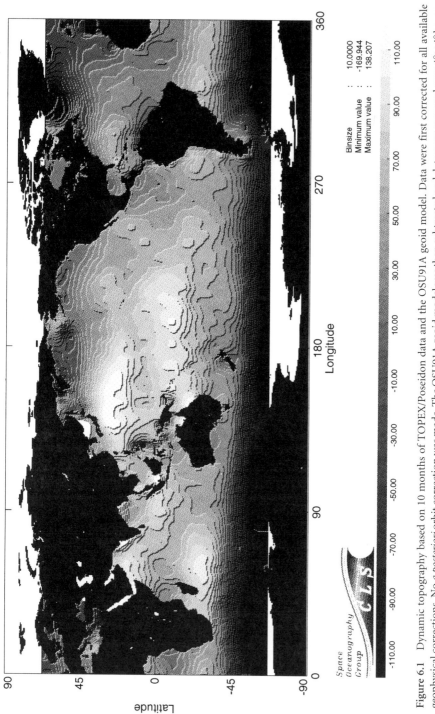

Figure 6.1 Dynamic topography based on 10 months of TOPEX/Poseidon data and the OSU91A geoid model. Data were first corrected for all available geophysical corrections. No *a posteriori* orbit correction was made. The OSU91A geoid model was then subtracted and data were averaged over $4° \times 4°$ boxes. Units are in centimeters.

June and a maximum in November. Maximum and minimum transport of the Gulf Stream occur when the Gulf Stream is north and south, respectively, of its mean position. Qiu et al.[76] have shown that during the first 2 years of the Geosat Exact Repeat Mission (ERM) (1987–1988) the Kuroshio extension as a whole shifted steadily northward at 0.1 km per day, while the surface height difference increased. This trend reversed during the following 6 months of 1989. Qiu et al.[76] suggest that the increase in surface height difference may have been caused by the intensification of the subtropical wind gyre with the 1986–1987 El Niño and the Southern Oscillation (ENSO) event. Zlotnicki[79] estimated the sea level differences across the Gulf Stream and the Kuroshio extension. The south minus north differences (over approximately 10° of latitude) in both regions are higher in the northern fall. Some 60% of this signal variance corresponds to an annual cycle with a maximum in late September–mid-October (9 cm for the Gulf Stream, 6.9 cm for the Kuroshio).

There have been few attempts, however, to study global large-scale variability from satellite altimetry data. The accuracy of past altimetric data did not really allow such investigations. Only TOPEX/Poseidon meets the requirements for such observations. Wunsch[80,81] combining Geosat data with tide gauge data analyzed the large space and timescale of the surface circulation. Although oceanic signals were clearly detectable, in particular, the sea level response to atmospheric forcing, the overall accuracy of the maps did not allow quantitative estimation. The methods he used are, however, well adapted to the retrieval of the large space and time scales of the surface circulation and could be applied to more accurate altimetric missions. Jacobs et al.[33] analyzed the global annual and semiannual harmonics of the oceanic circulation from the first 2 years of Geosat data. They removed the one-cycle-per-revolution orbit error by fitting a sinusoid over a complete revolution. However, as explained in Section 6.3, this procedure removes part of the large-scale oceanic signal. Jacobs et al.[33] first show the structure of the M_2 tidal error which for Geosat is aliased to a 317-day period signal not clearly separable from the annual signal. Given the spatio-temporal sampling of Geosat, this aliasing can produce a pattern similar to baroclinic Rossby waves. This problem does not occur for mesoscale studies when the tidal error orbit error is removed by a polynomial adjustment that also removes tidal error. Jacobs et al.[33] removed some of the tidal error by estimating M_2 in spherical caps three parallel Geosat tracks in diameter because the apparent phase of the tidal errors changes by about 70° for neighboring tracks. Their results then showed the dominance of annual signals in the equatorial regions and the presence of Rossby waves. They were also able to detect part of the seasonal cycle of the ocean due to heating and cooling. The analysis of the global annual cycle of sea level from Geosat data was also performed by Zlotnicki.[34] Zlotnicki[34] provides a comprehensive explanation of errors (tides, water vapor, electromagnetic bias) affecting this signal.

Much is expected from the analysis of TOPEX/Poseidon that should provide new insights on the large-scale oceanic circulation variability. Preliminary results based only on a few months of data illustrate the very high quality of TOPEX/Poseidon data (Le Traon et al.;[50] Nerem et al.[71]). For the first time, the large-scale sea level variability can be observed with satellite altimetry. Le Traon et al.[50] have, for example, analyzed the difference between TOPEX/Poseidon cycle 4 (October 22–November 1, 1992) and TOPEX/Poseidon cycle 22 (April 19–29, 1993) (Figure 6.2).* Data were averaged on 4° in longitude by 4° in latitude boxes. No *a posteriori* orbit error correction was made. The lowering of the sea level in the Northern Hemisphere is mostly due to the contraction of surface waters related to their cooling in fall and winter (e.g., Gill and Niiler[77]). The signal is stronger near the western boundary currents (Gulf Stream and Kuroshio) (–20 cm) partly because these warm currents exchange more heat with the atmosphere. The same phenomenon occurs in the Southern Hemisphere although the signal is much smaller than in the Northern Hemisphere. The lowering of the sea level in the Gulf Stream is also associated with a southward shift of the Gulf Stream mean axis.[71] In the tropical Pacific and Atlantic oceans, the sea level trough and crest between 5°N and 15°N correspond to the seasonal decreases in intensity of the North Equatorial Countercurrent and of the North Equatorial Current. Jacobs et al.[21] compared TOPEX/Poseidon data with Geosat data through a crossover analysis. They showed that, from the Geosat to the TOPEX/Poseidon period, the gyre-scale oceanic circulation intensified in agreement with the intensification of the wind field. Although these are preliminary results, they demonstrate that in the coming years satellite altimetry will provide us unique insights into the variations of the large-scale oceanic circulation.

6.7 Conclusion

As the reader realizes, the observation of the mean or variable basin-scale oceanic circulation from satellite altimetry requires a careful analysis and a good knowledge of errors affecting the altimetric measurement of sea level. The estimation of the mean and of the variable part of the basin-scale oceanic circulation are actually two different problems. As far as mean circulation is concerned, the main limitation comes now from the geoid. The present global geoids only allow a qualitative estimation that is not generally accurate enough to significantly improve our knowledge of the mean circulation. This estimation can be combined with *in situ* measurements to provide useful constraints on the mapping of the large-scale mean oceanic circulation but the contribution still needs to be assessed. This situation is unlikely to change until a dedicated gravimetric mission is flown. The large-scale oceanic variability signal can now, on the other hand, be quantitatively observed by satellite altimetry. The signal is small and poorly known (this is why altimetric data are so interesting), and we have much to learn in the coming years. Since the signal is small and does not decorrelate easily from altimetric measurement errors, one should always, however, question the data sets and look simultaneously at the

*Color plate follows p. 404.

corrections we applied to the data. This may not hold, of course, for all altimetric missions, but this is a general rule for a careful analysis.

Thus, the main conclusion of this chapter is that, provided careful analyses are performed, the understanding of the basin-scale circulation will be greatly improved with satellite altimetry. Satellite altimetry is a unique tool of the Global Ocean Observing System (GOOS). A dedicated and optimized series of altimetric missions are now needed to provide the long-term and uninterrupted monitoring of the oceans essential to climate and global change study (Koblinsky et al.[82]).

References

1. Luyten, J. R., Pedlosky, J., and Stommel, H., The ventilated thermocline, *J. Phys. Oceanogr.*, 13, 292, 1983.
2. Semtner, A. J. and Chervin, R. M., A simulation of the global ocean circulation with resolved eddies, *J. Geophys. Res.*, 93, 15,502, 1988.
3. Bryan, F. O. and Holland, W. R., A high resolution simulation of the wind and thermohaline driven circulation in the North Atlantic ocean, parametrization of small-scale processes, *Proc.'Aha Huliko'a, Hawaiian Winter Workshop*, Müller, P., and Henderson, D., Eds., University of Hawaii at Manoa, 1989, 99.
4. Colin de Verdière, A., On the interaction of wind and buoyancy driven flows, *J. Mar. Res.*, 47, 595, 1989.
5. Levitus, S., Climatological atlas of the world ocean, NOAA Tech. Paper, 3, 1982, 173 pp.
6. Wunsch, C. and Gaposchkin, E. M., On using satellite altimetry to determine the general circulation of the oceans with application to geoid improvement, *Rev. Geophys. Space Phys.*, 18, 725, 1980.
7. Stewart, R., Fu, L. L., and Lefebvre, M., Science opportunities from the TOPEX/Poseidon mission, JPL Publication 86-18, 1986, 58 pp.
8. Nerem, R. S., Tapley, B. D., and Shum, C. K., Determination of the ocean circulation using Geosat altimetry, *J. Geophys. Res.*, 95, 3163, 1990.
9. Rapp, R. H., Wang, Y. M., and Pavlis, N. K., The Ohio State 1991 Geopotential and Sea Surface Topography Harmonic Coefficient Models, report #410, Department of geodetic science and surveying, the Ohio State University, Columbus, OH, 1991.
10. Lerch, F. J. et al., Geopotential models of the earth from satellite tracking, altimeter and surface gravity observations: GEM-T3 and GEM-T3S, NASA Technical Memorandum 104555, 1992, 108 pp.
11. Fu, L. L., Recent progress in the application of satellite altimetry to observing the mesoscale variability and general circulation of the oceans, *Rev. Geophys. Space Phys.*, 21, 1657, 1983.
12. Cheney, R. E., Douglas, B. C., Agreen, R. W., Miller, L., Milbert, D., and Porter, D., The Geosat altimeter mission: a milestone in satellite oceanography, *Eos*, 67, 1354, 1986.
13. Douglas, B. C. and Cheney, R. E., Geosat: beginning a new era in satellite oceanography, *J. Geophys. Res.*, 95, 2833, 1990.
14. Le Traon, P. Y., Contribution of satellite altimetry to the observation of oceanic mesoscale circulation, *Oceanol. Acta*, 15, 441, 1992.
15. Nouel, F., Berthias, J. P., Deleuze, M., Guitart, A., Laudet, P., Piuzzi, A., Pradines, D., Valorge, C., Dejoic, C., Susini, M. F., and Taburiau, D., Precise CNES orbits for TOPEX/POSEIDON: is reaching 2 cm still a challenge ?, *J. Geophys. Res.*, 99(12), 24,405, 1994.
16. Glenn, S. M., Porter, D. L., and Robinson, A. R., A synthetic geoid validation of Geosat mesoscale dynamic topography in the Gulf Stream region, *J. Geophys. Res.*, 96, 7145, 1991.
17. Kelly, K. A. and Gille, S. T., Gulf stream surface transport and statistics at 69° W from the Geosat altimeter, *J. Geophys. Res.*, 95, 3149, 1990.

18. Fu, L. L. and Chelton, D. B., Observing the large-scale temporal variability of ocean currents by satellite altimetry with application to the Antarctic Circumpolar current, *J. Geophys. Res.*, 90, 4721, 1985.

19. Tai, C. K. and Fu, L. L., On crossover adjustment in satellite altimetry and its oceanographic implications, *J. Geophys. Res.*, 91, 2549, 1986.

20. Miller, L., Cheney, R. E., and Lillibridge, J. L., Blending ERS-1 altimetry and tide-gauge data, *Eos, Trans., AGU*, 74, 1993.

21. Jacobs, G. A., Mitchell, J. L., and Kindle, J. C., Linking the ERS-1 and TOPEX/Poseidon altimeter data sets through the Geosat exact repeat mission, Abstract European Geophysical Society Wiesbaden, *Ann. Geophys.*, 11, 103, 1993.

22. De Mey, P. and Menard, Y., Synoptic analysis and dynamical adjustment of Geos3 and Seasat altimeter eddy fields in the North West Atlantic, *J. Geophys. Res.*, 94, 6221, 1989.

23. Blanc, F. and Le Traon, P. Y., Note on the use of an altimeter mean sea surface for mesoscale variability studies, *Oceanol. Acta*, 15, 471, 1992.

24. Fu, L. L. and Glazman, R., The effect of the degree of wave development on the sea state bias in radar altimetry measurement, *J. Geophys. Res.*, 96, 829, 1991.

25. VanDam, T. M. and Wahr, J., The atmospheric load response of the ocean determined using Geosat altimeter data, *Geophys. J. Int.*, 113, 1, 1993.

26. Gaspar, P., Ogor, F., Le Traon, P. Y., and Zanife, O. Z., Joint estimation of the TOPEX and Poseidon sea-state biases, *J. Geophys. Res.*, 99(12), 24,981, 1994.

27. Fu, L. L. and Pihos, G., Determining sea level's response to atmospheric pressure forcing using TOPEX/Poseidon data, *J. Geophys. Res.*, 99(12), 24,633, 1994.

28. Ponte, R. M., The sea level response of a stratified ocean to barometric pressure forcing, *J. Phys. Oceanogr.*, 22, 109, 1992.

29. Tai, C. K., On the quasigeostrophic oceanic response to atmospheric pressure forcing: the inverted barometer pumping, NOAA Technical Memorandum NOS OES 005, 1993, 19 pp.

30. Schwiderski, E. W., Ocean tides. II. A hydrodynamic interpolation model, *Mar. Geodesy*, 3, 219, 1980.

31. Cartwright, D. E. and Ray, R. D., Oceanic tides from Geosat altimetry, *J. Geophys. Res.*, 95, 3069, 1990.

32. Parke, M. E., Stewart, R. H., Farless, D. L., and Cartwright, D. E., On the choice of orbits for an altimetric satellite to study ocean circulation and tides, *J. Geophys. Res.*, 92, 11,693, 1987.

33. Jacobs, G. A., Born, G. H., Parke, M. E., and Allen, P. C., The global structure of the annual and semiannual sea surface height variability from Geosat altimeter data, *J. Geophys. Res.*, 97, 17,813, 1992.

34. Zlotnicki, V., Measuring oceanographic phenomena with altimetric data, Lecture Notes in Earth Sciences, 50, in *Satellite Altimetry in Geodesy and Oceanography*, Rummel, R. and Sanso, F., Eds., Springer-Verlag, 1992, 479 pp.

34a. Le Provost, C., personal communication, 1993.

35. Kaula, W. M., *Theory of Satellite Geodesy*, Blaisdel Publishing, 1966, 124 pp.

36. Wagner, C. A., Radial variations of a satellite orbit due to gravitational errors : implications for satellite altimetry, *J. Geophys. Res.*, 90, 3027, 1985.

37. Engelis, T., Radial Orbit Error Reduction and Sea Surface Topography Determination Using Satellite Altimetry, Report #377, Department of Geodetic Science and Surveying, The Ohio State University, 1987.

38. Schrama, E. J. O., The role of orbit errors in processing satellite altimeter data, Report #33, *Neth. Geodetic Comm.*, 1989.

39. Balmino, G., Radial orbit error theory, Lecture Notes in Earth Sciences, 50, in *Satellite Altimetry in Geodesy and Oceanography*, Rummel, R. and Sanso, F., Eds., Springer-Verlag, New York, 1992, 479 pp.

40. Tapley, B. D., Fundamentals of orbit determination, Lecture Notes in Earth Sciences, 25, in *Theory of Satellite Geodesy and Gravity Field Determination*, Sanso, F. and Rummel, R., Eds., Springer-Verlag, New York, 1989, 235.

41. Denker, H. and Rapp, R., Geodetic and oceanographic results from the analysis of 1 year of Geosat data, *J. Geophys. Res.*, 95, 13,151, 1990.

42. Chelton, D. B. and Schlax, M. G., Spectral characteristics of time-dependent orbit errors in altimeter height measurements, *J. Geophys. Res.*, 12,579, 1993.

43. Tapley, B. D. and Rosborough, G. W., Geographically correlated orbit errors and its effect on satellite altimetry missions, *J. Geophys. Res.*, 90(11), 817, 1985.

44. Francis, O. and Bergé, M., Estimate of radial orbit error by complex demodulation, *J. Geophys. Res.*, 16,083, 1993.

45. Houry, S., Minster, J. F., Brossier, C., Dominh, K., Gennero, M. C., Cazenave, A., and Vincent, P., Radial orbit error reduction and marine geoid computation from the Geosat altimeter data, *J. Geophys. Res.*, in press.

46. Tai, C. K., Accuracy assessment of widely used orbit error approximations in satellite altimetry, *J. Atmos Ocean. Technol.*, 6, 1, 147, 1989.

47. Le Traon, P. Y., Boissier, C., and Gaspar, P., Analysis of errors due to polynomial adjustments of altimeter profiles, *J. Atmos. Ocean. Technol.*, 8, 385, 1991.

48. Tai, C. K., How to observe the gyre to global scale variability in satellite altimetry: signal attenuation by orbit error removal, *J. Atmos. Ocean. Technol.*, 8(2), 271, 1991.

49. Tai, C. K., Geosat crossover analysis in the tropical Pacific. I. Constrained sinusoidal crossover adjustment, *J. Geophys. Res.*, 93(10), 10621, 1988.

50. Le Traon, P. Y., Stum, J., Dorandeu, J., and Gaspar, P., Global statistical analysis of TOPEX/Poseidon data, *J. Geophys. Res.*, 99(12), 24,619, 1994.

51. Mazzega, P. and Houry S., An experiment to invert Seasat altimetry for the Mediterranean and Black Sea mean surfaces, *Geophys. J.* , 96, 259, 1989.

52. Blanc, F., Le Traon, P. Y., and Houry, S., A new method to extract mescocale variability from altimetry, *J. Atmos. Ocean. Technol.*, 12(1), 150, 1995.

53. Rapp, R. H., Use of altimeter data in estimating global gravity models, Lecture Notes in Earth Sciences, 50, in *Satellite Altimetry in Geodesy and Oceanography*, Rummel, R., and Sanso, F., Eds., Springer-Verlag, 1992, 479 pp.

54. Mather, R. S., Rizos, C., and Coleman, R., Remote sensing of surface ocean circulation with satellite altimetry, *Science*, 205, 11, 1979.

55. Lerch, F. J., Klosko, S. M., Laubscher, R. E., and Wagner, C. A., Gravity model improvement using GEOS 3 (GEM 9 and 10), *J. Geophys. Res.*, 84, 3897, 1979.

56. Lerch, F. J., Klosko, S. M., and Pastel, G. B., A refined gravity model from LAGEOS (GEM-L2), *Geophys. Res. Lett.*, 9, 1263, 1982.

57. Cheney, R. E. and Marsh, J. G., Global ocean circulation from satellite altimetry, *Eos Trans. AGU*, 63, 997, 1982.

58. Reigber, Ch. et al., GRIM-4 earth Gravity fields models in support of ERS1 and SPOT2, *Symp. G3, XXth IUGG*, Vienna, 1992.

59. Marsh, J. G., Koblinsky, C. J., Lerch, F., Klosko, S. M., Robbins, J. W., Williamson, R. G., and Patel, G. B., Dynamic sea surface topography, gravity, and improved orbit accuracies from the direct evaluation of Seasat altimeter data, *J. Geophys. Res.*, 95, 13,129, 1989.

60. Roemmich, D. and Wunsch, C., On combining satellite altimetry with hydrographic data, *J. Mar. Res.*, 40, 605, 1982.

61. Tapley, B. D., Nerem, R. S., Shum, C. K., Ries, J. C., and Yuan, D. N., Determination of the general ocean circulation from a joint gravity field solution, *Geophys. Res. Lett.*, 15, 1109, 1988.

62. Bode, A., Chen, Z., Reigber, C., and Schwintzer, P., Simultane globale schwerefeld und meerestopographie bestimmung mit Geosat altimeterdaten, GFZ Technical Rep., Potsdam, 1993.

63. Tai, C. K., On determining the large-scale ocean circulation from satellite altimetry, *J. Geophys. Res.*, 88, 10, 9553, 1983.

64. Tai, C. K. and Wunsch, C., An estimate of global absolute dynamic topography, *J. Phys. Oceanogr.*, 14, 457, 1984.

65. Engelis, T., Global Circulation from SEASAT Altimeter Data, *Mar. Geodesy*, 9, 45, 1986.
66. Tai, C. K., Estimating the basin-scale oceanic circulation from satellite altimetry. Part 1: Straightforward spherical harmonic expansion, *J. Phys. Oceanogr.*, 18, 10, 1398, 1988.
67. Rapp, R. H., The determination of geoid undulations and gravity anomalies from Seasat altimeter data, *J. Geophys. Res.*, 88, 1552, 1983.
68. Hwang, C., 1991. Orthogonal Functions Over the Oceans and Applications to the Determination of Orbit Error, Geoid and Sea Surface Topography from Satellite Altimetry, Report #414, Department of Geodetic Science and Surveying, The Ohio State University, Columbus, OH, 1991.
69. Martel, F. and Wunsch, C., Combined inversion of a finite difference model and altimetric sea surface topography, *Man. Geodesy*, 18, 219, 1993.
70. Fu, L. L., TOPEX/Poseidon Verification Workshop, a summary report, JPL D-10815, 1993.
71. Nerem, R. S., Schrama, E. J., Koblinsky, C. J., and Beckley, B. O., First comparison between TOPEX/Poseidon sea-level variations on the hemispheric scale and steric heights, *J. Geophys. Res.*, 99(12), 24,565, 1994.
72. Martel, F. and Wunsch, C., The North Atlantic circulation in the early 1980's — an estimate from inversion of a finite difference model, *J. Phys. Oceanogr.*, 23, 898, 1993.
73. Mercier, H., Ollitrault, M., and Le Traon, P. Y., An inverse model of the North Atlantic general circulation using Lagrangian float data, *J. Phys. Oceanogr.*, 23, 689, 1993.
74. Le Traon, P. Y. and Mercier, H., Estimating the North Atlantic mean surface topography by inversion of hydrographic and Lagrangian data, *Oceanol. Acta*, 15, 563, 1992.
75. Levitus, S., Interpentadal variability of steric sea level and geopotential thickness of the north Atlantic Ocean, 1970–1974 versus 1955–1959, *J. Geophys. Res.*, 95, 5233, 1992.
76. Qiu, B., Kelly, K. A., and Joyce T. M., Mean flow and variability in the Kuroshio extension from Geosat Altimetry data, *J. Geophys. Res.*, 96, 18,491, 1991.
77. Gill, A. E. and Niiler, P. P., The theory of the seasonal variability in the ocean, *Deep Sea Res.*, 20, 141, 1973.
78. Manabe, S., Stouffer, R. J., Spelman, M. J., and Bryan, K., Transient responses of a coupled ocean-atmosphere model to gradual changes of atmospheric CO_2. Part I. Annual mean response, *J. Climate*, 4, 785, 1991.
79. Zlotnicki, V., Sea level differences across the Gulf Stream and Kuroshio extension, *J. Phys. Oceanogr.*, 21, 599, 1991.
80. Wunsch, C., Global scale variability from combined altimetric and tide gauge measurements, *J. Geophys. Res.*, 96, 15,093, 1991.
81. Wunsch, C., Large-scale response of the ocean to atmospheric forcing at low frequencies, *J. Geophys. Res.*, 96, 15,083, 1991.
82. Koblinsky, C. J., Gaspar, P., and Lagerloef, G., Ed., The future of spaceborne altimetry: Oceans and Climate change, Joint Oceanographic Institutions Washington, D.C., 1992, 75 pp.

7

Sea Surface Temperature Observed by Satellites and Equatorial Dynamics

Masahisa Kubota

7.1 Introduction

An equatorial ocean is a very special area for geophysical fluids because the Coriolis parameter is zero at the equator, and as a result geostrophic balance is not established. El Niño, which is the most remarkable and intrinsic phenomenon in the equatorial Pacific Ocean, is closely related to these characteristics of the equatorial region. The history of El Niño research shows the development of the understanding of equatorial ocean dynamics. Initially, El Niño was considered to be a local phenomenon near the South American coast of Peru, but the relation between El Niño and the Southern Oscillation (ENSO) were noted.[1] The physical relation between the interannual oceanographic and meteorological variations in the tropical Pacific was proposed.[2] The importance of remote forcing was recognized in the 1970s.[3] Moreover, as the climate change problem became the center of public attention, ENSO was recognized as an important component of the global climate system.

To fully analyze the ENSO phenomenon, a rich set of data is needed in the tropical Pacific. However, the number of observations in the equatorial area has been limited for many years. Recently, *in situ* data observations in the equatorial ocean increased remarkably due to the Tropical Ocean and Global Atmosphere (TOGA) program established for the period between 1985 and 1995. Nevertheless, the data coverage is still sparse in the large spatial scale of the ocean. Also, simultaneous observations from the entire basin are needed because remote forcing is crucial for studying equatorial waves and equatorial ocean dynamics. Therefore, remotely sensed data and analysis techniques have become necessary for analyzing equatorial phenomena.

From the viewpoint of the influence on global climate, remotely sensed observations of sea surface temperature (SST) by infrared and microwave radiometers are extremely important because tropical SST is a critical driving force for atmospheric general circulation.

The purpose of this chapter is to describe scientific results obtained by analysis of satellite SST data from the equatorial oceans. Moreover, limitations of satellite SST data are also discussed. Because it is necessary to validate satellite SST data, results from comparisons between remotely sensed and *in situ* SST are described in the Section 7.2. Instability waves observed by remote sensors are presented in Section 7.3 with comparisons to findings based on other *in situ* observations and modeling studies. Other observed phenomena in an equatorial ocean are explained in Section 7.4. Conclusions are made in the last section.

7.2 Satellite Sea Surface Temperature in the Tropical Ocean

Estimates of SST can be derived using a variety of sensors: Advanced Very High-Resolution Radiometer (AVHRR), High Resolution Infrared Sounder and Microwave Sounding Unit (HIRS/MSU), Scanning Multichannel Microwave Radiometer (SMMR), and Visible Infrared Spin-Scan Radiometer Atmospheric Sounder (VAS). A brief introduction about each sensor is given in this section. AVHRR and SMMR are popular sensors to obtain SST data in the equatorial ocean. Thus, results of validating SST data from both sensors are also described.

AVHRR was first flown on Television Infrared Observing Satellite (TIROS)-N in 1978. AVHRR has been carried on the TIROS/National Oceanic and Atmospheric Administration (NOAA) series up to the present. The NOAA series has also carried HIRS/MSU. SMMR was flown on both Seasat and Nimbus-7. The Seasat/SMMR operated only for 3 months, while the Nimbus-7/SMMR operated from late 1978 through mid-1986. Prototypes of the VAS were first flown on Meteosat in 1977. Then, the Geostationary Operational Environmental Satellite (GOES) series carried the Visible Infrared Spin-Scan Radiometer (VISSR) having a single visible channel and a single IR channel. VAS with 12 IR channels instead of VISSR has been operated since 1980. Brief descriptions of the sensors and satellites appear in the

Appendixes of this book. These instruments employ different techniques for measuring SST, and each has unique advantages and limitations.[4] Evaluation of these data sets can be accomplished by comparison to other remotely sensed products (e.g., AVHRR, HIRS/MSU, and SMMR[5,6]) and to fields produced *in situ* data.[7-11] In the present, NOAA/AVHRR is the most popular sensor for the study of equatorial dynamics because of the fine resolution and the high accuracy, though it is strongly affected by clouds. Figure 7.1 shows distributions of ship and nighttime NOAA/AVHRR data during the period of August 8–14, 1993. A large number of satellite observations are clearly seen in the Figure 1 compared with the small number of the ship data. Therefore, the use of satellite data is quite useful, especially in regions of sparse *in situ* data. It should be noted that the AVHRR data can be operationally obtained by the continuity of the NOAA satellite.

It is important to evaluate these SST products in the tropical ocean for the study of equatorial dynamics. It should be noted that skin temperatures are measured by the NOAA/AVHRR, while bulk temperatures are measured by *in situ* observation. The skin temperature measured by the satellite is discussed in Chapter 10 for the global ocean. Here a focus is put on the equatorial area. The quality of the SST products from SMMR in the equatorial ocean (20° S–20° N) was assessed by ship and climatological SST comparison.[10] The standard deviation (SD) of the difference between SMMR and ship SSTs was smaller near the equator than in the midlatitudes for day and night. It was suggested that this was due partly to the equator being far from the day/night terminator and perhaps partly to the small average wind speed near the equator. On the other hand, AVHRR data for the period of 1982–1988 relative to *in situ* data from the Comprehensive Ocean-Atmosphere Data Set (COADS)[12] were evaluated.[11] Cross-correlations for anomaly time series in three so-called El Niño subregions of the equatorial Pacific are very high (0.85–0.92), and the effective degrees of freedom are very low. These statistics reflect the large, persistent nature of SST anomalies found here.

The 1986–1987 ENSO event is beautifully reflected in AVHRR data. Figure 7.2 shows the SST maps in the equatorial Pacific on October 1987 and 1988 observed by the NOAA/AVHRR. The high SSTs associated with El Niño can be clearly seen in the eastern equatorial Pacific on October 1987. On the other hand, the low SSTs there on October 1988 are remarkable. However, it should be noted that use of the satellite anomalies gives a distorted picture of the evolution of the 1982–1983 ENSO SST anomaly due to the effect of volcanic aerosol contamination on the remotely sensed SST signal.[11,13] The region 6° S–6° N, 160° E–150° W labeled region 4 is particularly important in modeling of air-sea interactions during ENSO, since atmospheric convection moves from the western Pacific to cover this region. SST in region 4 shows a large persistent negative bias relative to COADS for 1983–1984. It was concluded that such large biases in this region were unacceptable.[12] An unexplained cold bias appears at the end of the time series in this region and in two other regions as well. In the El Niño regions the signal-to-noise ratio is large and ranges from approximately 2 to greater than 4, indicating a denser coverage from the satellite data in these regions. Moreover, the aerosols from Mt. Pinatubo also

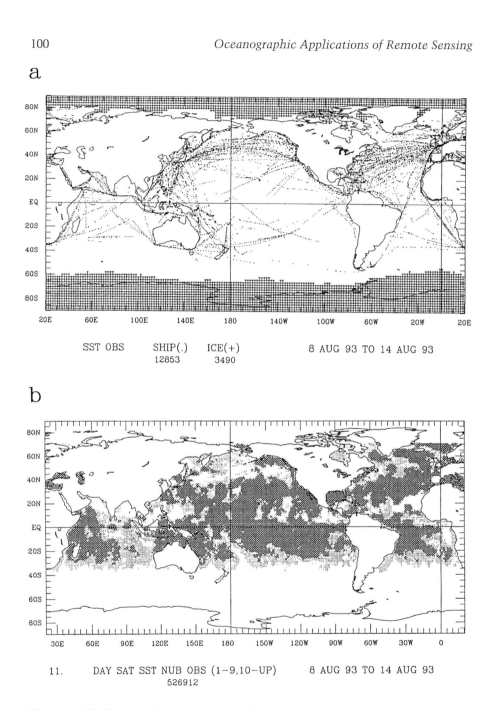

Figure 7.1   Distributions of (a) *in situ* and (b) satellite SST data during the period August 8–14, 1993. In the satellite data map the dots (·) indicate regions on a 1-degree grid with 1–9 observations, and the x's indicate 10 or more. In the *in situ* data map the dots (·) indicate observation points and the +'s indicate sea ice distribution. (From Reynolds, R.W., unpublished manuscript.)

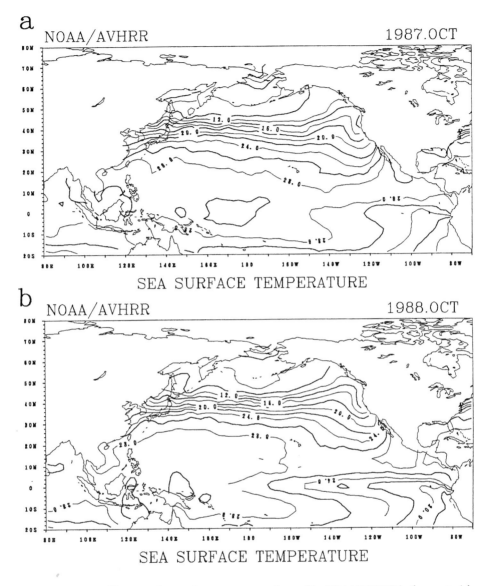

Figure 7.2 Monthly maps of sea surface temperature observed by NOAA/AVHRR in the equatorial Pacific Ocean for (a) October 1987 and (b) October 1988.

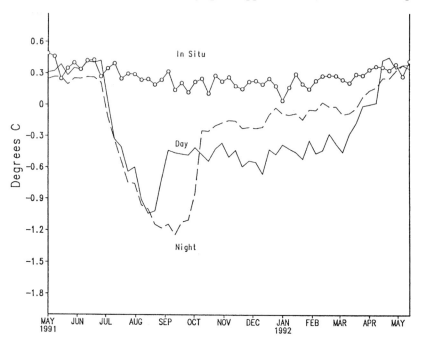

Figure 7.3   Time series of the *in situ,* daytime, and nighttime satellite SST anomalies between 20° N and 20° S for 55 weeks. The *in situ* curve is a solid line with circles, the daytime curve is a solid line, and the nighttime curve is dashed. On the abscissa the dates are labeled at the location of the first day of the month. The ordinate is in degrees Celsius. (Adapted from Reynolds, R.W. et al., *J. Climate,* 6(4), 768, 1993. With permission of the American Meteorological Society.)

caused negative biases in AVHRR data shown in Figure 7.3.[14] Therefore, it is difficult to interpret climate changes, especially in the equatorial region by satellite SST data alone. However, it is possible to correct the satellite data with *in situ* data as Reynolds and Marsico[15] tried. Probably an important issue is how to combine satellite data with *in situ* data.

7.3 Instability Waves

Detection of long equatorial waves by satellite SST data strongly demonstrated the usefulness of remotely sensed data. Figure 7.4 shows a map of SST observed by NOAA/AVHRR in the eastern equatorial Pacific for June 1984. The cusplike feature associated with the equatorial instability wave is clearly detected along the equator. In this section we first describe features of the long equatorial waves observed by satellites and other observing systems. Furthermore, generating and dissipating mechanisms of the waves are also given.

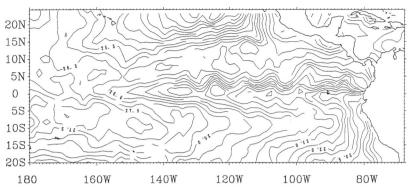

Figure 7.4 A map of sea surface temperature observed by NOAA/AVHRR in the eastern equatorial Pacific Ocean for June 1984.

Long equatorial waves were observed in both the equatorial Pacific[16–19] and the equatorial Atlantic Oceans.[20] These waves have been observed by both VISSR and AVHRR. However, the long waves are not detected by satellites in the western and central parts of the equatorial Pacific Ocean. They are characterized by westward propagating speeds of approximately 40 km/d, wavelengths of about 1000 km, and periods of 25 d. It is of interest that the long waves were absent in El Niño years such as 1976 and 1982.[16,17] The observed meridional extent between the crest and trough of the waves is about 300 km. Generally the long waves are found near 3° N corresponding to the strong shear region between the eastward North Equatorial Countercurrent (NECC) and the westward South Equatorial Current (SEC).

Moreover, instability waves were observed along the SST front off Panama during March 1985.[19] The waves moved southwestward along the Panama-Galapagos SST front at a phase speed of about 30 km/d, the wavelengths of 600 km, and the period of 20 d. They resemble the equatorial long waves previously described, but the relation between these two waves is unknown and should be the subject of future studies.

Equatorial long waves also have been identified by *in situ* observations. The temperature field along two 4000-km sections from 20° N to 17° S along 150° and 158° W was measured by airborne expendable bathythermographs (AXBTs) at approximately weekly intervals for 3 months (November 1977–January 1978).[21] The results suggest highly coherent fluctuations in the upper ocean, with zonal scales of at least 2000 km and timescales of several months. Data from 20 satellite-tracked drifting buoys in the eastern equatorial Pacific Ocean during the summer of 1979 were used to investigate the nature, effects, and energetics of the equatorial long waves.[22] These long waves are associated with a vigorous pattern of mesoscale eddies lying primarily between the equator and 7° N. The eddies cause an equatorward transport of heat that amounts to about two thirds of the poleward heat transport of the divergent Ekman transport in the near-surface waters.

Other observing systems can be used to analyze long waves. Long waves are linked to oscillations in sea level at Fanning Island and to the oscillatory trajectory of drifting buoys.[23] It is suggested by the drifting patterns of buoys that lateral oscillations of the Equatorial Countercurrent with a period of about 34 d are responsible for the observed variations of sea level. It should be noted that the long waves were observed in the central Pacific using *in situ* data, where the SST front is much weaker than in the eastern equatorial Pacific Ocean. Miller et al.[24] analyzed surface dynamic height monitored with inverted echo sounders at five sites from 0° to 9° N along 110° W. For the months of December–March (when the SST front is either absent or too weak to appear in the satellite observations), they noted the presence of 30-d oscillations of the dynamic height field. This result suggests that the dynamic height oscillations are not directly dependent on the presence of the front. Also, they explained that trochoidal features of the waves can be generated simply by having a mean pressure field and a series of pressure oscillations with amplitudes and meridional scales similar to those observed.

Long equatorial waves were observed by Seasat altimeter.[25] Synoptic maps of the sea surface topography derived from altimeter data show a set of eddies 600 km in diameter and 15–20 cm in amplitude moving westward with a velocity of about 40 km/d along 4.5° N.

Pullen et al.[18] determined equatorial long-wave characteristics using not only satellite SST but also *in situ* data. The wavelike structures in the SST maps are in agreement at the surface. Vertical expendable bathythermograph temperature sections made along the equator between 93° W and 125° W showed the phase of the waves tilting westward with increasing depth over the upper 75 m. Such a phase shift, if it extended 100–200 km meridionally in either direction from the equator, would induce an equatorward heat flux. Near-surface currents measured by moorings indicate a regular pattern of northward advection when wave cusps pass, followed by southwest flow during the passage of wave troughs. This is again consistent with an equatorward flux of heat, as well as with earlier drifter buoy findings.[22] Moreover, similar results have been reported in the Atlantic Ocean.[26]

Long waves in the eastern equatorial Atlantic during June and July of 1983 were observed by using AVHRR data.[20] The long waves observed in the Atlantic are similar to the waves in the eastern Pacific with several exceptions. The Atlantic long waves have a shorter duration (nearly three wave cycles at 15° W), and the wave amplitude is limited east of 10° W. The Atlantic equatorial front appears to have a minimum north-south displacement between 5° W and 10° W, while westward of 10° W the displacements exceed 400 km. At 15° W the time-dependent fluctuations of the SST front were shown to have a similar average period (24 d) to that measured by a current meter at 10-m depth on the equator.[27]

It is possible that the long equatorial waves are generated by the latitudinal shear of the NECC and the SEC. The principal source of energy for the fluctuations is the kinetic energy of the mean currents.[22] Theoretical studies on the instability of zonal equatorial currents by Philander[28,29] showed that the realistic meridional velocity profile is unstable and can support growing waves with a period of about 30 d and a wavelength of approximately 1100 km. The instability is primarily caused by the westward jet just north of the equator. A numerical study[30] indicated phase shifts 250 km off the equator, suggesting that baroclinic instability is responsible for the waves overcoming dissipation and radiation of wave energy. Such phase shifts are consistent with observational results.[18] Cox[31] showed that the instabilities draw both the kinetic and the potential energy of the mean flow. Moreover, he found that a portion of the mean potential energy is converted into eddy energy with the result being an equatorward heat transport in opposition to the poleward Ekman transport. The result of an equatorward heat transport is consistent with Hansen and Paul's[22] finding that equatorward heat flux is equal to two thirds of the poleward Ekman heat transport. The local meridional heat flux associated with the waves is on the order of 100 W m^{-2}, but their contribution to the net heat transport across the equator is small. The waves are highly nonstationary in time and inhomogeneous in space.[32]

On the other hand, there must be other processes that remove energy comparable to that produced by instabilities, because the amplitudes of the observed waves are fairly constant over long periods. Philander et al.[33] pointed out that one possible process is the large dissipation that is characteristic of equatorial regions;[34] and another is the availability of free waves at the period of the instability that can propagate into the deep ocean, thus draining energy from the surface layers.

There are periods when the long waves are exceptionally energetic. There are also periods, usually in the boreal spring and during El Niño events, when the waves are absent. This modulation of the wave amplitude is correlated with seasonal and interannual variations in the equatorial currents.[33]

7.4 Other Equatorial Phenomena Revealed by Satellite SST Data

Many theoretical studies for equatorial waves have been conducted after the pioneer work by Matsuno[35] and Blandford.[36] Equatorial Rossby waves, equatorial Kelvin waves, and Yanai waves (or mixed Rossby waves) are also stable equatorial-trapped waves that are possible solutions to the Navier-Stokes equations for the region.[37] Recently, many ENSO researchers indicate that these equatorial waves play an important role in equatorial ocean dynamics. However, there are very few examples of detecting these stable equatorial waves using satellite SST data except Tsai et al.[38] and Stumpf and Legeckis.[39] It should be noted that an example detecting equatorial Kelvin waves using satellite SST data cannot be found. The reason will be discussed in the next section.

In the Indian Ocean, the Yanai wave was first observed by Luyten and Roemnich[40] who found a prominent 26-d oscillation in the record of meridional velocities in the upper 200 m. They estimated the zonal wavelength of this oscillation to be approximately 800–1000 km, with meridional velocity of 0.15–0.3 m/sec. Reverdin and Luyten[41] examined the drifting buoy data and concluded that the meandering behavior exhibited by the drifting buoys was associated with motions of the Yanai waves. Tsai et al.[38] reported the 26-d oscillation observed in the satellite SST in the equatorial western Indian Ocean. They concluded that the Yanai wave is responsible for this signal and showed that the signal of the 26-d oscillation is trapped within 6° latitude of the equator and between 52° and 60° E. Also, the amplitude of this oscillation in the SST data is found to be the largest during the summer. These observed features are consistent with spatial and temporal variability in previous observational results[40] and modeling results.[42,43] Kindle and Thompson[42] explained that the excitation of the waves is caused by an instability of the Somali Current system in August and September. Subsequent generation (October–January) of the 26-d waves was theorized to be caused by meandering of the offshore extension of the East African Coastal Current within the equatorial waveguide. On the other hand, Moore and McCreary[44] demonstrated that a western boundary is an efficient source for these waves and that the southwest to northeast orientation of the western boundary is important for generation of the 26-d Yanai waves. Recently Kelly[45] demonstrated that a narrow band of waves with central period of 25 d occupies the interior of the domain because the high frequency, longer waves pass an observer quite rapidly and the low frequency, short waves were damped in and near the forced region. His results give an explanation for the existence of the 26-d waves in regions where the latitudinal shear in the mean zonal flow is small. Further observational investigations are necessary to elucidate the mechanism of the 26-d waves.

Mesoscale eddies off the Pacific coast of Central America have been observed by AVHRR. There are the two areas of prominent upwelling: the Gulf of Tehuantepec, Mexico, and the area near the Gulf of Papagayo, Costa Rica. These areas are located north of the Panama-Galapagos SST front along which instability waves were observed.[19] A series of upwellings in the Gulf of Tehuantepec during December 1973 were monitored and a large anticyclonic eddy was detected.[46] Stumpf and Legeckis[39] showed that these anticyclonic eddies, closely associated with wind-induced upwellings, propagate westward at an average speed of 13 km/d, which is approximately the speed of nondispersive baroclinic Rossby waves at latitude 12° N. Clarke[47] examined satellite SST data together with coastal wind data for the first 41 d of 1986 to see whether the expected wind path influenced SST near the Gulf of Tehuantepec and the Gulf of Papagayo. The time series of satellite images suggest that the initial development of cold surface water in a clockwise loop is due to the wind mixing the shallow thermocline surface water along its inertial path. In addition, stronger gradients of SST to the right of the expected wind flow are consistent with advection of SST by ocean surface Ekman flow. Hofmann et al.[48] used a model to point out the importance of the localized cyclonic wind stress curl for the generation of the Costa Rica dome. Umatani and Yamagata[49] also used a model and concluded that the Costa Rica dome may be classified into a nonlinear planetary mode, rather than a response to the local wind stress curl. Which mechanism is essential will be clarified by future observations. In another modeling study, Kubota and O'Brien[50] found that these eddies generated off Central America migrated southwestward as annual equatorial Rossby waves, as previously shown by using AVHRR data.[46]

Legeckis[51] examined the variability of satellite SST in the eastern equatorial Pacific Ocean for 1982–1986. The onset of the 1982–1983 ENSO event, the annual cycles of equatorial upwelling and currents, the westward propagation of instability waves, and the coastal upwelling along Central and South America were all evident in the data.

Most studies of satellite SST data are concerned with the equatorial Pacific Ocean, while fewer studies focus on the equatorial Atlantic and Indian oceans. The Somali Current is famous for its annual reversing of direction with change in monsoon season. Satellite and ship observations of SST during June–August of 1979 (the southwest monsoon) show the development of large wedge-shaped areas of cold water along the Somali coast at both 5° and 10° N during June and July. By late August the cold wedge at 5° N translated northeastward as far as 10° N at speeds of 15–30 cm s^{-1}, indicating a coalescence of the systems.[52] Legeckis[53] examined satellite SST observations of the Bay of Bengal and noted the existence during the latter part of February 1985 of two bands of warm water that resemble western boundary currents along the east coasts of India and Sri Lanka. February is a good season for examining the satellite observations because the SST in the Bay of Bengal reaches an annual low during this time. Thus, the SST gradients can be more easily detected by infrared sensors.

7.5 Conclusions

Estimates of SST can be made using many kinds of remote sensors. However, infrared radiometers such as AVHRR are best for the study of equatorial dynamics because of their fine resolution and high accuracy. However, the AVHRR data in the tropics in 1982–1983 and 1991–1992 were strongly affected by volcanic aerosols. Although the data could still indicate some of the ENSO warming there, it should be noted that the ENSO SSTs were distorted by the satellite data.

The satellite SST data have played important roles in research on equatorial dynamics and instability waves. From observations, long waves have westward propagating speed of 40 km/d, wavelengths of about 1000 km, and periods of 25 d. The waves have been observed in both the Pacific and Atlantic oceans. Initially, theory and observations supported the idea that the long waves were generated by barotropic instability because the waves were detected in the shear zone between the NECC and the SEC. Later observational and modeling results suggested that baroclinic instability cannot be ruled out for long wave generation.

There are very few indications of stable equatorial waves in satellite SST data except for the Yanai waves. This seems unexplainable to many because the roles of equatorial Rossby waves and equatorial Kelvin waves in annual and interannual variability have been widely recognized in many papers (e.g., Busalacchi and O'Brien;[54] Busalacchi et al.[55]). Perhaps one of the reasons for this lack of signature is the complicated mechanisms that determine SST. Horizontal and vertical advections, as well as heat flux from ocean-atmosphere interactions, all influence SST. Although these mechanisms are closely linked to equatorial Kelvin and Rossby waves, SST change is not determined solely by dynamics. Thermodynamics play a role as well. Therefore, it is difficult to identify those waves by satellite SST data alone.

Moreover, the limitation of satellite SST observation should be noted. Instability waves have generally been observed by satellite sensors only in the eastern Pacific Ocean in boreal spring to fall. However, observational evidence shows they exist in other parts of the Pacific as well. For example, Wyrtki[23] pointed out the existence of the instability waves by using sea level data in the central Pacific Ocean, and Miller et al.[24] also showed the existence of the instability waves by using surface dynamic height data from December to March when the SST front is either absent or too weak to appear in the satellite observations. These results indicate the limitation of satellite SST data, i.e., its weak sensitivity is not high enough where SST gradients are weak.

SST variability may be considered to be the result of complicated ocean dynamics and ocean-atmosphere interaction. Therefore, it is difficult to derive information about equatorial dynamics solely by SST data. Many satellites such as Geodetic Satellite (Geosat), Defense Meteorological Satellite Program (DMSP), European Remote Sensing Satellite (ERS-1), and TOPEX/Poseidon were recently launched for observation of physical variables related to the ocean. Moreover, Sea-Viewing Wide Field-of-View Sensor (SeaWiFS) and Advanced Earth Observation Satellite (ADEOS) will be launched in the near future. The combination of satellite SST data with other remotely sensed data on these satellites will become more useful than any single sensor data set in order to unlabel a complicated system of ocean dynamics and ocean-atmosphere interaction. For example, surface heat flux can be estimated by combining several satellite data including SST data. If effects of surface heat flux and advection on SST changes can be exactly evaluated by remote sensors, mechanisms of SST change will be easily understood. Also, assimilation of these types of data into numerical models is actively developed. Therefore, analyzing equatorial dynamics using numerical models and satellite SST combined with other observations will soon occur.

Acknowledgments

The author would like to thank Richard Reynolds for helpful comments on the manuscript and permission to use his figure. Ortis Brown is also thanked for his help in finding the image detecting the equatorial long wave. I am also grateful for the help of David Legler, Mark Luther, and Tommy Jensen. The NOAA/AVHRR data used for making Figures 7.2 and 7.4 were provided by the NASA Physical Oceanography Distributed Active Archive Center at the Jet Propulsion Laboratory/California Institute of Technology.

References

1. Walker, G.T., Correlation in seasonal variations of weather. IX. A further study of world weather, *Mem. Indian Meteorol. Dep.*, 24, 275, 1924.
2. Bjerknes, J., Atmospheric teleconnections from the equatorial Pacific, *Mon. Weather Rev.*, 97, 163, 1969.
3. Wyrtki, K., Sea level and seasonal fluctuations of the equatorial Pacific Ocean to atmospheric forcing, *J. Phys. Oceanogr.*, 5, 572, 1974.
4. Njoku, E.G., Barnett, T.P., Laurs, R.M., and Vastano, A.C., Advances in satellite sea surface temperature measurement and oceanographic applications, *J. Geophys. Res.*, 90, 11573, 1985.
5. Bernstein, R.L. and Chelton, D.B., Large-scale sea surface temperature variability from satellite and shipboard measurements, *J. Geophys. Res.*, 90, 11619, 1985.
6. Hilland, J.E., Chelton, D.B., and Njoku, E.G., Production of global sea surface temperature fields for the Jet Propulsion Laboratory Workshop comparisons, *J. Geophys. Res.*, 90, 11642, 1985.
7. McClain, E.P., Pichel, W.G., and Walton, C.C., Comparative performance of AVHRR-based multichannel sea surface temperatures, *J. Geophys. Res.*, 90, 11587, 1985.
8. Susskind, J. and Reuter, D., Sea surface temperature: observations from geostationary satellites, *J. Geophys. Res.*, 90, 11602, 1985.
9. Bates, J.J. and Smith, W.L., Sea surface temperature: observations from geostationary satellites, *J. Geophys. Res.*, 90, 11609, 1985.
10. Milman, A.S. and Wilheit, T.T., Sea surface temperature from the scanning multichannel microwave radiometer on Nimbus 7, *J. Geophys. Res.*, 90, 11631, 1985.
11. Bates, J.J. and Diaz, H.F., Evaluation of multichannel sea surface temperature product quality for climate monitoring: 1982–1988, *J. Geophys. Res.*, 90, 20613, 1991.
12. Woodruff, S.D., Slutz, R.J., Jenne, R.L., and Steurer, P.M., A Comprehensive Ocean-Atmosphere Data Sets, *Bull. Am. Meteorol. Soc.*, 68, 1239, 1987.
13. Reynolds, R.W., Folland, C.K., and Parher, D.E., Biases in satellite-derived sea-surface-temperature data, *Nature (London)*, 341, 728, 1989.
14. Reynolds, R.W., Impact of Mount Pinatubo aerosols on satellite-derived sea surface temperatures, *J. Climate*, 6, 768, 1993.
15. Reynolds, R.W. and Marsico, D.G., An improved real-time global sea surface temperature analysis, *J. Climate*, 6, 114, 1993.
16. Legeckis, R., Long waves in the eastern equatorial Pacific Ocean: a view from a geostationary satellite, *Science*, 197, 1179, 1977.
17. Legeckis, R., Pichel, W., and Nesterczuk, G., Equatorial long waves in geostationary satellite observations and in a multichannel sea surface temperature analysis, *Bull. Am. Meteorol. Soc.*, 64, 133, 1983.
18. Pullen, P.E., Bernstein, R.L., and Halpern, D., Equatorial long-wave characteristics determined from satellite sea surface temperature and *in situ* data, *J. Geophys. Res.*, 92, 742, 1987.
19. Legeckis, R., Upwelling off the Gulfs of Panama and Papagayo in the tropical Pacific during March 1985, *J. Geophys. Res.*, 93, 15485, 1988.
20. Legeckis, R. and Reverdin, G., Long waves in the equatorial Atlantic Ocean during 1983, *J. Geophys. Res.*, 92, 2835, 1987.
21. Barnett, T.P. and Patzert, W.C., Scales of thermal variability in the tropical Pacific, *J. Phys. Oceanogr.*, 10, 529, 1980.
22. Hansen, D.V. and Paul, C.A., Genesis and effects of long waves in the equatorial Pacific, *J. Geophys. Res.*, 89, 10431, 1984.
23. Wyrtki, K., Oscillations of dynamic topography in the eastern equatorial Pacific, *J. Phys. Oceanogr.*, 15, 1759, 1985.
24. Miller, L., Watts, D.R., and Wimbush, M., Oscillations of dynamic topography in the eastern equatorial Pacific, *J. Phys. Oceanogr.*, 15, 1759, 1985.

25. Malarde, J.-P., Mey, P.D., Périgaud, C., and Minster, J.-F., Observation of long equatorial waves in the Pacific Ocean by Seasat altimetry, *J. Phys. Oceanogr.,* 17, 2273, 1987.
26. McPhaden, M.J., Fieux, M., and Gonella, J., Meanders observed in surface currents and hydrography during an equatorial Atlantic transect, *Geophys. Res. Lett.,* 11, 757, 1984.
27. Weisberg, R.H., Instability waves observed on the equator in the Atlantic Ocean during 1983, *Geophys. Res. Lett.,* 11, 753, 1984.
28. Philander, S.G.H., Instabilities of zonal equatorial currents. I, *J. Geophys. Res.,* 81, 3725, 1976.
29. Philander, S.G.H., Instabilities of zonal equatorial currents. II, *J. Geophys. Res.,* 83, 3679, 1978.
30. Semtner, A.J. and Holland, W.R., Numerical simulation of equatorial ocean circulation. I. A basic case in turbulent equilibrium, *J. Phys. Oceanogr.,* 10, 667, 1980.
31. Cox, M.D., Generation and propagation of 30-day waves in a numerical model of the Pacific Ocean, *J. Phys. Oceanogr.,* 10, 1168, 1980.
32. Philander, S.G.H., Hurlin, W.J., and Pacanowski, R.C., Properties of long waves in models of the seasonal cycle in the tropical Atlantic and Pacific Oceans, *J. Geophys. Res.,* 91, 14207, 1986.
33. Philander, S.G.H., Halpern, D., Hansen, D., Legeckis, R., Miller, L., Paul, C., Watts, R., Wesberg, R., and Wimbush, M., Long waves in the equatorial Pacific Ocean, *Eos, Trans. Am. Geophys. Union,* 66, 154, 1985.
34. Crawford, W.R. and Osborn, T. R., Control of equatorial ocean currents by turbulent dissipation, *Science,* 212, 539, 1981.
35. Matsuno, T., Quasi-geostrophic motions in equatorial area, *J. Meteorol. Soc. Jpn.,* 2, 25, 1966.
36. Blandford, R., Mixed Rossby-gravity waves in the ocean, *Deep-Sea Res.,* 13, 941, 1966.
37. Gill, A.E., *Atmos.-Ocean Dynamics,* Academic Press, New York, 1982, chap. 11.
38. Tsai, P.T.H., O'Brien, J.J., and Luther, M.E., The 26-day oscillation observed in the satellite sea surface temperature measurements in the equatorial western Indian Ocean, *J. Geophys. Res.,* 97, 9605, 1992.
39. Stumpf, H.G. and Legeckis, R., Satellite observation mesoscale eddy dynamics in the eastern tropical Pacific Ocean, *J. Phys. Oceanogr.,* 7, 648, 1977.
40. Luyten, J.R. and Roemmich, D.H., Equatorial currents at semiannual period in the Indian Ocean, *J. Phys. Oceanogr.,* 12, 406, 1982.
41. Reverdin, G. and Luyten, J., Near-surface meanders in the equatorial Indian Ocean, *J. Phys. Oceanogr.,* 16, 1088, 1986.
42. Kindle, J.C. and Thompson, J.D., The 26- and 50-day oscillations in the western Indian Ocean: model results, *J. Geophys. Res.,* 94, 721, 1989.
43. Woodberry, K.E., Luther, M.E., and O'Brien, J.J., The wind-driven seasonal circulation in the southern tropical Indian Ocean, *J. Geophys. Res.,* 94, 17985, 1989.
44. Moore, D.W. and McCreary, J.P., Excitation of intermediate-frequency equatorial waves at a western ocean boundary: with application to observations from the Indian Ocean, *J. Geophys. Res.,* 95, 5219, 1990.
45. Kelly, B.G.J., On the generation and dispersion of Yanai waves, Technical Report, Florida State University, Tallahassee, FL, 1993.
46. Stumpf, H.G., Satellite detection of upwelling in the Gulf of Tehuantepec, Mexico, *J. Phys. Oceanogr.,* 5, 383, 1975.
47. Clarke, A.J., Inertial wind path and sea surface temperature patterns near the Gulf of Tehuantepec and Gulf of Papagayo, *J. Geophys. Res.,* 93, 15491, 1988.
48. Hofmann, E.E., Busalacchi, A.J., and O'Brien, J.J., Wind generation of the Costa Rica Dome, *Science,* 214, 552, 1981.
49. Umatani, S. and Yamagata, R., Response of the eastern tropical Pacific to meridional migration of the ITCZ: the generation of the Costa Rica Dome, *J. Phys. Oceanogr.,* 21, 346, 1991.
50. Kubota, M. and O'Brien, J.J., Seasonal variation in the upper layer thickness of the tropical Pacific Ocean model, *J. Oceanogr.,* 48, 59, 1992.

51. Legeckis, R., A satellite time series of sea surface temperatures in the eastern equatorial Pacific Ocean, 1982–1986, *J. Geophys. Res.*, 91, 12879, 1986.
52. Brown, O.B., Bruce, J.G., and Evans, R., Evolution of sea surface temperature in the Somali basin during the southwestern monsoon of 1979, *Science*, 209, 595, 1980.
53. Legeckis, R., Satellite observations of a western boundary current in the Bay of Bengal, *J. Geophys. Res.*, 92, 12974, 1987.
54. Busalacchi, A.J. and O'Brien, J.J., The seasonal variability in a model of the tropical Pacific, *J. Phys. Oceanogr.*, 10, 1929, 1980.
55. Busalacchi, A.J., Takeuchi, K., and O'Brien, J.J., Interannual variability of the equatorial Pacific-revisited, *J. Geophys. Res.*, 88, 7551, 1983.

8

Observations of Tropical Sea Level Variability from Altimeters

Gary T. Mitchum and Bernard J. Kilonsky

8.1 Introduction: The Special Value of Sea Level in the Tropics

Before considering the use of altimeters for studying sea surface height variations we should examine the reasons why sea level data, primarily from tide gauges, have long been recognized as an important source of information about oceanic variations.[1] Basically, the usefulness of these data follows from the fact that sea level is a direct measurement of the surface pressure variations of the ocean and that use of the geostrophic approximation allows horizontal sea level differences to be interpreted as surface current transport across the line connecting the two sea level stations. While these facts are true anywhere in the ocean, it can be argued that sea level is of particular interest in the tropical portions of the world oceans. Mitchum and Wyrtki[2] have provided a discussion along these lines for the tropical Pacific and only a brief summary is given here.

0-8493-4525-1/95/$0.00+$.50

First, in the tropics there is a rich variety of phenomena that have a sea level expression. For example, the equator serves as a waveguide supporting energy and phase propagation in both directions. These equatorial waves have clear sea level signatures and have been documented over a range of time and length scales; for example, Wunsch and Gill[3] described high-frequency inertial-gravity waves, Lukas et al.[4] described Kelvin and Rossby waves forced by bursts of westerly wind in the western Pacific, and Enfield[5] explained coastal signals in the eastern Pacific in terms of similarly forced Kelvin waves. Mitchum and Lukas[6] plotted sea level energy in the tropical Pacific as a function of frequency and latitude and identified signals due to all these phenomena and more. In addition to the energy in the equatorial waveguide there are well-known, large-scale signals associated with El Niño/Southern Oscillation (ENSO) events. In fact, the understanding of these events progressed hand in hand with the analysis and collection of sea level observations from the tropical Pacific by Wyrtki.[7-10]

Second, as an example of the indirect use of sea level, variations of the current structure of the tropical Pacific have also been described from sea level data by Wyrtki.[11,12] Also, due to the relatively simple vertical density structure found in the tropics, it is possible to use sea level variations as indirect measures of the variations of the entire warm upper layer of the tropical ocean. Basically, the sharp thermocline found in the tropics allows the use of a 1.5 layer approximation for the vertical structure, which makes sea level variations proportional to variations in the depth of the warm upper layer.[13] Wyrtki[14] has exploited this relationship to form estimates of the variation of the volume of warm water in the tropical Pacific using horizontal integral of the sea level variation. In a similar fashion Delcroix and Gautier[15] have added estimates of the temperature of the upper layer to estimate variations of the heat content of the tropical Pacific.

The purpose of this chapter is to discuss how satellite altimetric measurements of the height of the sea surface have extended, or promise to extend, the results from studies such as those cited above. The organization is as follows. We begin by describing the extensive efforts aimed at verifying altimetry data against *in situ* data and models. This discussion includes studies that reproduce previously known phenomena, as well as those that directly compare altimetry data to ground truth data. This section of the chapter concludes with a brief discussion of how various errors in the altimetry data do, or do not, limit its usefulness in the tropics. We then proceed to a discussion of a number of new scientific results from the analysis of altimetry data.

8.2 Validation of Altimetry Data in the Tropics

With a new observing system it is natural to evaluate it first by intercomparing its data with analogous data from an already accepted source. In the case of satellite altimetry the obvious sources of ground truth are sea level data from tide gauges. Other sources would be dynamic heights calculated from temperature or density data or simulations from numerical models that have themselves been checked against existing data. One could also check currents computed from the altimeter heights against *in situ* velocity measurements. In fact, all of these sorts of comparisons have been done.

The direct intercomparison of altimetric heights with *in situ* data has only been the first step in evaluating the usefulness of this relatively new measurement system. The next, and very logical, step has been to use the altimetric data to reproduce known signals. This is, of course, a much better test of the altimeter data in the mode that it must actually be used to study oceanic variability. We should repeat that it is not clear where this sort of validation ends and new science begins, and it is probable that other researchers would separate things differently.

Intercomparison with *in situ* Data and Models

Many of the direct intercomparisons of altimetry heights with *in situ* data have been done in the tropical Pacific. This is partly due to the fact that many researchers work in this region, but is probably mostly due to the greater number of island tide gauges in this area and the easy access to these data. Miller et al.[16] made an intercomparison of Seasat data with island tide gauges and concluded that the two data sources agreed to 5 cm on timescales longer than 1 month, but unfortunately there were only 3 months of data. Tai et al.[17] compared the first 17 months of Geosat data against island sea level and steric height data. They found that the data sets basically agreed, but it was necessary to use fairly heavy averaging in order to show this. In a separate study Cheney et al.[18] compared the Geosat data to island sea level and heights from a numerical model and concluded that the data sets agreed to about 4 cm for timescales longer than 1 month. Other intercomparisons have also shown good correlation between the altimetric heights and *in situ* data, albeit with somewhat higher estimates of the root mean square (rms) difference between the data sets.[19,20]

A simple example of an intercomparison of 2 years of height variability from Geosat with sea level variations observed at five stations in the tropical Pacific is shown in Figure 8.1. The Geosat time series were formed by simply taking data from the nearest ascending and descending arcs; no additional spatial averaging was done. The time series have been temporally smoothed to emphasize the agreement at periods greater than a month or so, and the correlation values are shown along with the rms difference. The means of the two time series are removed before the comparison is made. The correlations are good, as noted above, but the rms differences vary from 4 to nearly 10 cm. We note again that no spatial averaging has been used and also that the stations along the equator tend to agree better than the off-equatorial ones.

In a slightly different type of intercomparison Wyrtki and Mitchum[21] described the differences between the Geosat altimeter heights and the island sea levels that were spatially and temporally coherent. They concluded that the observations demonstrated the necessity for sea surface height data to be referred to a well-defined zero point, a datum, in order to reliably monitor interannual changes in the

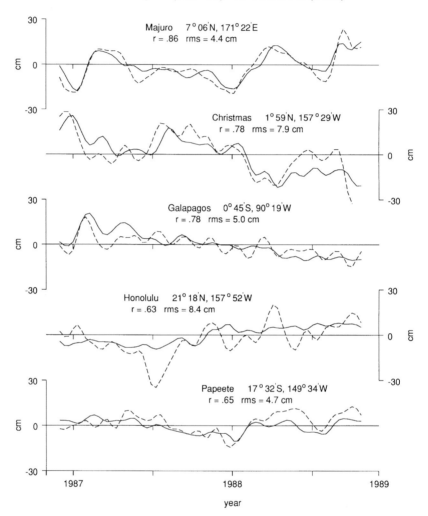

Figure 8.1   Intercomparison of time series of sea level from tide gauges and sea surface height from the Geosat altimeter. The sea level time series are obtained by low-pass filtering hourly observations. The Geosat time series are obtained from applying a low-pass filter with similar characteristics to the set of points obtained from an interpolation of the nearest ascending and descending passes of the altimeter to daily values. The low-pass filter has a half amplitude point of 45 days. The means of each series are removed. Units are in centimeters.

sea surface topography. In yet another approach Picaut et al.[22] computed the variability of geostrophic currents from the Geosat heights, compared these to direct measurements in the equatorial eastern and central Pacific, and found that the computed and observed current variations correlated well, but with an rms difference of order 20 cm/s. One advantage of this approach is that it provides a much more stringent test for the altimeter due to the necessity of differentiating the sea surface heights.

To this point all of the intercomparisons discussed have been in the Pacific. Comparisons done on a global scale[23] find that the tropical correlations are generally good and rms differences are about 6 cm. In the Atlantic, Arnault and collaborators[24,25] used steric height data, which provide better coverage than sea level from the sparse set of Atlantic islands, and have obtained results similar to those cited above for the Pacific. In the Indian Ocean relatively little direct intercomparison has been done, probably because of the paucity of *in situ* data. Comparisons between Seasat data and steric heights have been done, however, with the heights from altimetry crossover; and those estimated from the expendable bathythermograph (XBT) sections have 7-cm rms differences.[26] In recent years more Indian Ocean sea level data have become available and will be useful for the evaluation of the TOPEX/Poseidon and ERS-1 altimeters.

We end this section by considering comparisons of altimetric heights to the output from numerical models. One advantage of doing these sorts of comparisons is that the points of comparison are not limited by the availability of islands or hydrographic data. Typically models do reasonably well at reproducing tropical sea level changes at seasonal and longer timescales, which makes these intercomparisons particularly easy to justify. Only a few examples will be given, however. In the tropical Pacific a recent article by Chao et al.[27] compare Geosat altimeter heights, island sea level, and heights from an ocean model driven by monthly mean wind observations. It was found that all three data sets correlate well, 0.69 for Geosat and island sea level and 0.71 for the ocean model; and the rms differences are similar to those found in the comparisons to *in situ* data discussed above. In the Atlantic, Didden and Schott[28] found consistent spatial pattern agreement between seasonal circulation changes in the North Equatorial Countercurrent and the North Brazil Current computed from Geosat and those taken from the WOCE numerical model with a rms difference between Geosat and model velocities in the North Equatorial Countercurrent (NECC) core of 7/cm s^{-1}. In the Indian Ocean, Perigaud and Delecluse[29] also found good agreement for seasonal variations observed by Geosat and those computed from a reduced-gravity model driven by monthly mean winds, with correlations between the observed and simulated amplitudes of 0.79 in time and 0.58 in space. The observed and simulated phases had correlations of 0.79 and 0.78, respectively.

Reproducing Known Phenomena

We turn now to the second sort of validation mentioned above, which involves evaluating the ability of the altimetric data to recover known features of the oceanic variability. An example of this type of validation concerns Kelvin waves in the equatorial Pacific that are forced by westerly wind events in the western Pacific. Lukas et al.,[30] using *in situ* sea level and wind data, described these waves and identified possible Rossby waves generated by reflection at the Pacific eastern boundary. The link to westerly winds was clear, and the possible importance of these waves to the ENSO phenomenon was discussed. Using Geosat data, Miller et al.[31] provided an excellent description of a Kelvin wave forced by a westerly wind event in May 1986 preceding the ENSO of 1986–1987. This wave was also observed by *in situ* data,[32] but the ability of Geosat to observe and describe this event greatly increased confidence in the altimetric data.

Attempts to reproduce known ocean features and variability from altimetry have been made simultaneously with the intercomparisons to *in situ* data. Even the rather limited Seasat and GEOS-3 data sets provide examples of this type of application. Three examples from the tropics will be given. First, Menard[33] composited both data sets in the Atlantic to obtain a description of seasonal variations between 10° S and 20° N that was comparable to that measured by dynamic heights computed from the historical hydrographic data and demonstrated that altimetry was able to detect a low-frequency signal with amplitude of the order of 10 cm. Second, Perigaud et al.[34] used Seasat data to describe the variations of the zonal slope of sea level across the Indian Ocean. Finally, Malarde et al.[35] and Musman[36] succeeded in recovering sea level signals in the eastern Pacific that are consistent with Legeckis waves, with periods of about 25 days and wavelengths of about 1000 km.

Geosat has been even more successful at reproducing known features of the tropical variability, largely because of its longer time series and consistent sampling during the Exact Repeat Mission. For example, Delcroix et al.[37] used the time series from Geosat to detail Kelvin and Rossby wave variations in the equatorial Pacific. They were able to recover propagation speeds and also to detail Rossby wave reflection characteristics in the eastern Pacific. Chao et al.[38] showed that the Geosat data in the tropical Pacific was able to capture variability consistent with seasonal changes, the basin-wide ENSO of 1986–1987 (see Figure 8.2), as well as the May 1986 Kelvin wave event discussed earlier. In the tropical Atlantic, Arnault et al.[39] found interannual variability during 1986 and 1987 that agreed with the observed dynamic topography to within 4 cm rms and confirmed earlier hypotheses concerning interannual changes in the strength of the 1987 Gulf of Guinea upwelling. As a final example we note that Jacobs et al.,[40] in a global description of the annual changes observed with Geosat, recovered signals in the tropics consistent with expectations, (e.g., strong variability in the Intertropical Convergence Zone (ITCZ), phase reversal between the hemispheres, westward propagating waves, and phase shifts between the major zonal currents).

Outstanding Questions and Problems

In the preceding sections we have provided examples of intercomparisons of altimetric heights and *in situ* data and models that demonstrate good correlations and reasonable, but not negligible, rms differences between the data sets. We have also given examples of tropical variability that are successfully recovered using altimetric heights, which are even more effective in convincing us of the practicality

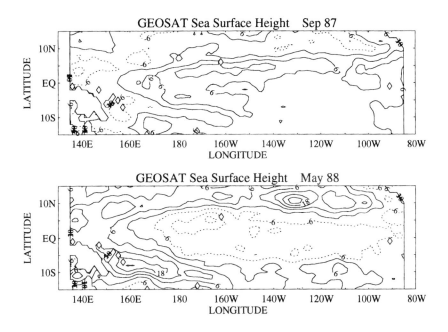

Figure 8.2 Maps of sea surface height anomalies derived from the Geosat altimeter data during peak phases of (top) the 1987 El Niño and (bottom) the 1988 El Niña. The contour interval is 6 cm. (From Chao, Y. et al., *J. Geophys. Res.*, 98, 6947, 1993. With permission of the American Geophysical Union.)

of using altimeters as an important source of information about variations in the height of the tropical ocean. However, the question naturally arises as to whether all the problems involved in using the altimetry data are solved. A good discussion of these issues in the context of planning future altimetry missions has recently been given by Koblinsky et al.[41]

There are reasons to believe that several types of error that are potentially quite serious are actually not as critical for the use of altimetry data in studies of tropical variability. For example, orbit error, which is discussed elsewhere in this volume, has been a continuing concern. We, however, do not consider this to be critical now. First, the computed orbits for Geosat have improved greatly in the past several years and techniques for removing orbit error at the one cycle per orbit period, which generally use some version of the sinusoidal fitting method devised by Tai,[42] work well at reducing the orbit error even further. Second, the orbits being computed for TOPEX/Posiden are accurate to better than 5 cm. This standard should be achievable by future altimetric missions as well.

With the advent of more accurate orbits, those errors associated with tide models are receiving more attention. Although models that use altimetric data[43] and other methods[44] have greatly improved the data set, our comparison of the first 300 days of the TOPEX sea surface heights and the *in situ* tide gauge sea level has shown that the removal of a 60-day harmonic, which corrects for aliasing from both the M_2 and

S_2 tidal components, reduces the total variance of the differences between the sea surface heights and sea level by 35%. This implies that the tidal model errors are probably the largest current source of uncertainty.

Other errors, such as those associated with ionospheric range delay, atmospheric pressure variations, and sea state, are also not as worrisome as in the past. In the case of the ionospheric range delay, dual-frequency measurements have basically removed the problem. It should be noted that future missions can be assured of avoiding this error only by using dual-frequency altimeters. Atmospheric pressure is not a problem simply because its variability in the tropics is small.[45]

The error associated with uncertainty in our knowledge of the atmospheric water vapor changes is still a point of worry for Geosat and for any future altimetry missions that do not carry radiometers. In these cases water vapor must be derived indirectly, often from atmospheric models, and significant errors arise.[46,47] It has been argued that studies of mesoscale variability do not suffer from these errors,[48-50] but even this is not obviously true if one considers the annual modulations of the mesoscale energy.[51] It is often assumed that water vapor errors are reduced by low-pass temporal and spatial smoothing, but this is not clearly the case. The study of Wyrtki and Mitchum[52] referred to earlier is relevant on this point. These authors leveled the Geosat altimeter heights and the tide gauge sea levels by forcing the mean difference over the first year of the exact repeat mission to be zero. The mean difference over the second year was then examined and used to identify low-frequency drift between the two data sets. The differences are large scale in nature and are also consistent with the water vapor errors associated with the 1986–1987 ENSO event, as inferred by Zimbelman and Busalacchi.[53] It has been argued, however, that these errors are primarily due to unresolved orbit errors.[54] We consider this to be an issue needing more attention.

8.3 New Scientific Results from Altimetry

In the preceding sections we have reviewed a great deal of work establishing altimetry as a reliable source of information about sea surface height variations. While this work is crucial, the goal is not to validate the data set or to reproduce known phenomena, but to extend our knowledge and understanding of tropical sea level variability. It is only natural that validation has received most of the attention to this point, but the success of Geosat and the launch of TOPEX/Poseidon will mark the beginning of a time focused on developing new scientific results. In this section we will review a few examples of new scientific results that have already been obtained by exploiting the unprecedented spatial resolution and coverage afforded by the satellite measurements as compared to the existing *in situ* networks.

In the Pacific, Perigaud[55] has given a detailed analysis of shear waves occurring between the North Equatorial Countercurrent and the North Equatorial Current primarily in the eastern portion of the basin. She has exploited the spatial resolution of the altimetric data to describe waves of 50–90 day period and 630–950 km wavelength (Figure 8.3) that modulate temporally in conjunction with the strength of the shear between the two currents. In the eastern Pacific, Hansen and Maul[56] have identified an eddy field off the coast of Central America and presented evidence that these eddies are generated by the North Equatorial Countercurrent as it turns northward upon encountering the boundary. In the waveguide, du Penhoat et al.[57] have done a thorough analysis of Kelvin and Rossby waves and have addressed a number of interesting issues regarding the reflection processes at the eastern Pacific boundary. They find that wind stress near the boundary plays an important role in determining the strength of the reflected Rossby waves generated by upwelling vs. downwelling Kelvin waves. In the western Pacific, White et al.[58] have used the spatial resolution available from Geosat to detail the reflection of annual Rossby waves from the western Pacific boundary, which they argue is important to the understanding of the ENSO phenomenon. Finally, on a larger scale, Miller and Cheney[59] have expanded on the work of Wyrtki,[60] who argued that upper layer volume fluctuations of the tropical Pacific represented a loss of mass to higher latitudes. They present evidence that during ENSO cycles there is also a significant recirculation of mass within the tropics.

Outside of the Pacific, Carton[61] used Geosat data to add significant detail to the description of the seasonal variation of the Atlantic North Equatorial Countercurrent and North Equatorial Current. This work was extended by Carton and Katz[62] in a study describing the origin of the North Equatorial Countercurrent in the North Brazil Current retroflection and its weakening as it moves eastward due to recirculation toward the south. Also in the Atlantic, Musman[63] provides evidence for the existence of near equatorial shear waves with periods of about 25 days. It is unlikely that this work in the Atlantic would have been possible without Geosat because of the lack of island sea level stations and the scarcity of hydrographic data. In the Indian Ocean another interesting application of Geosat data is described by Harangozo.[64] In his work a coastal flooding event in the Maldives is explained as being due to destruction of sea defenses by surface gravity waves generated by wind in the southern Indian Ocean and not symptomatic of sea level rise through global warming. This conclusion would have been difficult to reach without the spatial coverage provided by the satellite measurements.

8.4 The Future

It has been established that altimetric sea surface heights in the tropics correlate well with *in situ* data with typical differences at a given point in the range of 4–10 cm, depending on the amount of temporal and spatial smoothing done. It is also clear that the data from altimeters can reproduce a variety of phenomena in the tropical oceans and that new scientific results can be obtained by exploiting the excellent spatial resolution and coverage of these data.

Of course these conclusions have been reached for altimetry as defined by Geosat, and there is every reason to believe that the situation will improve even more with TOPEX/Poseidon and other future altimetry missions. One area in

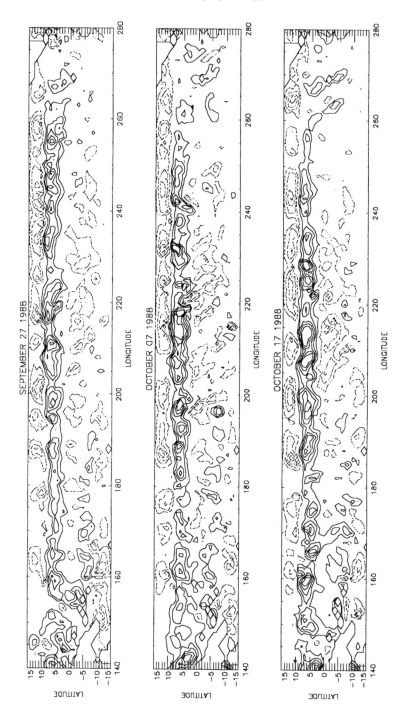

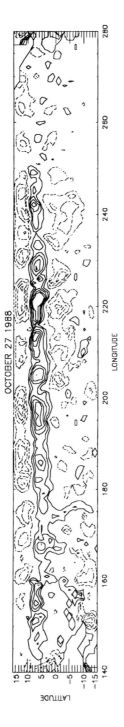

Figure 8.3 Maps of sea surface height anomalies derived from the Geosat altimeter data (anomalies are residuals relative to the 24-month mean). Isocontours are 5 cm with solid lines for positive values and dotted lines for negative. The zero line is not plotted for clarity. (From Perigaud, C., *J. Geophys. Res.*, 95, 7239, 1990. With permission of the American Geophysical Union.)

which we hope to see significant improvement is in the study of the large-scale, low-frequency variability of the ocean. Unfortunately, many of the errors, particularly those associated with the orbit determination, water vapor corrections, and tide corrections, conspire against this type of research. Also, the length of the time series available for study are short. Despite these problems there is reason to believe that progress can be made in this area. Early on Tai[65,66] suggested that large-scale structure could be studied by using spherical harmonic expansions of the altimetry data and demonstrated some success in this area. Also, Haines et al.[67] have attempted an estimate of the decadal variation of the ocean by differencing the Geosat and the Seasat data sets. This is an interesting calculation, although the difference is found to be dominated in the tropics by the signals due to the 1986–1987 ENSO event; for example, the difference is dominated by an aliased signal and does not actually reflect decadal scale variability.

In order to accurately study the large-scale, low-frequency changes in the ocean from altimetry it will be necessary to use the altimetry data and the *in situ* data simultaneously, as suggested by Wyrtki and Mitchum;[68] and progress is being made along these lines. Wunsch[69] has tried a technique that expands the altimetry data using spherical harmonics with the tide gauge data included as well. Unfortunately the error estimates used for the *in situ* data were somewhat large, and the analysis was only weakly constrained by the addition of the *in situ* data. Preliminary results from another approach, based on empirical orthogonal function expansions of the month-to-month changes in the large-scale height fields, were described by Mitchum et al.[70] These calculations, however, were done using the TOPEX/Poseidon interim data set and were contaminated by tide errors. A distinct but related method has been used by Miller et al.[71] to obtain blended maps of the height field in the western tropical Pacific. At this time it appears that all of these approaches are limited to regions having a relatively dense set of *in situ* observations, and other methods will have to be devised and tested.

Finally, we note that altimetric data, as well as *in situ* data, are beginning to be used for assimilation into ocean numerical models. It may well be that this is ultimately the best way to blend the data sets into a single description of the sea surface variations. A general review of these efforts has been given elsewhere,[72] and we note only that a few efforts specifically aimed at the tropical variability have already appeared.[73-75]

References

1. Chelton, D. and Enfield, D., Ocean signals in tide gauge records, *J. Geophys. Res.*, 91, 9081, 1986.
2. Mitchum, G. T. and Wyrtki, K., Overview of Pacific sea level variability, *Mar. Geodesy*, 12, 235, 1988.
3. Wunsch, C. and Gill, A., Observations of equatorially trapped waves in Pacific sea level variations, *Deep-Sea Res.*, 23, 371, 1976.
4. Lukas, R., Hayes, S., and Wyrtki, K., Equatorial sea level response during 1982–83 El Niño, *J. Geophys. Res.*, 89, 10425, 1984.

5. Enfield, D., The intraseasonal oscillation in eastern Pacific sea levels: how is it forced?, *J. Phys. Oceanogr.*, 10, 557, 1987.
6. Mitchum, G. and Lukas, R., The latitude-frequency structure of Pacific sea level variance, *J. Phys. Oceanogr.*, 17, 2362, 1987.
7. Wyrtki, K., Sea level during the 1972 El Niño, *J. Phys. Oceanogr.*, 7, 779, 1977.
8. Wyrtki, K., The response of sea surface topography to the 1976 El Nino, *J. Phys. Oceanogr.*, 9, 1223, 1979.
9. Wyrtki, K., The slope of sea level along the equator during the 1982/83 El Niño, *J. Geophys. Res.*, 89, 10419, 1984.
10. Wyrtki, K., Sea level fluctuations in the Pacific during the 1982/83 El Niño, *Geophys. Res. Lett.*, 12, 125, 1985.
11. Wyrtki, K., Sea level and the seasonal fluctuations of the equatorial currents in the western Pacific Ocean, *J. Phys. Oceanogr.*, 4, 91, 1974.
12. Wyrtki, K., Equatorial currents in the Pacific 1950 to 1970 and their relations to the trade winds, *J. Phys. Oceanogr.*, 4, 372, 1974.
13. Rebert, J., Donguy, R. J., Eldin, G., and Wyrtki K., Relations between sea level, thermocline depth, heat content and dynamic height in the tropical Pacific Ocean, *J. Geophys. Res.*, 90, 11719, 1985.
14. Wyrtki, K., Water displacements in the Pacific and the genesis of El Niño cycles, *J. Geophys. Res.*, 90, 7129, 1985.
15. Delcroix, T. and Gautier, C., Estimates of heat content variations from sea level measurements in the central and western tropical Pacific from 1979 to 1985, *J. Phys. Oceanogr.*, 17, 725, 1987.
16. Miller, L., Cheney, R., and Milbert, D., Sea level time series in the equatorial Pacific from satellite altimetry, *Geophys. Res. Lett.*, 13, 475, 1986
17. Tai, C.-K., White, W. B., and Pazan, S. E., GEOSAT crossover analysis in the tropical Pacific. II. Verification analysis of altimetric sea level maps with expendable bathythermograph and island sea level data, *J. Geophys. Res.*, 94, 897, 1989
18. Cheney, R. E., Douglas, B. C., and Miller, L., Evaluation of GEOSAT altimeter data with application to tropical Pacific sea level variability, *J. Geophys. Res.*, 94, 4737, 1989
19. Shibata, A. and Kitamura, Y., GEOSAT sea level variability in the tropical Pacific in the period from November 1986 to February 1989, obtained by collinear method, *Oceanogr. Mag.*, 40, 1, 1990.
20. Maul, G. A., Hanson, D. V., and Bravo, N. J., A note on sea level variability at Clipperton island from GEOSAT and *in situ* observations, sea level changes: determination and effects, *Geophys. Monogr. 69 (IUGG)*, 11, 145, 1992.
21. Wyrtki, K. and Mitchum, G., Interannual differences of Geosat altimeter heights and sea level: the importance of a datum, *J. Geophys. Res.*, 95, 2969, 1990.
22. Picaut, J., Busalacchi, A. J., McPhaden, M. J., and Camusat, B., Validation of the geostrophic method for estimating zonal currents at the equator from GEOSAT altimeter data, *J. Geophys. Res.*, 95, 3015, 1990.
23. Koblinsky, C. J., Nerem, R. S., Williamson, R. G., and Klosko, S. M., Global scale variations in sea surface topography determined from satellite altimetry, sea level changes: determination and effects, *Geophys. Monogr. 69 (IUGG)*, 11, 155, 1992.
24. Arnault, S., Menard, Y., and Merle, J., Observing the tropical Atlantic Ocean in 1986–1987 from altimetry, *J. Geophys. Res.*, 95, 921, 1990.
25. Arnault, S., Gourdeau, L., and Menard, Y., Comparison of the altimetric signal with *in situ* measurements in the tropical Atlantic Ocean, *Deep-Sea Res.*, 39, 481, 1992.
26. Perigaud, C. and Minster, J.-F., Variability of the Somali Current as observed from SEASAT altimetry, *J. Phys. Oceanogr.*, 18, 25, 1988.
27. Chao, Y., Halpern, D., and Perigaud, C., Sea surface height variability during 1986–1988 in the tropical Pacific Ocean, *J. Geophys. Res.*, 98, 6947, 1993.

28. Didden, N. and Schott, F., Seasonal variations in the western tropical Atlantic: surface circulation from GEOSAT altimetry and WOCE model results, *J. Geophys. Res.*, 97, 3529, 1992.

29. Perigaud, C. and Delecluse, P., Annual sea level variations in the southern tropical Indian Ocean from GEOSAT and shallow-water simulation, *J. Geophys. Res.*, 97, 20169, 1992.

30. Lukas, R., Hayes, S., and Wyrtki, K., Equatorial sea level response during 1982–83 El Niño, *J. Geophys. Res.*, 89, 10425, 1984.

31. Miller, L., Cheney, R. E., and Douglas, B. C., GEOSAT altimeter observations of Kelvin waves and the 1986–87 El Niño, *Science*, 239, 52, 1988.

32. McPhaden, M. J., Freitag, H. P., Hayes, S. P., Taft, B. A., Chen, Z., and Wyrtki, K., The response of the equatorial Pacific Ocean to a westerly wind burst in May 1986, *J. Geophys. Res.*, 93, 10589, 1988.

33. Menard, Y., Observing the seasonal variability in the tropical Atlantic from altimetry, *J. Geophys. Res.*, 93, 13967, 1988.

34. Perigaud, C., Minster, J.-F., and Reverdin, G., Zonal slope variability of the tropical Indian Ocean studied from SEASAT altimetry, *Mar. Geodesy*, 10, 53, 1986.

35. Malarde, J.-P., de Mey, P., Perigaud, C., and Minster, J.-F., Observation of long equatorial waves in the Pacific Ocean by SEASAT altimetry, *J. Phys. Oceanogr.*, 17, 2273, 1987.

36. Musman, S., Sea height wave form in equatorial waves and its interpretation, *J. Geophys. Res.*, 94, 3303, 1989.

37. Delcroix, T., Picaut, J., and Eldin, G., Equatorial Kelvin and Rossby waves evidenced in the Pacific Ocean through GEOSAT sea level and surface current anomalies, *J. Geophys. Res.*, 96, 3249, 1991.

38. Chao, Y., Halpern, D., and Perigaud, C., Sea surface height variability during 1986–1988 in the tropical Pacific Ocean, *J. Geophys. Res.*, 98, 6947, 1993.

39. Arnault, S., Menard, Y., and Merle, J., Observing the tropical Atlantic Ocean in 1986–1987 from altimetry, *J. Geophys. Res.*, 95, 17921, 1990.

40. Jacobs, G. A., Born, G. H., Parke, M. E., and Allen, P. C., The global structure of the annual and semiannual sea surface height variability from GEOSAT altimeter data, *J. Geophys. Res.*, 97, 17813, 1992.

41. Koblinsky, C. J., Gaspar, P., and Lagerloef, G., The future of spaceborne altimetry: oceans and climate change, Joint Oceanographic Institutions, Inc., Washington, D.C., 1992, 75 pp.

42. Tai, C.-K., GEOSAT crossover analysis in the tropical Pacific. I. Constrained sinusoidal crossover adjustment, *J. Geophys. Res.*, 93, 10621, 1988.

43. Cartwright, D. E. and Ray, R. D., Oceanic tides from GEOSAT altimetry, *J. Geophys. Res.*, 95, 3069, 1990.

44. Le Provost, C., Lyard, F., and Molines, J.-M., Improving ocean tide predictions by using additional semidiurnal constituents from spline interpolation in the frequency domain, *Geophys. Res. Lett.*, 18, 845, 1991.

45. Luther, D. S., Evidence of a 4–6 day barotropic, planetary oscillation of the Pacific Ocean, *J. Phys. Oceanogr.*, 12, 644, 1982.

46. Tapley, B. D., Lundberg, J. B., and Born, G. H., The SEASAT altimeter wet tropospheric range correction, *J. Geophys. Res.*, 87, 3213, 1982.

47. Emery, W. J., Born, G. H., Baldwin, D. G., and Norris, C. L., Satellite-derived water vapor corrections for Geosat altimetry, *J. Geophys. Res.*, 95, 2953, 1990.

48. Fu, L.-L., On the wave number spectrum of oceanic mesoscale variability observed with the SEASAT altimeter, *J. Geophys. Res.*, 88, 4331, 1983.

49. Monaldo, F., Path length variations caused by atmospheric water vapor and their effects on the measurement of mesoscale ocean circulation features by a radar altimeter, *J. Geophys. Res.*, 95, 2923, 1990.

50. Jourdan, D., Boissier, C., Braun, A., and Minster, J.-F., Influence of wet tropospheric correction on mesoscale dynamic topography as derived from satellite altimetry, *J. Geophys. Res.*, 95, 17993, 1990.

51. Zlotnicki, V., Fu, L.-L., and Patzert, W., Seasonal variability in global sea level observed with GEOSAT altimetry, *J. Geophys. Res.*, 94, 17959, 1989.

52. Wyrtki, K. and Mitchum, G., Interannual differences of Geosat altimeter heights and sea level: the importance of a datum, *J. Geophys. Res.*, 95, 2969, 1990.

53. Zimbelman, D. F. and Busalacchi, A. J., The wet tropospheric range correction: product intercomparisons and the simulated effect for tropical Pacific altimeter retrievals, *J. Geophys. Res.*, 95, 2899, 1990.

54. Cheney, R. E. and Miller, L., Recovery of the sea level signal in the western tropical Pacific from GEOSAT altimetry, *J. Geophys. Res.*, 95, 2977, 1990.

55. Perigaud, C., Sea level oscillations observed with GEOSAT along the two shear fronts of the Pacific North Equatorial Countercurrent, *J. Geophys. Res.*, 95, 7239, 1990.

56. Hansen, D. V. and Maul, G. A., Anticyclonic current rings in the eastern tropical Pacific Ocean, *J. Geophys. Res.*, 96, 6965, 1991.

57. du Penhoat, Y., Delcroix, T., and Picaut, J., Interpretation of Kelvin/Rossby waves in the equatorial Pacific from model-GEOSAT data intercomparison during the 1986–1987 El Niño, *Oceanol. Acta*, 15, 545, 1992.

58. White, W. B., Graham, N., and Tai, C.-K., Reflection of annual Rossby waves at the maritime western boundary of the tropical Pacific, *J. Geophys. Res.*, 95, 3101, 1990.

59. Miller, L. and Cheney, R., Large-scale meridional transport in the tropical Pacific Ocean during the 1986–1987 El Niño from GEOSAT, *J. Geophys. Res.*, 95, 17905, 1990.

60. Wyrtki, K., Water displacements in the Pacific and the genesis of El Niño cycles, *J. Geophys. Res.*, 90, 7129, 1985.

61. Carton, J. A., Estimates of sea level in the tropical Atlantic Ocean using GEOSAT altimetry, *J. Geophys. Res.*, 94, 8029, 1989.

62. Carton, J. A. and Katz, E. J., Estimates of the zonal slope and seasonal transport of the Atlantic North Equatorial Countercurrent, *J. Geophys. Res.*, 95, 3091, 1990.

63. Musman, S., GEOSAT altimeter observations of long waves in the equatorial Atlantic, *J. Geophys. Res.*, 97, 3573, 1992.

64. Harangozo, S. A., Flooding in the Maldives and its implications for the global sea level rise debate, sea level changes: determination and effects, *Geophys. Monogr. 69 (IUGG)*, 11, 95, 1992.

65. Tai, C.-K., Determining the large-scale ocean circulation from satellite altimetry, *J. Geophys. Res.*, 88, 9553, 1983.

66. Tai, C.-K., Estimating the basin-scale ocean circulation from satellite altimetry. I. Straightforward spherical harmonic expansion, *J. Phys. Oceanogr.*, 18, 1398, 1988.

67. Haines, B., Born, G., and Koblinsky, C., An estimate of decadal changes in sea surface topography from SEASAT and GEOSAT altimetry, sea level changes: determination and effects, *Geophys. Monogr. 69 (IUGG)*, 11, 181, 1992.

68. Wyrtki, K. and Mitchum, G., Interannual differences of Geosat altimeter heights and sea level: the importance of a datum, *J. Geophys. Res.*, 95, 2969, 1990.

69. Wunsch, C., Global-scale sea surface variability from combined altimetric and tide gauge measurements, *J. Geophys. Res.*, 96, 15053, 1991.

70. Mitchum, G., Busalacchi, A., Lukas, R., and Wyrtki, K., Using tide gauge data for verification and data blending: initial results, *Trans. Am. Geophys. U. (EOS)*, 10127, 1992.

71. Miller, L., Cheney, R., and Lillibridge, J., Blending ERS-1 altimetry and tide-gauge data, *Trans. Am. Geophys. U. (EOS)*, 74, 185, 1993.

72. Arnault, S. and Perigaud, C., Altimetry and models in the tropical oceans: a review, *Oceanol. Acta*, 15, 411, 1992.

73. Miller, R. N. and Cane, M. A., A Kalman filter analysis of sea level height in the tropical Pacific, *J. Phys. Oceanogr.*, 19, 773, 1989.

74. Bourles, B., Arnault, S., and Provost, C., Toward altimetric data assimilation in a tropical Atlantic model, *J. Geophys. Res.*, 97, 20271, 1992.

75. Gourdeau, L., Arnault, S., Menard, Y., and Merle, J., GEOSAT sea level assimilation in a tropical Atlantic model using Kalman filter, *Oceanol. Acta*, 15, 567, 1992.

Part II

Water Properties

9

Sea Surface Temperatures from the Along-Track Scanning Radiometer

Peter J. Minnett

9.1 Introduction

Over the last 15 years satellite remote sensing has developed into an integral part of oceanographic and atmospheric research. It provides data that are complementary to those from conventional sources and intrinsically more attuned to the requirements of numerical models, both diagnostic and prognostic, of oceanic, atmospheric, and climatological processes. Some variables, including sea surface temperature, can be measured from satellites to comparable or greater accuracy than by conventional *in situ* sensors. Furthermore, the measurements from satellites offer a globally consistent data set that is particularly advantageous for research into

0-8493-4525-1/95/$0.00+$.50
© 1995 by CRC Press, Inc.

climate change. However, to achieve appropriate levels of accuracy, care in the design and application of the retrieval algorithms is needed, especially in how they compensate for interfering effects.

The sea surface temperature (SST) plays a crucial role in the global climate system and in coupling the atmosphere and ocean. The skin of the ocean provides the lower boundary condition for the upwelling infrared radiation in the marine atmosphere. The air-sea temperature difference is a controlling factor in the exchange of heat and moisture (evaporation), and also determines the marine boundary layer stability, which has consequences on the air-sea fluxes of heat and momentum.[1] The temperature dependence of the solubility of gases such as CO_2 requires accurate knowledge of the surface temperature for, e.g., the estimation of the global carbon inventory.[2] In fact the surface skin effect,* which is not considered in the calculation of the global carbon budget using conventional SST climatologies, can account for ~0.7 Gt C year^{-1} of extra carbon flux to the ocean, which is up to about 50% of the discrepancy between estimates of net oceanic CO_2 uptake derived from models calibrated by tracers, and from integrating the air-sea exchanges over the global oceans.[2] The latter approach is sensitive to SST. The measurement of the surface skin temperature of the ocean requires accurate radiometers mounted on ships,[3] aircraft,[4,5] or satellite.[6,7] Of these, only satellites provide sufficient data to have any impact on global studies. The other platforms have a role in process studies and in the validation of the satellite data.

SSTs from conventional sources are limited in their geographic range to the main shipping routes;[8,9] and their inadequate sampling, which is concentrated in the Northern Hemisphere, can give rise to significant uncertainties when they are used to generate climatologically meaningful maps.[9] In contrast, there are typically over an order of magnitude more satellite SST determinations more uniformly distributed around the globe.[10] Accurate satellite measurements provide the possibility of generating a consistent global climatology of SST that can be used as time progresses to diagnose the rate of climate change. To quote J. T. Houghton:**

> The questions are often asked: How can climate change be measured? Are measurements available to show how climate changed over the last 100 years and can we observe changes in the future? Regarding temperature change, our current assessments are based on measurements at a few sites where careful records have been kept over a long period. Such sites are limited to a few countries at mid-latitudes in the Northern Hemisphere and therefore represent a very inadequate sample so far as the whole globe is concerned. Consistent measurements of the temperature of the sea surface from satellites would overcome the sampling problem very well and would be very suitable for looking at climatic trends provided that they could be made with sufficient accuracy. An accuracy of say better than 0.2 K would be required.

*The surface skin of the ocean is usually colder, by a few tenths of a degree, than the underlying bulk temperature of the water. This is due to the flow of heat from the ocean to atmosphere by molecular processes through the upper millimeter or so of the water where turbulent transfer is suppressed by the density difference across the interface. See Chapter 10.

**From the discussion following the presentation of Harries, J.E. et al., Observations of sea-surface temperature for climate research, *Philos. Trans. R. Soc. London*, A309, 395, 1984. With permission.

The Along-Track Scanning Radiometer (ATSR) was designed and built to reduce, as far as is practicable, the uncertainties in the measurement of sea surface temperature from space. In particular, the problems of onboard calibration were resolved by using two blackbody calibration targets in thermally controlled enclosures, and a very careful prelaunch calibration was conducted. The greatest innovation in the ATSR is the use of a conical scan mechanism to sample the emitted infrared radiation from the sea surface through two atmospheric paths of different length. In addition, the detector noise is reduced by having the focal plane assembly of the ATSR cooled to liquid-nitrogen temperatures by a novel mechanical refrigerator. The details of the instrument are given in Appendix G, and this chapter deals with the correction of the atmospheric effect and the sea surface temperature measurements.

9.2 Limits on the Accuracy of Infrared Measurements of Sea Surface Temperature

The largest source of uncertainty in the retrieval of SST from spaceborne infrared radiometric measurements is in the correction for the effects of the intervening atmosphere. The presence of clouds poses a great problem for the measurement of SST by infrared radiometry, as they obscure the sea surface. Those parts of the data stream (image) that are contaminated by the effects of clouds must be identified and excluded from the SST retrieval procedure. The largest component of the clear sky atmospheric effect is due to water vapor, which is highly variable in space and time. In the spectral region where these measurements are made, at ~3.7 and ~10–13 μm, the effect is dominated by the so-called water vapor continuum. The physical basis of the continuum absorption is poorly understood,[11–13] but it is more pronounced toward longer wavelengths. Other factors that contribute significantly to the atmospheric effect and are also very variable in their properties and distributions are aerosols, both in the troposphere and stratosphere. Stratospheric aerosols are particularly abundant after some powerful volcanic eruptions, such as El Chichon and Mt. Pinatubo.

9.3 Sea Surface Temperature Retrieval Algorithms

The spectral gradient of the water vapor continuum across the 10–13 μm atmospheric window, and between this and the 3.7 μm window, permits a correction of the gaseous atmospheric effect using the coregistered multichannel measurements alone without recourse to external measurements. The effects of aerosols are usually to reduce the channel radiances measured in space that can result in an erroneous atmospheric correction.[14] The multichannel atmospheric correction is the conventional approach that has been used for more than a decade with the measurements of the Advanced Very High-Resolution Radiometer (AVHRR); however, since the launch of the ATSR, it has been possible to measure the effect of the atmosphere, rather than infer it from its spectral properties.

Multichannel Atmospheric Corrections

For independent, collocated measurements in nearby spectral intervals, the radiative transfer equation, the Planck function, and the spectral dependence of the sea surface emissivity can be linearized to good approximation. The atmospheric effect can then be accounted for by combining collocated measurements at different spectral intervals (channels) using a simple expression:[15]

$$SST = a_0 + a_{3.7\mu m} T_{3.7\mu m} + a_{11\mu m} T_{11\mu m} + a_{12\mu m} T_{12\mu m} \tag{1}$$

where $T_{3.7\,\mu m}$, $T_{11\,\mu m}$, and $T_{12\,\mu m}$ are the equivalent blackbody brightness temperatures (temperatures derived from the measured radiances by inverting the Planck function) measured in the channels at 3.7, 11, and 12 µm. This is the approach used to extract SSTs from the measurements of the AVHRR; and although more complex formulations exist, such as the cross-product SST algorithm (CPSST[16]), they all rely on an inference of the effect of the atmosphere based on multispectral measurements.

Dual-Scan Angle Atmospheric Corrections

An alternative technique is to measure the same area of the sea surface through two atmospheric paths of differing lengths. For example, a measurement at nadir is through one atmospheric mass, while a measurement at an angle of 60° is through about two atmospheric masses. The difference between these two measurements of the same area of the sea surface (assuming the atmosphere is horizontally homogeneous on the scale required to accommodate both the vertical and slant paths) is a direct measure of the effect of the atmosphere:

$$SST = b_0 + b_{i,n} T_{i,n} + c_{i,s} T_{i,s} \tag{2}$$

(where i represents any of the atmospheric window channels, n refers to the nadir measurements, and s refers to measurements through a slanted atmospheric path), or in combination with the multispectral information:

$$\begin{aligned} SST = c_0 &+ c_{3.7\mu m,n} T_{3.7\mu m,n} + c_{11\mu m,n} T_{11\mu m,n} + c_{12\mu m,n} T_{12\mu m,n} \\ &+ c_{3.7\mu m,s} T_{3.7\mu m,s} + c_{11\mu m,s} T_{11\mu m,s} + c_{12\mu m,s} T_{12\mu m,s} \end{aligned} \tag{3}$$

This is the measurement strategy of the ATSR.

With ATSR data it is possible to compare the performance of the conventional multichannel algorithm with the more direct corrections that make use of the dual-angle information.

Blended Analyses

A different approach used, for example, at the National Meteorological Center, is to blend satellite and *in situ* fields. The spatial information in the satellite-derived field is coupled to the absolute accuracy of the *in situ* field.[10,17] However, in so doing the resultant values are bulk temperatures and the possibility of determining the skin temperatures is lost, along with the chance, for example, to improve the specification of the air-sea CO_2 exchange. For this the skin temperature is needed, and this can only be realistically achieved using satellite data with algorithms designed to give a direct measure of the skin SST.

Skin and Bulk Temperatures

The success of the atmospheric correction algorithm hinges on the coefficients in Equations 1–3 being correct and appropriate for the conditions under which the satellite data were taken.[18,19] There are two ways of deriving them: regression of satellite measurements against collocated *in situ* temperatures, and using a radiation transfer model to simulate satellite data. The former has been used to derive the coefficients used in the National Oceanic and Atmospheric Administration (NOAA) operational retrievals,[20] and the second has been used in the research community.[21] The latter approach results in an algorithm that is appropriate for the determination of the oceanic skin temperature, because that is the temperature used in the model as the source of upwelling radiation at the base of the atmosphere. The former approach results in bulk temperature estimates as this type of measurement is used in the regressions, but in this case the skin temperature gradient masquerades as part of the atmospheric effect.

Fuller discussions of the differences between bulk and skin temperature are to be found in Chapter 10 and elsewhere.[6,22]

9.4 Cloud Screening

A vital operation in the procedure to derive SSTs from satellite radiance measurements is the identification of those pixels in which the sea surface is obscured, maybe only in part, by the presence of clouds. The cloud screening implemented in the ATSR data processing stream[23] is a development of that described by Saunders and Kriebel[24] that involves the successive application of a series of tests to identify cloud-free pixels. These tests are based on the physical differences between the sea surface and cloud tops and include: a simple temperature test (clouds are usually colder than the underlying sea surface); a test for the amount of reflected sunlight in the short wavelength channel because during the day, clouds are bright and the ocean surface away from where there is sun glitter is dark; and a test for the spatial variability of temperature over a small region (say 3×3 pixels) because away from strong surface fronts, the sea surface is more uniform in temperature than are cloud tops. A further set of tests, based on the different brightness temperatures measured in the different channels, which result from differing spectral properties of the emissivities of the sea surface and clouds and the shorter atmospheric path length above the clouds, has been elaborated for use with ATSR data to include the differences in the path lengths between the forward and nadir views. Atmospheric radiative transfer modeling has been used to define the specific threshold values for use in the series of tests for application to ATSR data.[23]

9.5 High-Resolution Data

An example of ATSR data at full 1-km resolution is given in Figure 9.1, which shows a 512 × 512 pixel scene of the Alboran Sea in the western Mediterranean, derived from the measurements for the 12-μm channel nadir swath. The good radiometric resolution can be seen in the smoothness of the gray tones defining the surface temperature features. The temperature range is from about 10 to 15°C. This image encompasses the track of the R/V Alliance between 19:46 on October 6 and 17:52 on October 7, and the brightness temperatures extracted from the image along the position of the ship's track for the two long-wavelength channels in both the nadir and forward swaths are shown as functions of longitude in Figure 9.2 together with the bulk temperature measured from the ship at a depth of about 3 m. The relative effects of the spectral brightness temperature gradient and the atmospheric path length difference are apparent and agree with physical expectations. The brightness temperature differences caused by the spectral and path length changes are dependent on the state of the atmosphere at the time of the measurement; for example, in other cases the atmosphere is such that the nadir measurement in the 12-μm channel is very close to the 10.8-μm channel values in the forward swath.

SSTs were calculated from these brightness temperatures using prelaunch atmospheric coefficients[23,25] (Figure 9.3), for both the nadir swath multichannel split-window algorithm and the dual-angle multichannel algorithm (Equations 1 and 3, but using only 11- and 12-μm measurements). The resulting SST traces along the ship's track are shown in Figure 9.4 with the 3-m depth SST for comparison. There is a difference between the two SST retrievals, which amounts to 0.5 K in places (away from cloud contamination). Since the retrieval coefficients for the two algorithms were derived in an identical manner, these differences result from the inclusion of the additional information from the dual-angle measurements. The improved accuracy of the ATSR SST product (i.e., the dual-angle, multispectral retrieval) has been demonstrated by Barton et al.[26] and Mutlow et al.,[27] who argue that the design goals of absolute SST retrieval accuracy of ~0.3 K have been met.

The thorough validation of the ATSR measurement scheme and the dual-angle, multispectral atmospheric correction is the subject of continuing research.

9.6 50-km Resolution Data

For some applications, such as numerical weather forecasting, a near-real time, spatially averaged SST field is more appropriate than high-resolution images. An extensive comparison between the nadir-only and dual-path SST retrievals conducted at the U.K.

Figure 9.1   An image of 12-μm channel brightness temperatures from the nadir swath of ATSR taken on October 6, 1991 at 21:59. The scene is 512 km square and covers part of the western Mediterranean Sea. The data are © ESA and SERC. (From Minnett, P.J. and Stansfield, K.L., The validation of ATSR measurements with *in situ* sea temperatures, in *Proc. 2nd ERS-1 Symp., Space at the Service of Our Environment*, European Space Agency Publications Division, Noordwijk, The Netherlands, 1994. With permission.)

Meteorological Office using global, cloud-screened SSTs averaged over 50×50 km has revealed significant discrepancies. An example is shown in Figure 9.5, which indicates that the largest problems occur in the tropics, with the nadir-only (i.e., AVHRR-like) algorithm producing SSTs up to 1 K colder than the dual-angle multichannel algorithm.[28] The residual difference is a manifestation of an effect of the atmosphere because the coefficients used in the algorithms were derived using the same radiative transfer model, and are therefore mutually consistent. The difference in this case is influenced by stratospheric aerosols from the Mt. Pinatubo eruption, which corrupted the

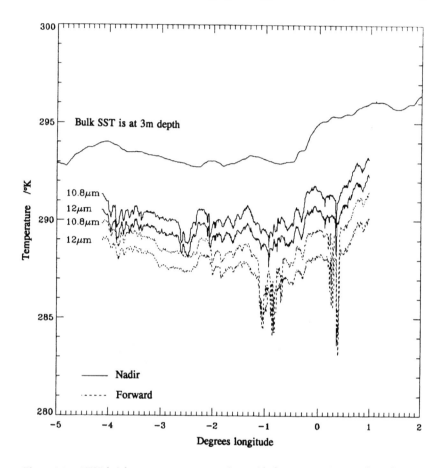

Figure 9.2   ATSR brightness temperature sections with the −3 m *in situ* SST along the track of R/V Alliance for the image taken on October 6, 1991. The displacements to cold temperatures are the effects of intervening clouds. (From Minnett, P.J. and Stansfield, K.L., The validation of ATSR measurements with *in situ* sea temperatures, in *Proc. 2nd ERS-1 Symp., Space at the Service of our Environment,* European Space Agency Publications Division, Noordwijk, The Netherlands, 1994. With permission.)

nadir-only multichannel retrievals;[29] and examination of comparable data later in 1992, when the Mt. Pinatubo effects were diminished, showed smaller differences, with maximum value of 0.6 K.[14] This discrepancy is consistent with the results of Emery et al.[30] who attribute the poor performance of the multichannel retrievals in the tropics to an inability to compensate adequately for the effects of large atmospheric water vapor burdens.

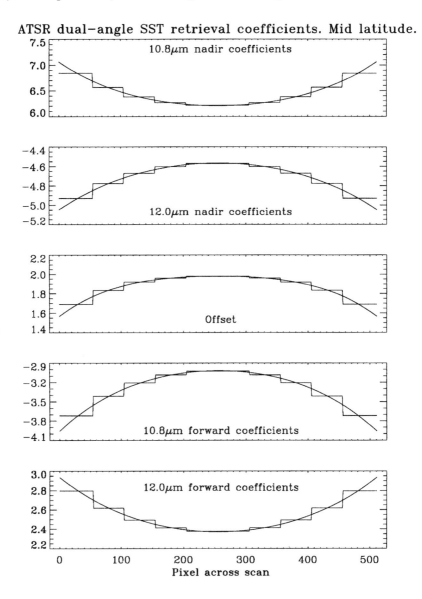

Figure 9.3   Coefficients for the retrieval of SST from the ATSR 10.8- and 12-μm brightness temperatures derived from radiative transfer modeling. The coefficients are for the retrieval of skin temperature through atmospheric conditions characteristic of midlatitude conditions, and are shown as functions of pixel number across the swath. The changing values across the ATSR swath account for the changing atmospheric path lengths and surface emissivity. (The smooth curves are derived by fitting a fourth-order polynomial to the stepped values courtesy of A. M. Závody.)

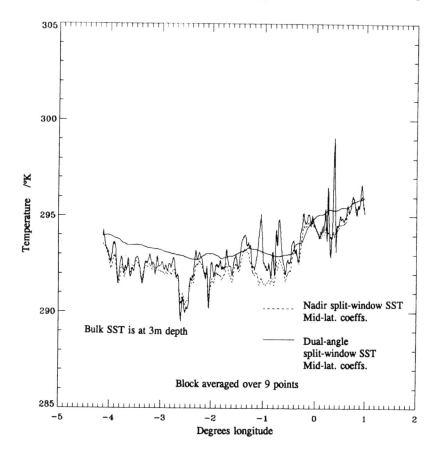

Figure 9.4   ATSR SST retrievals along the path of R/V Alliance in the western Mediterranean Sea during early October 1991. The differences between the satellite SSTs and the *in situ* measurement are shown. It can be seen that the dual angle information leads to a better atmospheric correction. For example, between 2.5° and 3.5° W, the dual-angle SST is about 0.5 K below that of the *in situ* sensor, while the nadir-only SST is about 1 K cooler. Given the expected size of the skin effect, i.e., a few tenths of a degree, it is believed that the dual-angle correction is more appropriate. Note that the inclusion of more channels of information in the SST algorithm leads to a noisier trace; some of this may be due to the geometric spreading of the field of view in the slant measurements causing a mismatch in the size of pixels in the two swaths. Apart from applying the atmospheric correction algorithm, the traces are uncorrected for any other effect, such as cloud contamination. At the time of the satellite overpass, the longitude of the ship's position was 3.53° W. (From Minnett, P.J. and Stansfield, K.L., The validation of ATSR measurements with *in situ* sea temperatures, in *Space at the Service of Our Environment*, Proc. 2nd ERS-1 Symp., European Space Agency Publications Division, Noordwijk, The Netherlands, 1994. With permission.)

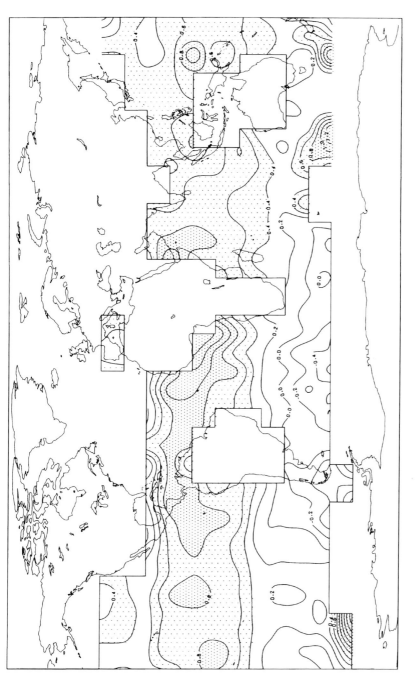

Figure 9.5　The effects of a direct dual-angle measurement atmospheric correction compared to the conventional multichannel correction, revealed as the differences in ATSR SST fields calculated with and without the dual-angle measurements. The figure shows dual-angle SST minus multichannel SST for February 1992. Contours are at intervals of 0.2 K. (From Harrison, D., Validation of ATSR near-real-time products. Presented at the ATSR Science Team Meeting: Validation Results, Abingdon, U.K., March 30, 1993. With permission.)

9.7 Summary

The novel innovations incorporated in the ATSR to improve the accuracy of SST measurement from space appear to reduce the residual errors in the correction for the effect of the atmosphere, although at these levels of uncertainty it is difficult to demonstrate this conclusively by comparison with conventional *in situ* measurements.[31] Nevertheless, initial results are very promising,[27] and efforts to thoroughly validate the instrument and its products are continuing. Aircraft measurements in the tropics show the dual-angle, multichannel measurements to be more robust than the nadir-only multichannel (AVHRR-like) retrievals in the presence of extraneous contamination, such as stratospheric aerosols.[14]

Acknowledgments

The author is indebted to colleagues referenced in this chapter for making available recent information and illustrations. The ATSR data (©1991 by ESA and SERC) were supplied by the ATSR Project Team at the Rutherford Appleton Laboratory, U.K. This was prepared under the auspices of the U.S. Department of Energy under Contract No. DE-AC02–76CH00016, with partial support from the National Oceanic and Atmospheric Administration, Grant NA26GP0266–01.

References

1. Smith, S.D., Coefficients for sea surface wind stress, heat flux, and wind profiles as a function of wind speed and temperature, *J. Geophys. Res.*, 93, 15,467, 1988.
2. Robertson, J.E. and Watson, A.J., Thermal skin effect of the surface ocean and its implications for CO_2 uptake, *Nature (London)*, 358, 738, 1992.
3. Hepplewhite, C.L., Remote observation of the sea surface and atmosphere: the oceanic skin effect, *Int. J. Remote Sensing*, 10, 801, 1989.
4. Saunders, P.M., Aerial measurements of sea-surface temperatures in the infrared, *J. Geophys. Res.*, 72, 4109, 1967.
5. Saunders, R.W. and Minnett, P.J., The measurement of sea surface temperature from the C-130. MRF Internal Note No. 52. Meteorological Research Flight, Royal Aerospace Establishment, Farnborough, Hampshire, U.K., 1990.
6. Schluessel, P., Emery, W.J., Grassl, H., and Mammen, T., On the bulk-skin temperature difference and its impact on satellite remote sensing of sea surface temperatures, *J. Geophys. Res.*, 95, 13,341, 1990.
7. Wick, G.A., Emery, W.J., and Schluessel, P., A comprehensive comparison between satellite-measured skin and multichannel sea surface temperature, *J. Geophys. Res.*, 97, 5569, 1992.
8. Hilland, J.E., Chelton, D.B., and Njoku, E.G., Production of global sea surface temperature fields for the Jet Propulsion Laboratory Workshop, *J. Geophys. Res.*, 90, 11,642, 1985.
9. Folland, C.K., Reynolds, R.W., Gordon, M., and Parker, D.E., A study of six operational sea surface temperature analyses, *J. Climate*, 6, 96, 1993.
10. Reynolds, R.W. and Marisco, D.C., An improved real-time global sea surface temperature analysis, *J. Climate*, 6, 114, 1993.
11. Grant, W.B., Water vapor absorption coefficients in the 8–13 μm spectral region: a critical review, *Appl. Optics*, 26, 451, 1990.

12. Kilsby, C.G., Edwards, D.P., Saunders, R.W., and Foot, J.S., Water-vapor continuum absorption in the tropics: aircraft measurements and model comparisons, *Q. J. R. Meteorol. Soc.,* 118, 715, 1992.

13. Rudman, S.D., Saunders, R. W., Kilsby, C.G., and Minnett, P.J., Water vapor continuum absorption in mid-latitudes; aircraft measurements and model comparisons, *Q. J. R. Meteorol. Soc.,* 120, 795, 1994.

14. Smith, A.H., Saunders, R.W., and Závody, A.M., The validation of ATSR using aircraft radiometer data over the tropical Atlantic, *J. Atmos. Oceanogr. Technol.,* (in press).

15. McMillin, L., Estimation of sea-surface temperatures from two infrared window measurements with different absorption, *J. Geophys. Res.,* 80, 5113, 1975.

16. Walton, C., Nonlinear multichannel algorithms for estimating sea surface temperature with AVHRR satellite data, *J. Appl. Meteorol.,* 27, 115, 1988.

17. Reynolds, R.W., A real-time global sea-surface temperature analysis, *J. Climate,* 1, 75, 1988.

18. Minnett, P.J., A numerical study of the effects of anomalous North Atlantic Atmospheric conditions on the infrared measurement of sea-surface temperature from space, *J. Geophys. Res.,* 91, 8509, 1986.

19. Minnett, P.J., The regional optimization of infrared measurements of sea-surface temperature from space, *J. Geophys. Res.,* 95, 13,497, 1990.

20. Strong, A.E. and McClain, E.P., Improved ocean surface temperatures from space — comparisons with drifting buoys, *Bull. Am. Meteorol. Soc.,* 85, 138, 1984.

21. Llewellyn-Jones, D.T., Minnett, P.J., Saunders, R.W., and Závody, A.M., Satellite multichannel infrared measurements of sea-surface temperature of the N.E. Atlantic Ocean using AVHRR/2, *Q. J. R. Meteorol. Soc.,* 110, 613, 1984.

22. Robinson, I.S., Wells, N.C., and Charnock, H., The sea surface thermal boundary layers and its relevance to the measurement of sea surface temperature by airborne and spaceborne radiometers, *Int. J. Remote Sensing,* 5, 19, 1984.

23. Závody, A.M., Gorman, M.R., Lee, D.J., Bailey, P., Eccles, D., Mutlow, C.T., and Llewellyn-Jones, D.T., The ATSR data processing scheme developed for EODC, *Int. J. Remote Sensing,* 15, 827, 1994.

24. Saunders, R.W. and Kriebel, K.T., An improved method for detecting clear sky and cloudy radiances from AVHRR data, *Int. J. Remote Sensing,* 9, 123, 1988.

25. Závody, A.M., Mutlow, C.T., and Llewellyn-Jones, D.T., A radiative transfer model for sea surface temperature retrieval for the Along-Track Scanning Radiometer, *J. Geophys. Res.,* 100, 937, 1995.

26. Barton, I.J., Prata, A.J., and Llewellyn-Jones, D.T., The Along Track Scanning Radiometer — an Analysis of coincident ship and satellite measurements, *Adv. Space Res.,* 13(5), 69, 1993.

27. Mutlow, C.T., Závody, A.M., Barton, I.J., and Llewellyn-Jones, D.T., Sea surface temperature measurements by the Along-Track Scanning Radiometer on the ERS-1 Satellite: early results, *J. Geophys. Res.,* 99, 22,575, 1994.

28. Harrison, D., Validation of ATSR near-real-time products, Presented at the ATSR Science Team Meeting: Validation results, Abingdon, U.K., March 30, 1993.

29. Harrison, D., personal communication, 1993.

30. Emery, W., Yu, Y., Wick, G., Schluessel, P., and Reynolds, R., Correcting infrared satellite estimates of sea surface temperature for atmospheric water vapor contamination, *J. Geophys. Res.,* 99, 5219, 1994.

31. Minnett, P.J., Consequences of sea surface temperature variability on the validation and applications of satellite measurements, *J. Geophys. Res.,* 96, 18,475, 1991.

10

Skin and Bulk Sea Surface Temperatures: Satellite Measurement and Corrections

William J. Emery, Gary A. Wick, and Peter Schluessel

Historically, sea surface temperatures were measured from bucket samples or engine cooling water samples that measure the temperature of the upper meter of the ocean. Known as the bulk sea surface temperature, this measurement has been used to compute air-sea heat fluxes using empirical formulas. Infrared satellite sensors, however, collect radiation emitted from the millimeter thin skin of the ocean. This skin temperature differs from the bulk temperature by about 0.3 K on average. Depending on atmospheric and oceanic conditions, the bulk-skin temperature differences can be as large as +1.2 or −1.4 K. Since the commonly desired bulk temperature and measured skin temperature are different, it is necessary to know the relationship between the two. These differences are studied using both coincident measurements of skin and bulk temperature from a ship and simulated temperatures from a one-dimensional numerical model. Differences between satellite-measured sea surface temperature algorithms designed to give either the skin or bulk temperature are discussed. The newest skin sea surface temperature algorithm uses microwave satellite data to estimate atmospheric water vapor and correct Advanced High-Resolution Radiometer (AVHRR) infrared estimates of sea surface temperature.

10.1 Introduction

Due to its ease of measurement, sea surface temperature (SST) is one of the most widely observed ocean variables. At the interface between the ocean and atmosphere, the SST is an important variable in controlling the fluxes of heat and moisture between the atmosphere and the ocean, which plays a critical role in dictating the climate. Traditionally, SST has been measured from ships initially using bucket samples and then later using the temperature of the seawater taken in to cool the ship's engines. The bulk air-sea heat flux formulas that were developed over the years utilized these bulk SST measurements in their empirical formulations.

The era of meteorological satellites introduced the capability of mapping SST from space using infrared radiometers on the satellites. Starting in the late 1970s, data from the scanning radiometer on the National Oceanic and Atmospheric Administration (NOAA) polar orbiting spacecraft were used to compute SST on an operational basis.[1] The algorithms used for the computation of SST from the satellite infrared data were calibrated against ship SST measurements. Later, it was recognized that the ship SST data had a warm bias[2,3] due to the heating of the engine room where the SST was measured. As the SST algorithms improved with the introduction of new spaceborne radiometers, the calibration shifted to using data from freely drifting buoys that reported their SST measurements through the same NOAA satellites.[4,5]

These drifting buoys again measure the bulk SST through thermistors that are located on the lower part of the buoy hull. Early buoys were large spar buoys often made of aluminum or steel. Here the thermistor was attached to the inner part of the hull, and the ocean temperature was sensed through the hull. As buoy designs changed to glass spheres and other small fiberglass floats, the thermistor was installed on the lower part of the hull, often extending out into the seawater. While the hull size is quite small, these buoys do not ride on the surface skin of the ocean, but rather are often submerged below the surface depending on the activity of the wave field. The result is that the buoys also measure the bulk SST anywhere from 0.5 to 1.5 m below the surface.

The most widely used satellite sensor for monitoring SST is the AVHRR flying on the NOAA polar orbiters. The AVHRR initially had four channels, but this number was later increased to five to provide two thermal infrared channels for improved SST retrievals.[4] The differential absorption by atmospheric water vapor in these two infrared channels provided added information to correct the thermal infrared data for signal attenuation by atmospheric moisture. Known as the split-window algorithm, the SST was computed as a combination of the two thermal infrared channels of the AVHRR with the final algorithm coefficients set by comparisons with SST data from drifting buoys.[4]

Due to the high emissivity of seawater in the infrared, infrared radiometers in space only receive radiation from the upper few micrometers of the ocean. The drifting buoys are incapable of measuring the temperature of this radiative skin layer of the ocean. Even if a buoy temperature sensor could be placed in contact with the ocean skin, this very contact would destroy the skin layer and erase the skin temperature effect. Early studies[6-11] clearly demonstrated the existence of this thin skin layer that is the top of the molecular boundary layer between the turbulent ocean and turbulent atmosphere. This molecular boundary layer affects the heat transport between these two turbulent regimes. Studies also have shown that even after a breaking wave has momentarily destroyed the skin layer, this skin layer will reestablish itself within 9–12 s after its destruction.[8,9] Thus, the skin layer is almost always present in the ocean.

Since researchers are most often interested in the bulk temperature and satellites only give measurements of the skin temperature, it is necessary to understand the relationship between the two temperatures. This chapter addresses the relationship by examining physical measurements and simulations of the temperature difference and discussing differences between algorithms designed to determine bulk and skin SSTs. The basic character of the oceanic skin layer will be introduced and its variability discussed. Direct shipboard measurements of the skin temperature with an infrared radiometer will be used with coincident measurements of the bulk temperature to define the relationship between the skin and bulk SSTs. Global maps of biweekly skin and buoy calibrated satellite SSTs will be compared to show differences stratified by latitude and season. Simulations from a one-dimensional numerical model showing a clear skin layer that responds to changes in the surface air-sea heat exchange will also be shown. Finally, a new skin-calibrated satellite SST algorithm with an explicit correction for atmospheric water vapor will be presented.

10.2 The Skin Layer

The nature of the skin layer has been studied by a number of investigators.[6-12] The thickness of the skin layer and the temperature gradient across this skin layer are determined by the surface heat exchange processes. The outgoing longwave radiative, latent, and sensible heat fluxes emerge from the skin of the ocean thereby cooling the skin. Longwave radiation is emitted only from the upper few micrometers of the ocean, and the latent and sensible heat fluxes occur through molecular processes across the skin layer. During the day, solar radiation enters the ocean, but only a small portion is absorbed right at the surface. The remainder is absorbed at

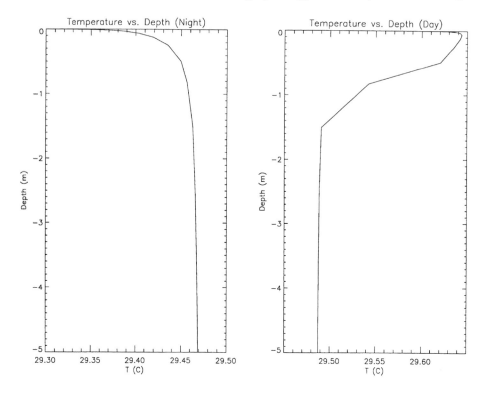

Figure 10.1   Temperature profiles of the shallow upper layer of the ocean during the (left) night and (right) day.

greater depths. As a result, the skin of the ocean is generally cooler than the water immediately below the surface even during the daytime.

Examples of possible near surface temperature profiles are shown in Figure 10.1 for night and day conditions. Because of the difficulty in obtaining accurate temperature measurements right at the surface, the profiles are taken from a model simulation. The model will be described in greater detail later. During the night (Figure 10.1a), with the only fluxes acting to cool the surface, there is the clear presence of a cool skin. During the day, absorption of shortwave solar radiation leads to heating of the upper layer of the ocean and the formation of a diurnal thermocline (Figure 10.1b). If the bulk temperature is measured at a depth of 2 m or lower, the skin temperature can be greater than the bulk temperature depending on the exact depth of the diurnal thermocline. Note from the figure though, that the temperature still increases with depth right near the surface. The penetration depth of shortwave radiation depends on the amount of suspended material in the ocean. These profiles are typical of clear, open-ocean conditions. Differences between the skin temperature and the temperature 1 m below the surface range between +1 and −1 K, with the skin temperature being about 0.3 K cooler on the average.[13]

10.3 Measuring Skin Temperature

The primary reason that people use drifting buoy temperatures in the determination of satellite SST algorithms is their availability. Regressions against *in situ* data are performed to determine the appropriate algorithm coefficients. Strictly speaking, the regressions should be performed using radiative skin SST measurements near sea level where there is no need for atmospheric corrections.

Using a Barnes PRT-5 radiometer mounted on a ship, measurements of skin SST were collected simultaneously with bulk SST measurements at 0.1, 2, and 4 m below the surface. The first of these sets of measurements was made in the North Atlantic in the fall of 1984.[13,14] The PRT-5 radiometer was calibrated by alternately viewing the ocean and a well-mixed bucket of seawater for 1-m intervals. Resistance thermometers were used to measure the temperatures 2 and 4 m below the surface, and a small float was deployed when the ship was on station to measure the temperature at 0.1 m.

The regular calibration of the PRT-5 was deemed necessary to achieve the required accuracy in the skin measurements. The calibration allowed the compensation for such effects as the nonblackness of the sea surface, variations in surface reflectivity with viewing angle, reflected sky and cloud radiation, sea spray contamination of the optics, temperature drift of the internal reference blackbody, and drift in the instrument electronics.[15] The reference bucket was kept well stirred by continuously overturning water pumped into the bucket to keep a skin layer from forming in the bucket. The temperature of the bucket was monitored with a platinum resistance thermometer accurate to 0.0125 K.

Based on all of the calibration measurements, the skin temperature was determined from the radiometer counts. The relationship between the reference temperature and radiometer counts was found to be linear. For each ocean measurement, the slope was taken from a regression of all calibration temperature and count measurements, and the offset was taken as the mean of the offsets of the nearest two calibration cycles. By considering the nearest calibration cycles, the effect of reflected radiation from moving fields of scattered clouds was minimized. This calibration procedure produced a final accuracy of 0.05 K for the skin temperature measurements.

10.4 Measuring Bulk Temperature

Many different methods are used to measure bulk SST. In fact, there is no general agreement as to the definition of bulk SST and where/when it should be measured. In the past SST has referred mainly to a ship-based *in situ* measure of the surface temperature that is more representative of the bulk temperature. Today bulk temperature not only is measured from ships but also is measured from buoys both moored and drifting. In fact, there is general agreement that buoys provide better measurements of SST than do ships. This view reflects the problems of using ship injection temperatures that are known to be biased up from the real SST due to heating in the engine room. The problem with the concept that buoys more accurately measure SST is that there is a wide variety of buoy types instrumented with an equally wide range of SST sensors. In 1978 Tabata[16] examined a variety of ship-based methods for measuring SST and concluded that mean offsets range from 0.08 to 0.6°C with standard deviations of about 0.2–0.5°C. Similar comparisons by Kent et al.[15] for volunteer observing ships (VOS) suggested that hull mounted sensors could measure bulk SST as accurately as specially designed buckets for measuring SST.

The advantage of using moored buoys to measure SST is that temperature sensors can be located at various levels in the water column. An example is showed here in Figure 10.2 that is a surface buoy deployed near Bermuda.[16] Here the thermistors are installed along one of the legs of the three-legged mooring and located at depths of 26, 48, 67, 89, 110, 131, 152, and 173 cm. Now buoy motion over its mooring watch circle and vertical mooring motion will result in modest variations of these depths; this system should provide SST measurements very near to the ocean surface. The data should be able to resolve changes in this shallow upper layer as heating and cooling occurs through the ocean surface.

The sample data series for a 3-d period in 1987 demonstrates this capability (Figure 10.3) where the sensors all agree well at night (actually in the early morning before sunrise) and then separate during the day when air-sea processes (heating/cooling, rainfall, etc.) cause changes in the vertical stratification. In this example, cool rain occurred on September 10 resulting first in the shallow sensors cooling over those below; later in the day, surface heating changes this relationship to where the near surface temperatures exceed those below. On September 11 there was no rain event, and the solar insolation was much stronger resulting in an even more dramatic vertical stratification and separation of temperature lines. A similar but slightly smaller amplitude pattern is found for September 12. Each day shows a convergence to the same line at night.

10.5 Difference Between Skin and Bulk Temperatures

Along with the well-calibrated skin SST measurements, the simultaneous values of the parameters needed to compute the air-sea heat fluxes with the traditional bulk formulas were also measured. This made it possible to formulate an expression for the difference between the skin and bulk temperatures based on the heat exchange between the ocean and the atmosphere. An example of a series of measured bulk-skin (ΔT) temperature differences is shown in Figure 10.4 along with bulk temperature as represented by the temperature measured at 2 m below the surface. Note the marked diurnal changes with the ΔT going negative during midday when the skin temperature is higher than the 2 m temperature due to solar heating of the shallow, near-surface layer. During most of the rest of this 2-d time series, the ΔT is slightly positive indicating the presence of a cool skin layer. It is interesting that the bulk-skin temperature difference is positive in the middle of the time series when the

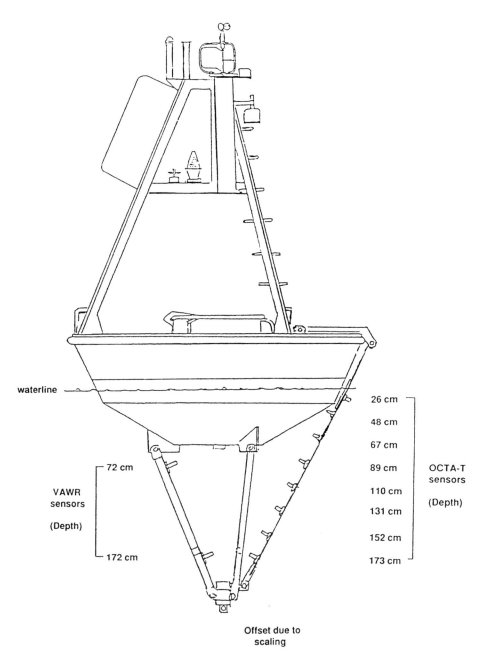

Figure 10.2 Line drawing of moored surface buoy. The Vector Averaging Wind Recorder (VAWR) installed on the buoy collects temperatures from thermistors at 72 and 172 cm below the mean waterline. A similar instrument (OCTA-T) using the same type of thermistors was also mounted on the hull; this instrument recorded data from eight thermistors mounted in an array deployed next to the buoy hull at the depths indicated on the drawing.

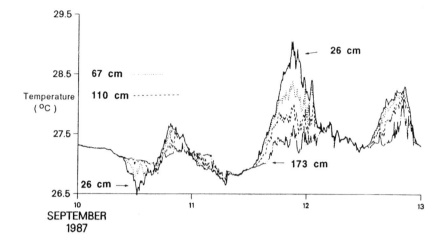

Figure 10.3 Three days of data from four of the sensors of the OCTA-T. On September 10, 1987, cool rain resulted in the sensors nearest the surface being cooler while solar heating on each day restratified the upper ocean layer and warmest temperatures are found near the surface. At night, when it was not raining, the temperature was uniform over the depths between 26 and 173 cm.

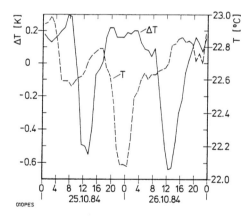

Figure 10.4 Two-day time series of the temperature difference (ΔT) between the ship radiometer skin temperature and the temperature measured at 2 m.

2-m temperature is at a minimum. This indicates that the skin layer was cooler than even this minimum temperature value at 2 m.

The end goal is to be able to predict the value of ΔT from available measurements. The most common method has been to parameterize ΔT in terms of the surface heat fluxes. This was done initially by Saunders[17] and later by Hasse[11] and Grassl.[9] Due to differences in the skin layer with day and night conditions, the best

Table 10.1   Mean ΔT Values for Various Cloud, Daytime, and Wind Speed Conditions

N (OCTAS)	u (m/sec)	Day	Night
0–5	>5	0.23 K	0.33 K
0–5	<5	0.17 K	0.18 K
6–8	>5	0.16 K	0.28 K
6–8	<5	−0.07 K	0.22 K

results were obtained by using separate expressions for day and night. The nighttime expression was

$$\Delta T = a_0 + a_1 u\left(T_s - T_a\right) + a_2\left(q_s - q_a\right) + a_3 L \tag{1}$$

while the daytime expression was

$$\Delta T = a_0 + a_1 S/u + a_2\left(q_s - q_a\right) + a_3 L \tag{2}$$

Here, T_s is the bulk sea surface temperature, T_a is the overlying air temperature, q_s and q_a are the water vapor mixing ratios for air and the sea surface, and L is the latent heat of evaporation. The term S represents the net solar radiative flux, and u is the surface wind speed.

The nighttime coefficients are $a_0 = -0.285$ K, $a_1 = 0.0115$ s/m, $a_2 = 37.255$ K, and $a_3 = -0.00212$ km^2/W. These were optimized to yield a net accuracy (standard error) with respect to the entire set of cruise data, of 0.11 K. The daytime values are $a_0 = -0.415$ K, $a_1 = -0.0037$ km^3/sW, $a_2 = 48.043$ K, and $a_3 = -0.00355$ km^2/W. These yielded a standard error of 0.17 K.[14]

The complete set of the bulk-skin temperature differences was used together with the surface heat flux variables to determine the dependence of ΔT on wind, cloud, and heat flux magnitudes. In these results, the 2-m temperature is taken as the bulk temperature. Comparisons between the ΔT relative to 2-m and 0.1-m temperatures, when available, showed similar results. A comparison of temperature differences computed with various measured cloud conditions showed no definite trends. A separation into day and night conditions revealed only positive ΔT values at night, consistent with the cool skin conditions that must prevail. During the day, some negative values were obtained showing that solar insolation can once again create skin temperatures greater than those at 2 m.

One very surprising result is the one shown here in Table 10.1, which gives the ΔT values for various cloud and wind speed conditions. The surprise is that the difference between the bulk and skin temperature increases with increasing wind. From traditional studies of the upper mixed layer, one would expect the skin and bulk temperatures to be more alike as the wind speed increases due to the turbulent stirring of the upper layer. Instead the outgoing radiation due to sensible and latent

Table 10.2.   Mean Correlations Between T_s and T_2 and
Their Standard Deviations (SD)

Δx (km)	Correlation	SD
20	0.5	0.36
40	0.62	0.33
80	0.74	0.29
160	0.85	0.21
320	0.92	0.02
640	0.92	0.01

fluxes increases, which cools the skin layer and widens the difference between the skin and bulk temperatures. These flux terms are related to the square of the wind speed, and thus a small change in wind speed can cause a strong increase in the outgoing fluxes that cool the skin. Likewise, when the wind is weak, the outgoing heat fluxes are reduced and the skin SST warms up to become equal to or greater than the subsurface bulk SST. The overturning of the mixed layer, due to stronger winds, is effective at stirring conditions below the skin layer but does not change the temperature of the skin layer. Once wave breaking occurs, the skin layer will be destroyed and the observed ΔT will decrease. These results are consistent with the results of initial numerical simulations as will be discussed later.

Subsequent analysis of this same data set demonstrated that the increase of ΔT with increasing wind speed only occurred during part of the overall 1984 cruise on the Meteor. Similar data from other cruises clearly demonstrated the uniqueness of the 1984 data set since all of them showed a decrease of ΔT with increasing wind speed. Examining the Meteor data set closer revealed that the increase of ΔT with wind speed only occurred at ocean stations where the ship was not steaming. Data from cruise segments where the ship was steaming showed a modest decrease of ΔT for the higher wind speeds.

Another surprising result was found by correlating the skin and bulk temperatures collected as time series while the ship was steaming. The mean correlation coefficients for all legs of the Atlantic cruise are given here in Table 10.2. Note the increase in correlation as the spatial lag increases, and the eventual leveling out of this correlation at lag scales longer than 300 km. Wave number spectra were computed for each of the cruise sections, and an example is shown in Figure 10.5. The spectrum peaks at a wavelength of around 200 km with a coherence greater than 0.9; at shorter wavelengths, the spectral value decreases to a minimum below 0.2 at a wavelength shorter than 60 km. While the shapes of the coherence spectra varied considerably with the individual cruise legs, all of the wave number spectra showed this same decrease in energy with decreasing wavelength.

Thus, skin and bulk temperatures vary together (are well correlated) for spatial scales longer than 80 km while they are not related for scales shorter than 80 km. The heat flux mechanisms that create the differences between skin and bulk temperatures dominate at the shorter scales while the ocean circulation processes that are related to the large-scale distribution of temperature dominate at the larger

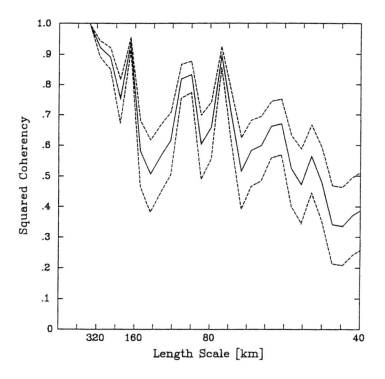

Figure 10.5 Wave number coherence spectra between coincident skin and bulk (2 m) temperature measured along the ship track.

scales. If one is interested in the variability of the large-scale circulation, it does not matter if one uses a skin temperature computation or a bulk temperature formulation to process satellite infrared data. If one is interested in the heat flux at the sea surface, however, it is necessary to use the skin temperature to represent the SST changes related to the heat fluxes.

10.6 Satellite Image Derived Global SST Maps

The motivation for measuring the skin temperature from a ship was to validate the computation of skin SST from satellite infrared radiometers.[14] To illustrate the effect that the skin layer can have on the remote sensing of SST, algorithms designed to give skin and bulk temperatures from satellite data have been compared. The most widely used algorithm for satellite-measured SST in the past has been the multichannel SST (MCSST). By deriving the algorithm coefficients from regressions between satellite measurements and drifting buoys,[20] this algorithm attempts to give a measure of the bulk temperature. This algorithm was compared with a skin SST algorithm derived by Schluessel et al.[14]

The algorithms were compared by computing coincident global maps of satellite SST from global area coverage (GAC) data from the AVHRR in 1984 and 1985. The

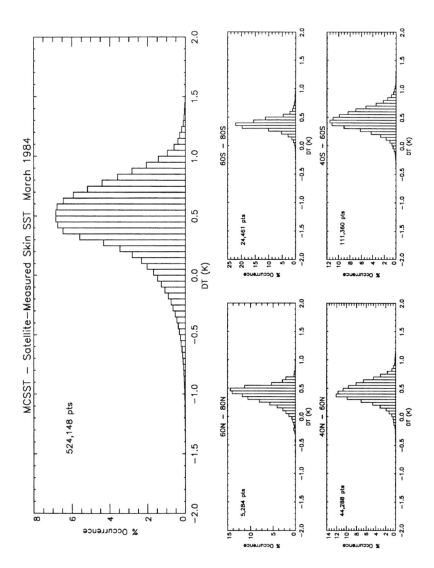

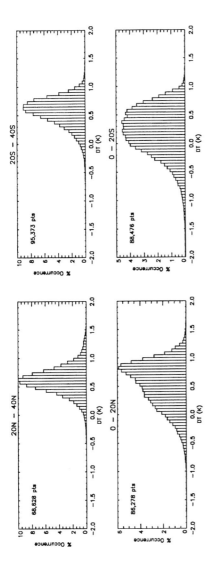

Figure 10.6 Histograms of the differences between the MCSST and the skin SST algorithm computed from satellite infrared imagery.

comparisons were performed with 2-week data sets from four different seasons.[18] Histograms, both global and stratified by latitude, of the bulk-skin temperature difference for March 1984 are shown in Figure 10.6. All of the histograms have a bias of about 0.5 K. This is consistent with the bulk temperature being greater than the skin temperature and demonstrates that the skin layer can have a significant effect on satellite-measured SST. Nearly all of the histograms are Gaussian in shape with some tendency for a longer tail on the negative side, particularly in the lower latitudes. The histograms are fairly narrow at high latitudes and are wider at low latitudes. The wider histograms in the low latitudes are most likely due to increased atmospheric absorption. The greater resulting differences between the brightness temperatures of the radiometer channels are amplified differently by the algorithms.

The histograms were generally the same for all four seasons with a mean value near 0.5 K and wider distributions at the lower latitudes. A comparison from March 1985 also showed similar results indicating few interannual changes. The persistence of the algorithm differences indicates that the bulk-skin difference does affect satellite SST measurements. Previous results show that the bulk-skin difference is variable, and thus using a measure of skin temperature to estimate the bulk temperature can lead to errors if only the mean ΔT is considered. Some knowledge of the nature of the temperature difference needs to be incorporated.

10.7 Numerical Model Results

One promising method for connecting the skin and bulk temperatures uses one-dimensional, primitive equation, numerical ocean models. Although more complicated than the simple parameterizations presented earlier, primitive equation models are generally more physically based. One model based on second moment turbulent closure hypotheses[19,20] has been used to simulate the near-surface temperature profile with millimeter resolution. The model has been run with both simulated and real data taken from cruise measurements.

The response of the simulated temperature profiles to changes in wind speed was studied using data representative of tropical conditions. Runs were performed for wind speeds of 1.5, 3, and 7 m/sec while holding all other forcing parameters constant. The results are shown in Figure 10.7. The plots on the left show the temperature profile in the upper 10 m while the plots on the right show the upper 20 cm in greater detail. Though these are for daytime conditions, all of the profiles show the presence of a cool skin at the surface of the ocean. As the wind speed increases from 1.5 m/sec in Figure 10.7a to 7 m/sec in Figure 10.7c, the increase in the outgoing latent and sensible heat fluxes is seen to lead to greater cooling at the surface. The resulting increase of the bulk-skin temperature difference is consistent with the previous observations from the cruise. Greater mixing leads to smaller temperature gradients through the upper 10 m, but does not decrease the bulk-skin temperature difference.

While the above results show the model to behave correctly qualitatively, additional runs were performed to observe whether the predicted temperatures agreed with actual measurements. Using input data from the Atlantic cruise, the model was run for a 1-d period. The comparisons between the actual and modeled temperatures at the surface and a depth of 2 m are shown in Figure 10.8. During the morning hours, the model only slightly underestimates the skin and 2-m temperature. In the middle of the day, the model begins to overestimate the 2-m temperature but drastically underestimates observed warming of the skin. The precise cause of this warming is still under investigation. During the evening, the model does an excellent job of reproducing both the skin and 2-m temperature.

These and other results show the model to perform very well during the night and under lower wind speeds. Additional work must go into the treatment of the absorption of solar insolation if better results are to be obtained during the day. At higher wind speeds (>5 m/sec), modeled bulk-skin temperature differences are not as great as observations. One possible source of this discrepancy is the wind energy that goes into the local surface wave field rather than into vertical temperature mixing. If these problems can be resolved, the model will provide a powerful tool to relate satellite measurements of the skin temperature to estimates of the bulk temperature.

10.8 Water Vapor Corrections for Skin Sea Surface Temperature Observations

Recently, a new satellite-measured skin SST algorithm that explicitly includes a term for atmospheric water vapor content was developed. The skin algorithms discussed previously — the MCSST and its newer operational replacement, the cross-product SST (CPSST)[21] — all correct for water vapor attenuation through the differential absorption of radiation in the two different thermal infrared channels of the AVHRR. The new water vapor SST (WVSST) was derived to see whether any improvement was possible by treating the water vapor content explicitly.

The WVSST algorithm can be written as:

$$T_s = a_1 + a_2 T_4 + a_3 \left(T_4 - T_5\right) + a_4 w\left(T_4 - T_5\right) \qquad (3)$$

where T_s is the skin SST, T_4 and T_5 are the brightness temperatures from channels 4 and 5 of the AVHRR, and w is the total column atmospheric water vapor. The coefficients were derived from atmospheric simulations[22] using a suite of marine radiosound data. The total column atmospheric water vapor was calculated from the Special Sensor Microwave/Imager (SSM/I) using the algorithm developed by Schluessel and Emery.[23] Since the SSM/I swaths do not overlap, the equatorial gaps must be filled by compositing over time. Two-day composites were used to formulate the atmospheric water vapor fields for the correction term.

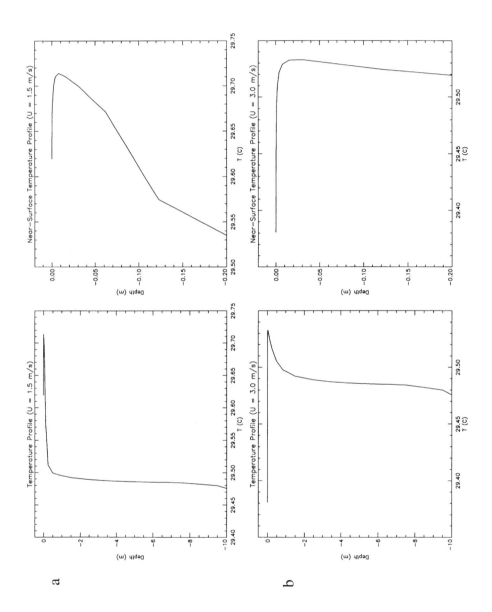

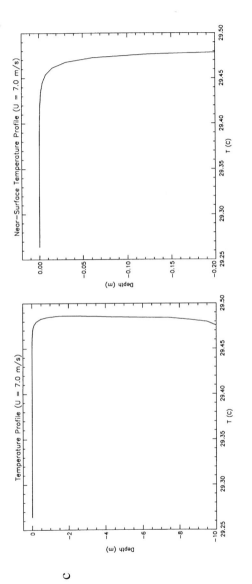

Figure 10.7   Modeled temperature profiles for wind speeds of (a) 1.5 m/sec, (b) 3.0 m/sec, and (c) 7.0 m/sec.

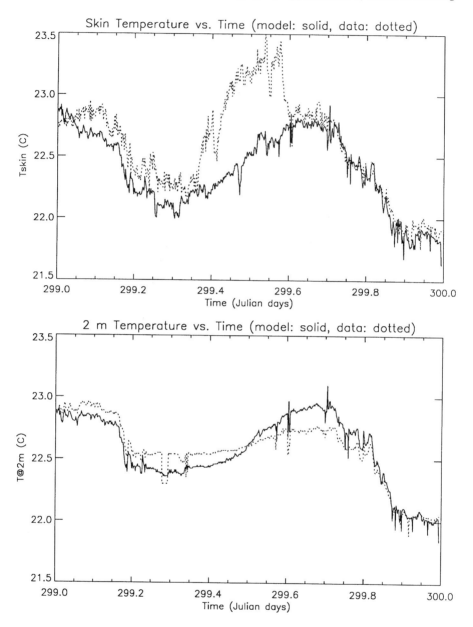

Figure 10.8   Time series of measured and simulated temperatures at the surface and a depth of 2 m.

Table 10.3. Differences Between AVHRR-SST Estimates and *In Situ* SST Measurements

	WV	CP	MC'	MC	WV	CP	MC'	MC
	Skin SST				Bulk SST			
Arctic Ocean								
2/26/91 11:00 (day)								
Data points: 19								
Temp. range: 5.0–7.0°C								
Mean difference (K)	−0.18	−0.97	0.52	−0.39	−0.29	−1.09	0.41	0.50
rms (K)	0.25	1.00	0.58	0.45	0.34	1.11	0.48	0.55
3/06/91 02:40 (night)								
Data points: 160								
Temp. range: 4.5–7.0°C								
Mean difference (K)	−0.24	−1.14	0.71	−0.88	−0.20	−1.11	0.71	−0.85
rms (K)	0.36	1.18	0.80	0.94	0.29	1.13	0.82	0.89
3/08/91 10:48 (day)								
Data points: 76								
Temp. range: 5.0–7.0°C								
Mean difference (K)	−0.05	−0.66	0.28	−0.57	0.03	−0.57	0.36	−0.49
rms (K)	0.22	0.70	0.39	0.63	0.33	0.67	0.49	0.59
3/9/91 02:06 (night)								
Data points: 175								
Temp. range: 5.0–7.0°C								
Mean difference (K)	−0.07	−0.92	0.72	−0.78	−0.29	−1.14	0.50	−1.00
rms (K)	0.17	0.93	0.79	0.82	0.32	1.15	0.59	1.03
South Pacific								
3/27/90 01:49 (day)								
Data points: 106								
Temp. range: 18.5–22.0°C								
Mean difference (K)	−0.01	−0.07	0.81	0.56	−0.35	−0.41	0.47	0.22
rms (K)	0.39	0.53	0.95	0.73	0.59	0.72	0.74	0.54
3/28/90 01:38 (day)								
Data points: 78								
Temp. range: 19.0–22.0°C								
Mean difference (K)	0.00	−0.05	0.71	0.24	0.08	0.03	0.79	0.32
rms (K)	0.29	0.28	0.76	0.36	0.24	0.24	0.82	0.39

To evaluate and compare the new algorithm, similar simulations were conducted to determine coefficients for the traditional MCSST algorithm. This algorithm, referred to as MCSST' below, is roughly equivalent to the skin SST algorithm discussed previously since the simulations and regressions were based on skin temperature. Because of its nonlinear character, new coefficients were not determined for the CPSST algorithm. Comparisons were performed between the WVSST, MCSST', and operational MCSST and CPSST algorithms. The algorithms were applied to AVHRR images from the South Pacific and northern North Atlantic (Arctic) for which coincident *in situ* measurements of bulk and skin SST were available from research vessels.

The comparisons between the algorithms and the skin and bulk measurements are presented in Table 10.3. For each scene, the date, time, location, and number of comparison points are given along with the mean and root mean square (rms) differences. In almost all cases, the number of comparisons is sufficiently large to give stable and significant statistics. A quick review of the table indicates that the WVSST algorithm has the best overall performance relative to both the bulk and skin *in situ* measurements. The differences between the WVSST and MCSST algorithms are due to the different treatment of atmospheric absorption. Clearly, the WVSST offers an important improvement. The differences between the WVSST and the operational algorithms are also primarily due to the different atmospheric corrections, but they also incorporate errors due to the variability of the bulk-skin temperature difference. Thus, both good atmospheric corrections and knowledge of the bulk-skin temperature difference are needed to accurately determine SST from satellites.

Since data was used from both the Arctic and midlatitudes, the comparisons are not restricted to a single uniform environment. The South Pacific data were from an area of significant atmospheric moisture with some important fronts, while in the Arctic there were very limited water vapor and weak moisture fronts. It is interesting that both of the operational algorithms performed better in regions with more moisture. In the driest conditions, the MCSST performed better than the newer CPSST algorithm.

10.9 Summary and Conclusions

The skin of the ocean is the radiatively active layer that emits the infrared radiation sensed by environmental satellites. This skin temperature is generally cooler than the water immediately below, but can be higher than the bulk temperature during midday when solar heating becomes important. To reconcile satellite measurements of the skin temperature with estimates of the bulk temperature, the bulk-skin temperature difference must be predicted. The temperature difference can be parameterized in terms of the constituent heat flux measurement parameters and, with an increasing degree of success, simulated using one-dimensional numerical ocean models. This bulk-skin temperature difference was observed to increase with increasing wind speed, and correlations between the skin and bulk temperature were found to be high for large space scales while small for smaller scales.

Differences between skin and bulk corrected satellite SST algorithms of similar forms were found to have a mean difference of 0.5 K. This difference was consistent with the observed cool skin of the ocean. A more accurate skin temperature algorithm was obtained by including an explicit correction for atmospheric water vapor attenuation. Together, these results show that good atmospheric corrections and proper ground truth validation are critical to satellite SST algorithm development.

References

1. Brower, R. L., Gohrband, H. S., Pichel, W. G., Signore, T. L., and Walton, C. C., Satellite derived sea surface temperatures from NOAA spacecraft, *NOAA Tech. Memo. NESS 78*, National Environmental Satellite Service, Washington, D.C., 1976.

2. Saur, J. F. T., A study of the quality of sea water temperatures reported in logs of ship's weather observations, *J. Appl. Meteorol.*, 2, 417, 1963.

3. Tabata, S., On the accuracy of sea-surface temperatures and salinities observed in the Northern Pacific Ocean, *Atmos.-Ocean*, 16, 237, 1978.

4. McClain, E. P., Pichel, W. G., and Walton, C. C., Comparative performance of AVHRR-based multichannel sea surface temperatures, *J. Geophys. Res.*, 90, 11,587, 1985.

5. McClain, E. P., Pichel, W. G., Walton, C. C., Ahmad, Z., and Sutton, J., Multi-channel improvements to satellite-derived global sea surface temperatures, *Adv. Space Res.*, 2, 43, 1983.1374, 1960.

6. Clauss, E., Hinzpeter, H., and Mueller-Glewe, J., Messungen der Temperaturstruktur im Wasser and der Brenzflaeche Ozean-Atmosphaere, *Meteor-Forschungsergeb., Reihe B*, 5, 90, 1970.

7. Ewing, G. and McAlister, E. D., On the thermal boundary layer of the ocean, *Science*, 131, 1374, 1960.

8. Grassl, H. and Hinzpeter, H., The cool skin of the ocean, GATE-Report 14, 1, WMO/ICSU, Geneva, , 1975, 229.

9. Grassl, H., The dependence of the measured cool skin of the ocean on wind stress and total heat flux, *Boundary Layer Meteorol.*, 10, 465, 1976.

10. Katsaros, K. B., The aqueous thermal boundary layer, *Boundary Layer Meteorol.*, 18, 107, 1980.

11. Hasse, L., The sea surface temperature deviation and the heat flow at the sea-air interface, *Boundary Layer Meteorol.*, 1, 368, 1971.

12. Paulson, C. A. and Simpson, J. J., The temperature difference across the cool skin of the ocean, *J. Geophys. Res.*, 86, 11,044, 1981.

13. Schluessel, P., Shin, H. -Y., and Emery, W. J., Comparison of satellite derived sea surface temperatures with *in situ* skin measurements, *J. Geophys. Res.*, 92, 2859, 1987.

14. Schluessel, P., Emery, W. J., Grassl, H., and Mammen, T., On the bulk-skin temperature difference and its impact on satellite remote sensing of sea surface temperature, *J. Geophys. Res.*, 95, 13,341, 1990.

15. Kent, E. C. Truscott, B. S., Taylor, P. K., and Hopkins, J. S., The Accuracy of ahips' meteorological observations: the results of the VSOP/NA, WMO/TD No. 455, 1991.

16. Weller, R. A. and Emery, W. J., *In situ* measurements of sea surface temperature, in *Sea Surface Temperature*, Joint Oceanographic Institute, July, 1992.

17. Saunders, P. M., The temperature at the ocean-air interface, *J. Atmos., Sci.*, 24, 269, 1967.

18. Walton, C. C., McClain, E. P., and Sapper, J. F., Recent changes in satellite-based multi-channel sea surface temperature algorithms, paper presented at Science and Technology for a New Oceans Decade, MTS 90, Marine Technological Soc., Washington, D.C., September 1990.

19. Wick, G. A., Emery, W. J., and Schluessel, P.,A comprehensive comparison between satellite-measured skin and multichannel sea surface temperatures, *J. Geophys. Res.*, 97, 5569, 1992.

20. Mellor, G. L. and Yamada, T., A hierarchy of turbulence closure models for planetary boundary layers, *J. Atmos. Sci.*, 31, 1791, 1974.

21. Mellor, G. L. and Yamada, T., Development of a turbulence closure model for geophysical fluid problems, *Rev. Geophys. Space Phys.*, 20, 851, 1993.

22. Emery, W. J., Wick, G. A., Schluessel, P., and Reynolds, R. W., Improving satellite infrared sea surface temperature estimates by including independent water-vapor observations, *J. Geophys. Res.*, 99, 5219, 1994.

23. Schluessel, P. and Emery, W. J., Atmospheric water vapor over oceans from SSM/I measurements, *Int. J. Remote Sensing*, 11, 753, 1990.

11

Coastal Zone Color Scanner on Nimbus and Sea-Viewing Wide Field-of-View Sensor on Seastar

Marlon R. Lewis

11.1 Introduction

The many-hued nature of the ocean surface results from the presence and activities of microscopic organisms that make up the base of the marine food chain. These organisms, the phytoplankton, alter the ocean optical properties — the absorption coefficient, the scattering coefficient, and the volume scattering function — in a manner that gives rise to changes in the color of the ocean surface as seen from above.[1,2] They also alter the effective penetration of visible light.

Variations in the concentration of phytoplankton in the upper ocean and consequent variations in the penetration of visible light have a fundamental impact on prediction of biological, physical, and geochemical oceanographic processes. The phytoplankton absorb solar energy and convert it to organic matter, thus providing the basis for the world fisheries production. In the process, phytoplankton remove inorganic carbon from the water, and reduce the partial pressure of carbon dioxide in the surface layers. Under some circumstances, the partial pressure may be reduced below that in the overlying atmosphere; a net flux of carbon dioxide from atmosphere to ocean results. Solar irradiance absorbed by the phytoplankton also contributes to the local heating rate and thus influences the development of the thermal structure and dynamics of the upper ocean. Under conditions of discharge of nutrients from either agricultural runoff or domestic sewage discharge, phytoplankton concentrations increase, often dramatically and occasionally with toxic consequences. Finally, the underwater visibility is also affected by variations in the concentration of phytoplankton, an important consideration for many applied interests.

For the past century, biological oceanographers have attempted to describe the range of variability in the marine phytoplankton in both time and space, with a view toward developing sufficient understanding to enable prediction, both of their concentration, and their influence on marine fisheries, on global biogeochemical cycles, and on the penetration of light in the ocean. This has proved to be a daunting task from traditional ship-based platforms; the concentrations of phytoplankton exhibit such strong and rapid variations in both time and space that shipboard observational programs attempting to build a large-scale view of their variations are all but precluded.

The concentrations of phytoplankton in the surface ocean can, however, be determined over the required synoptic scales from space-based remote ocean color observing platforms. Based on the 8 years of life of the first ocean color observation satellite — the Coastal Zone Color Scanner (CZCS, see Appendix B) — it can now be unequivocally shown that this is not only true, but that the sensitivity of the methodology is surprisingly good. Significant advances have subsequently been made in all of the scientific and applied areas touched by the phytoplankton; with the planned launch of several ocean color satellites over the next decade, one can be optimistic that many of the challenges that have plagued biological oceanographers for the past century will now be put to rest.

In what follows, the physical and biological bases for the observation of ocean color from space are briefly described. Next, a range of applications, based primarily on CZCS observations are discussed. Finally, the challenges that await new generations of both ocean observing visible radiometers and new generations of scientists that will make use of the data are outlined.

11.2 Physical and Biological Bases

The color of the ocean as seen from above results from solar energy which is backscattered from the ocean surface and interior. The deep blue of the open, oligotrophic waters results from the selective absorption and scattering of pure seawater, unadulterated with phytoplankton or other optically active substances. As one moves closer to shore, nutrient inputs generally increase. The consequent development of higher concentrations of phytoplankton changes the color from

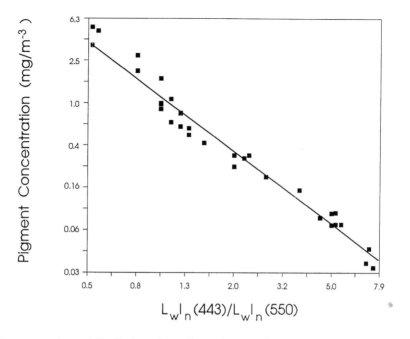

Figure 11.1 Spectral distribution of upwelling radiance (relative scale) just below the sea surface from expendable optical buoy measurements. The four lines indicate dramatically different water types from the relatively oligotrophic equatorial Pacific at one end to an intense bloom of dinoflagellates (red tide) that reached concentrations of 100 mg m^{-3} of chlorophyll.[30]

blue to green. As one moves even closer to shore, the situation becomes much more optically complicated, with contributions not only from higher concentrations of phytoplankton, but also from sediments and dissolved materials from terrestrial and inland sources. The color often approaches yellow-brown and in some circumstances, red.

The color as perceived by human observers can be quantified through the measurement of the spectral distribution of the upwelling, water-leaving radiances. For example, four optically very different water types are depicted in Figure 11.1, where the radiance distribution (as measured from expendable buoys,[3]) varies from the equatorial Pacific where the maximum in upwelling radiance is in the blue, to the increasingly greener waters of Monterey Bay, with an open connection to the sea, to Bedford Basin, an enclosed marine inlet on Canada's eastern shore. A massive bloom of dinoflagellates in Bedford Basin is reflected in the fourth upwelling spectrum: a red tide. Although these spectra were obtained from in-water measurements, their character would be preserved as one observed the oceans from spaceborne optical spectrometers.

However, the situation is somewhat more complicated, and it is necessary to consider not only oceanographic, but also atmospheric processes in the interpretation of observations of the oceans from space. As much as 95% (or even greater) of the radiances measured by the satellite sensor derive from light scattered out of the atmosphere as the solar energy transits to the earth's surface. The radiance observed by the satellite can be partitioned as:

$$L_t(\lambda) = L_r(\lambda) + L_a(\lambda) + t(\lambda, \theta) L_g(\lambda) + t(\lambda, \theta) L_w(\lambda) \qquad (1)$$

where L_t is the radiance observed at the satellite at a given waveband λ, L_r represents radiance molecularly (Rayleigh) scattered from the atmosphere, L_a is radiance scattered by atmospheric aerosols, L_g is that specularly reflected from the sea surface, and L_w is the desired water-leaving radiances derived from the interior of the oceans.[4,5] Both the specularly scattered and water-leaving radiances are reduced by the diffuse transmission back through the atmosphere to the satellite (t) observing the ocean at some angle θ to the vertical.

The so-called atmospheric correction involves estimation of the atmospherically scattered radiances which reach the spacecraft sensor in order to extract the small portion of the signal that provides useful information about the ocean. This exercise has met with considerable success. The procedure involves computations of the two sources of scattered light. The molecular or Rayleigh scattering typically makes the largest contribution;[6-10] and it has been found that multiple scattering terms must be included, particularly for high latitudes.[11] The aerosol scattering term is more problematic; formally, it is necessary to consider the vertical distribution of aerosols, their effective size distribution, and their optical properties.[12] Operationally, the clear-water approach used for the CZCS[13] involves determination of relatively pigment-free pixels in the image (typically <0.25 mg m^{-3} pigment) where the ocean can be assumed to be optically black in the near infrared region; any radiances observed over these wave bands are assumed to derive from the atmospheric scattering. Using an assumed spectral dependency for the aerosol scattering term, and assuming that the aerosol concentration and type are uniform over some region of the image, the other channels can then be corrected over the image. Future atmospheric correction models may use iterative schemes, which convolute the in-water radiance models with the atmospheric models to provide a pixel-by-pixel correction.

Assuming that the atmospheric contribution to the signal can be accounted for, it remains to interpret the water-leaving radiances in terms of the optical characteristics of the upper ocean, or alternatively, in terms of the variations in the concentrate and types of dissolved and particulate matter that contribute to variations in the bulk optical properties. The computations to achieve this are collectively referred to as the bio-optical algorithms.[14]

A convenient place to start is with the definition of the normalized water-leaving radiances, $L_w|_n$. These radiances are what would be expected to exit the ocean with a zenith sun, and with no intervening atmosphere. They then provide a means to compare different ocean regions without consideration of the amplitude or geometric distribution of the downwelling irradiance. As will be shown below, the normalized radiances are strictly a function of the optical properties of the ocean:[13,15]

$$L_w\big|_n = \frac{(1-\rho)(1-\bar{\rho})RF_0}{n^2 Q(1-rR)} \qquad (2)$$

where ρ is the Fresnel reflectance for upwelling radiance at normal incidence (assumed constant at 0.021); $\bar{\rho}$ is the albedo for irradiance incident at the sea surface from sun and sky; R is the irradiance reflectance just below the sea-surface ($\equiv E_U^{0-}/E_D^{0-}$, where E_U^{0-} is the upwelling irradiance just below the sea surface and E_D^{0-} is the downwelling irradiance, again just below the sea surface); n is the index of refraction; and $Q \equiv (E_U^{0-}/L_U^{0-})$ is the distribution function that is equal to π for a completely diffuse field, and normally equal to ≈ 4.6. L_U^{0-} is the upwelling radiance just below the sea surface. The term $1 - rR$ accounts for the internal reflectance below the sea surface. Most of the above terms have a strong spectral dependence that has been dropped for clarity.

The normalized, water-leaving radiances can be related to the inherent optical properties of the upper ocean, and from them, to the concentration and type of optically active constituents of the water, including the seawater medium itself, phytoplankton pigment (C, mg m^{-3}), dissolved organic material or yellow substances, and others.[4,13,16,17] For most open ocean areas, variations in the optical properties can be parameterized to a good approximation by variations in the pigment concentrations[17,18] (Case I waters); this is not necessarily a good approximation for many coastal and inland waters because of the presence of other dissolved and particulate materials that do not covary with pigment concentrations (Case II). The water-leaving radiances, therefore, provide the link between remote observations of the electromagnetic signature of the upper ocean and the inherent optical properties of the upper ocean, and thus the concentrations of dissolved and particulate matter of interest.

The irradiance reflectance R and the distribution function Q contain the necessary information; they are the variable portions of Equation 4, and occur as a ratio. Gordon et al.[16] have suggested, based on Monte Carlo techniques with a specified phase function and for Case I waters, that to a very good approximation this ratio can be cast in terms of inherent optical properties of the water:

$$\frac{R}{Q} = \sum_{i=1}^{2} l_i \left(\frac{b_b}{a+b_b}\right)^i \qquad (3)$$

where b_b is the backscattering coefficient and a is the absorption coefficient, both of which are inherent optical properties; and l_i are empirically determined coefficients. They are difficult to measure at sea; more commonly, the attenuation coefficient, K_D, is measured as the e-folding optical length scale for downwelling irradiance in the ocean. Although the attenuation coefficient varies with respect to solar zenith to some degree, K_D can be considered a quasi-inherent optical property for solar zenith angles less than 60°, and R/Q can also be related to K_D to the first order:

$$\frac{R}{Q} = 0.110 \frac{b_b}{K_D} \tag{4}$$

While these relationships are derived based on Case I waters, the functional form should also be valid for Case II waters. However, the coefficients may vary somewhat depending on the specific volume scattering phase function assumed that will almost certainly vary as water types make the transition from Case I to II.

The challenge now is to parameterize variations in K_D and b_b in terms of the biological quantities that are desired, with a view toward inverting Equations 1–6 to retrieve the biological quantities. In order to do this, the dependence of both K_D and b_b on the concentration of chlorophyll pigment is required. The first is easier than the second.

The assumption is made that the various constituents contribute to both the attenuation and backscattering in an additive fashion:

$$K_D = K_w + K_p + K_d \tag{5}$$

$$b_b = \left(b_b\right)_w + \left(b_b\right)_p \tag{6}$$

The seawater medium itself provides a constant background attenuation (K_w) and backscattering ($(b_b)_w$), with a well-known spectral dependence.[19] Morel[17] has compiled an extensive database to estimate the component contribution to the attenuation coefficient; for wavelengths shorter than 575 nm, the explained variance in the log-log relationship between K_D and the concentration of pigment is in excess of 80% without any explicit consideration of the influence of attenuation of dissolved organic matter, K_d. For these Case I waters then, Equation 7 only has two terms:

$$K_D(\lambda) = K_w(\lambda) + \phi(\lambda)Ce^{\psi(\lambda)} \tag{7}$$

where the empirically derived coefficients ϕ and ψ, have been tabulated with respect to wavelength,[17] as have the spectral variations in K_w.

For the backscattering, it is first necessary to outline the dependence of the total scattering coefficient, b (i.e., over all solid angles, m^{-1}) on pigment. For a relatively smaller data set than that used to derive the relationship between pigment and attenuation (and also a worse fit), it has been shown that:[4,16]

$$b_{550} = 0.3\, C^{0.62} \tag{8}$$

where as above, C is the concentration of pigment. The relationship between pigment and scattering is not nearly as good as that for attenuation.[20-24] The uncertainty is due to relatively high natural variability in the scattering relationship, in the relative contributions of detrital to viable phytoplankton, and in variations in the size structure of the marine particulate populations. Furthermore, certain species of phytoplankton, notably coccolithophorids, result in extraordinarily high scattering coefficients.[25,26] At 550 nm, scattering by particulate matter in the euphotic zone is generally much greater than the scattering due to pure seawater. The backscattering probability for particulates generally in any event is low, and varies inversely with wavelength. It is also depressed in the spectral regions with strong absorption. Purely empirical relationships have been developed to relate the backscattering probability to pigment concentration,[16,17] which can then be used with Equation 8 to estimate the particulate contribution to the backscattering coefficient. An alternative approach would be to estimate the size dependence of the resident particle population, and with a knowledge of their size-specific index of refraction and scattering theory, compute directly the backscattering contribution.[24] The difficulty here is in the routine estimation of the size structure and the size-dependent scattering efficiency.

The retrieval of pigment concentration then reduces to inverting the above to obtain the estimate given measurements of the normalized, water-leaving radiances. A significant improvement in the skill in estimation can be had by taking ratios of two wavelengths; the relative lack of spectral structure in the backscattering coefficient reduces its influence. The relationships developed for use with the CZCS archive are empirical representations of these relationships between upwelled radiance and concentration of phytoplankton pigment in the upper ocean, for example:

$$C = 1.15 \left(\frac{L_w|_n(443)}{L_w|_n(560)} \right)^{-1.42} ;\ C < 1mg\ m^{-3} \tag{9}$$

$$C = 3.64 \left(\frac{L_w|_n(500)}{L_w|_n(560)} \right)^{-2.62} ;\ C > 1mg\ m^{-3} \tag{10}$$

where the coefficient of determination is greater than 0.95 and the relative error is approximately 20% for Equation 10 and 30% for Equation 11.[15] Results from this relationship are seen in Figure 11.2. It should be noted that these relationships are based on a limited data set used in the calibration of the CZCS algorithms; regional

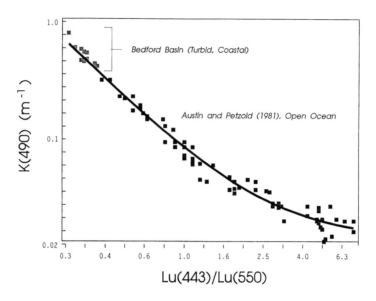

Figure 11.2   Ocean surface pigment concentrations as a function of the ratio between normalized upwelling radiances at 443 and 550 nm. Figure redrawn from Reference 16. Note that both axes are logarithmic.

relationships may depart substantially from them,[27] particularly in optically complex coastal regions.[28]

In the case of prediction of the penetration of visible light, the situation is somewhat more straightforward, since K_D is what is being determined directly. Consequently, the errors are diminished. Austin and Petzold[29] compiled a large body of data to develop an empirical relationship between upwelling radiance just below the surface (not normalized radiance) and the attenuation coefficient at 490 nm, which has proved robust, even in some coastal waters (Figure 11.3). This relationship can be combined with that in Equation 9 to provide a generalized means to recover the spectral attenuation coefficient:[30]

$$K_D(\lambda) = \phi(\lambda) \cdot 1.421^{\Psi(\lambda)} \left(\frac{L_{u,443}}{L_{u,550}} \right)^{\Psi(\lambda) - 2.124} + K_w(\lambda) \qquad (11)$$

11.3 Applications

The Coastal Zone Color Scanner operated intermittently from 1978 to 1986 (see Appendix B). During that time, over 60,000 clear scenes of 2-min duration were collected over the global ocean. Coverage was not uniform in either time or space,

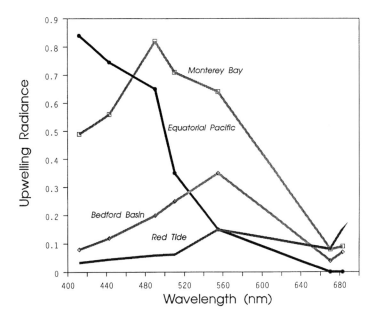

Figure 11.3 Attenuation coefficient at 490 nm as a function of the ratio of upwelling radiances at 443 and 550 nm. Data are taken from Reference 29 (open ocean) and unpublished data by the author. Note that the ordinate is logarithmic.

but the rich data set that resulted has served to stimulate much research, and perhaps as expected, generated as many questions as answers. The resulting data have been used in research on biogeochemical cycles, fisheries, upper ocean thermal dynamics, and a number of other areas; they are briefly reviewed below.

Global Biogeochemical Cycles

The primary application of ocean color data has been, and will likely continue to be, in the large-scale estimation of the role of the phytoplankton in global fluxes of carbon and other biogeochemically important compounds in the context of large international programs such as the Joint Global Ocean Flux Study. Excellent recent reviews and analyses of ocean color observations in this context are available.[31-35]

The marine phytoplankton carry out photosynthesis that utilizes sun energy to provide the organic matter necessary to sustain the marine food chain. In this process, they remove inorganic carbon, as well as other important plant nutrients such as nitrate and silicate, from the upper layers and release oxygen. Photosynthesis by phytoplankton therefore is a key process in controlling the biogeochemical cycles of carbon, nitrogen, and oxygen in the ocean; on a global scale, however, it varies by a factor of 10 or greater in both space and time. The role of the ocean phytoplankton remains controversial, in large part because of the current uncertainties in defining the amplitudes and scales of variability in a manner that permits accurate integration of the pigment biomass and productivity over the ocean surface free of alias.

The accumulated data from CZCS have now just started to be analyzed with a view toward defining the large-scale temporal and spatial variability, although the nonoperational nature of the CZCS operations makes this difficult in many regions.[36-40] Although the temporal and spatial patterns observed are consistent to the first order with previous ship-based studies,[41] the increased coverage has provided a view much more consonant with the large-scale ocean models currently under development.[42] For example, although the spring bloom of the phytoplankton in the North Atlantic has a rich historic observational basis,[43] much of this was from coastal waters. Few oceanographers anticipated the observed coherence and phasing of the North Atlantic spring bloom across the entire basin, or the implications that these dramatically increased levels might have for the carbon dynamics of the upper ocean.[44,45] Likewise, the increased levels of phytoplankton in the equatorial Pacific were known from the relatively few ship observations, but few expected that the equatorial parallel could be well defined from space from discolorations of the surface waters.[39]

Another bioproduct of phytoplankton photosynthesis also has potential geochemical implications. It has been argued that the release of dimethylsulfide from marine phytoplankton into the atmosphere may be a major source of cloud condensation nucleii;[46,47] and efforts are underway to utilize the synoptic ocean color data set, in conjunction with physiological models, to estimate its global significance.

The use of the CZCS data to address large-scale biogeochemical dynamics and to estimate the role of the ocean biota in prediction of future climate change is currently an area of great interest and activity.[31,34,35,48,49] The continued large-scale observations that will result from future satellite ocean color measurements are a spur to this effort, as well as a key source of data for use in the validation of future predictions.

Fisheries Oceanography

The marine phytoplankton are the principal source of organic matter that sustains the marine food chain. The productivity of the global commercial fisheries clearly depends on the phytoplankton for sustenance, and this has motivated many to make use of ocean pigment distributions in the estimation of the spatial distribution, and in the feeding success of several fisheries. Envisioned applications include both tactically directed fisheries, as well as use of the data to understand the food chain dynamics that may contribute to variability in recruitment success. An excellent review of the application of remote sensing, in general, and ocean color observations, in particular, to fisheries can be found in Pettersson et al.[50]

For example, the distribution of tuna in the northern Pacific,[51,52] the distribution of anchovy spawning,[53] and the distribution of shrimp in the Gulf of Mexico[54] have all been shown to be dependent on the spatial and temporal structure and dynamics of the ocean color field. During the brief existence of the CZCS, several organizations provided near-to-real-time ocean color maps to the fishermen; it is likely that operational use of data from future ocean color sensors will provide substantial benefits to both the commercial fishing industry, as well as the scientific management of economically important fish stocks.

Thermal Structure and Dynamics

Variations in the color of the sea reflect variations in the penetration of visible light. Roughly half of the incoming solar energy flux is associated with visible wavelengths, and the penetration of this energy represents in many cases a significant flux in the vertical. The consequences for the development of upper ocean thermal structure and dynamics are surprisingly large.

One of the earliest clues came from the mixed-layer modeling work of Denman[55] where the attenuation coefficient, which is a strong function of the phytoplankton concentration, was shown to be one of the most sensitive parameters. Lewis et al.[56] used the CZCS data over the equatorial Pacific to show that, contrary to what was thought previously, the upper mixed layer over most of the basin was not heated in the mean. The observations of net surface heat flux at the surface that led to the previous assumption were correct; however, the flux of visible light out of the base of the layer, which was ignored, was approximately equivalent to that entering the layer. Studies in the Arabian Sea confirmed the importance of consideration of the variable phytoplankton concentrations in the seasonal evolution of the thermal structure of the upper ocean.[57] The satellite data are particularly useful in evaluation of the role of optical variability in upper ocean heat budgets since the retrieval for the required attenuation coefficient is much better than that for pigment.

Other Applications

There are several other applications for which satellite ocean color data have proved useful. These include the remote determination of ocean organic pollution through its influence on the development of phytoplankton blooms; the determination of fronts, eddies, and ocean current features; and the use of ocean color data as a navigational aid. One of the more active areas of interest concerns the use of ocean color data in the determination of optical visibility for national defense requirements.

11.4 Future Prospects, Potential, and Challenges

With the proposed launch during the 1990s of at least five new instruments capable of accurate imaging of the color of the ocean surface, the future looks promising. The near term is dominated by the launch in 1994 of the Sea-Viewing, Wide Field-of-View Sensor (SeaWiFS) on the Seastar satellite by Orbital Sciences Corporation; it represents a combined research and commercial ocean color data source. Next in line is the Japanese Ocean Color and Temperature Sensor (OCTS) to fly on the Advanced Earth Observation Satellite (ADEOS) spacecraft in 1996 (see Appendix

I). Toward the end of the decade, the EOS/AM-1 satellite will contain the moderate resolution imaging spectrometer (MODIS) and the European Envisat will carry the MERIS ocean color satellite. Finally, the National Aeronautics and Space Administration (NASA) has plans to launch a free-flyer follow-on to SeaWiFS (EOS Color). The combined data volume will be enormous and will represent a rich resource for research into global change as well as substantial commercial and operational opportunities.

There is much that needs to be done to make full use of these data. The atmospheric models require improvement to take advantage of the improved sensor capabilities on the new generation of ocean color imagers. The bio-optical models developed for use with CZCS in the retrieval of ocean pigment concentration will require further work to evaluate the temporal and spatial scales over which the global models can be relied on, and to develop site-specific algorithms for regional studies where they cannot. One of the largest challenges will be in the evaluation of the rates of primary production solely from satellite data, both with respect to the total rate, as well as the rate at which carbon is exported from the upper ocean. As interest grows in applications in the coastal and inshore regions of the world, new models will have to be developed to account for the complex optical nature of these important waters. Finally, but far from the least important, the data rate and volume from the new generation of ocean observing satellite systems threaten to overwhelm the current capability for ingesting, processing, distributing, and archiving satellite data streams; improvement in this area will be required in order that the benefits of the improved global ocean color coverage can be realized.

Acknowledgments

This work was supported by the Naval Research Laboratory (U.S.), NASA, the U.S. Office of Naval Research, and the Natural Sciences and Engineering Research Council (Canada).

References

1. Yentsch, C. S. The influence of phytoplankton pigments on the color of seawater, *Deep-Sea Res.*, 7, 1, 1959.
2. Clarke, G. L., Ewing, G. C., and Lorenzen, C. J., Spectra of backscattered light from the sea obtained from aircraft as a measure of chlorophyll concentration, *Science*, 167, 1119, 1970.
3. McLean, S. D. and Lewis, M. R., *Expendable Spectral Radiometer Buoy*, in Proc. IEEE Oceans 91 Symp., IEEE Press, 1991.
4. Gordon, H. R. and Morel, A. Y. *Remote Assessment of Ocean Color for Interpretation of Satellite Visible Imagery: A Review*, Springer-Verlag, New York, 1983, 114 pp.
5. Gordon, H. R., Removal of atmospheric effects from satellite imagery of the oceans, *Appl. Optics*, 17, 1631, 1978.
6. Deschamps, P. Y., Herman, M., and Tanre, D., Modeling of the atmospheric effects and its application to the remote sensing of ocean color, *Appl. Optics*, 22, 3751, 1983.
7. Gordon, H. R., Brown, J. W., and Evans, R. H., Exact Rayleigh scattering calculations for use with the Nimbus-7 Coastal Zone Color Scanner, *Appl. Optics*, 37, 862, 1988.

8. Andre, J. M. and Morel, A., Simulated effects of barometric pressure and ozone content upon the estimate of marine phytoplankton from space, *J. Geophys. Res.*, 94, 1029, 1989.

9. Andre, J. M. and Morel, A., Atmospheric corrections and interpretation of marine radiances in CZCS imagery revisited, *Oceanol. Acta*, 14, 3, 1991.

10. Bricaud, A. and Morel, A., Atmospheric corrections and interpretation of marine radiances in CZCS imagery: use of a reflectance model, *Oceanol. Acta*, 7, 33, 1987.

11. Gordon, H. R. and Castano, D. J., Coastal Zone Color Scanner atmospheric correction algorithm: multiple scattering effects, *Appl. Optics*, 26, 2111, 1987.

12. Gordon, H. R. and Castano, D. J., Aerosol analysis with the Coastal Zone Color Scanner: a simple method for including multiple scattering effects, *Appl. Optics*, 28, 1320, 1989.

13. Gordon, H. R. and Clark, D. K., Clear water radiances for atmospheric correction of coastal zone color scanner imagery, *Appl. Optics*, 20, 4175, 1981.

14. Smith, R. C. and Baker, K. S., The bio-optical state of ocean waters and remote sensing, *Limnol. Oceanogr.*, 23, 247, 1978.

15. Gordon, H. R., Ocean color remote sensing systems: radiometric requirements, in Recent Advances in Sensors, Radiometry and Data Processing for Remote Sensing, Society of Photo-Optical Instrumentation Engineers, 924, 1988, 151.

16. Gordon, H. R., Brown, O. B., Evans, R. H., Brown, J. W., Smith R. C., Baker, K. S., and Clark, D. K., A semianalytic radiance model of ocean color, *J. Geophys. Res.*, 93, 10909, 1988.

17. Morel, A., Optical modeling of the upper ocean in relation to its biogenous matter content (Case I waters), *J. Geophys. Res.*, 93, 10749, 1988.

18. Morel, A. and Prieur, L., Analysis of variations in ocean color, *Limnol. Oceanogr.*, 22, 709, 1977.

19. Morel, A., Optical properties of pure water and pure sea-water, in *Optical Aspects of Oceanography*, Jerlov, N. G. and Nielsen, E. S., Eds., 1974, 1.

20. Bricaud, A. and Morel, A., Light attenuation and scattering by phytoplanktonic cells: a theoretical modeling, *Appl. Optics*, 25, 571, 1986.

21. Bricaud, A., Morel, A., and Prieur, L., Optical effciency factors of some phytoplankters, *Limnol. Oceanogr.*, 28, 816, 1983.

22. Prieur, L. and Sathyendranath, S., An optical classification of coastal and oceanic waters based on the specific absorption curves of phytoplankton pigments, dissolved organic matter, and other particulate materials, *Limnol. Oceanogr.*, 26, 671, 1981.

23. Morel, A., Chlorophyll-specific scattering coefficient of phytoplankton. A simplified theoretical approach, *Deep-Sea Res.*, 34, 1093, 1987.

24. Bricaud, A., Bedhomme, A.-L., and Morel, A., Optical properties of diverse phytoplankton species: experimental results and theoretical interpretation, *J. Plankton Res.*, 10, 851, 1988.

25. Holligan, P., Viollier, M., Harbour, D. S., Camus, P., and Champagne-Philippe, M., Satellite and ship studies of coccolithophore production along the continental shelf edge, *Nature (London)*, 304, 339, 1983.

26. Balch, W. M., Eppley, R. W., Abbott, M. R., and Reid, F. M. H., Bias in satellite-derived pigment measurements due to coccolithophores and dinoflagellates *J. Plankton Res.*, 11, 575, 1989.

27. Muller-Karger, F. E., McClain, C. R., Sambrotto, R. N., and Ray, G. C., A comparison of ship and CZCS-mapped distributions of phytoplankton in the Southeastern Bering Sea, *J. Geophys. Res.*, 95, 11483, 1990.

28. Carder, K. L., Steward, R. G., Harvey, G. R., and Ortner, P. B., Marine humic and fulvic acids: their effects on remote sensing of ocean chlorophyll, *Limnol. Oceanogr.*, 34, 68, 1989.

29. Austin, R. W. and Petzold, T. J., Remote sensing of the diffuse attenuation coefficient of sea water using the coastal zone color scanner, in *Oceanography from Space*, Gower, J. R. F., Ed., 1981, 239 pp.

30. Lewis, M. R. and McLean, S. D., Air-launched expendable drifting buoys in JGOFS EQPAC, in preparation.

31. Morel, A., Light and marine photosynthesis: a spectral model with geochemical and climatological implications, *Prog. Oceanogr.*, 26, 263, 1991.

32. Abbott, M. R. and Chelton, D. B., Advances in passive remote sensing of the ocean, U.S. National Report to International Union of Geodesy and Geophysics 1987–1990, Contributions in Oceanography, American Geophysical Union, 1991, 571.

33. Balch, W. M., Evans, R., Brown, J., Feldman, G., McClain, C., and Esaias, W., The remote sensing of ocean primary productivity: use of a new data compilation to test satellite algorithms, *J. Geophys. Res.*, 97, 2279, 1992.

34. Platt, T. and Sathyendranath, S., Oceanic primary production: estimation by remote sensing at local and regional scales, *Science*, 241, 1613, 1988.

35. Balch, W. M., Abbott, M. R., and Eppley, R. W., Remote sensing of primary production. I. A comparison of empirical and semi-analytical algorithms, *Deep-Sea Res.*, 36, 281, 1989.

36. Brown, O. B., Evans, R. H., Brown, J. W., Gordon, H. R., Smith, R. C., and Baker, K. S., Phytoplankton blooming off the U.S. East Coast: a satellite description, *Science*, 229, 163, 1985.

37. Eslinger, D. L., O'Brien, J. J., and Iverson, R. L., Empirical orthogonal function analysis of cloud-containing Coastal Zone Color Scanner images of northeastern North American coastal waters, *J. Geophys. Res.*, 94, 10884, 1989.

38. Maynard, N. G. and Clark, D. K., Satellite color observations of spring blooming in Bering Sea shelf waters during the ice edge retreat in 1980, *J. Geophys. Res.*, 92, 7127, 1987.

39. Feldman, G., Clark, D., and Halpern, D., Satellite color observations of phytoplankton distribution in the eastern Equatorial Pacific during the 1982–1983 El Nino, *Science*, 226, 1069, 1984.

40. Feldman, G. C., Kuring, N., Ng, C., Esaias, W., McClain, C., Elrod, J., Maynard, N., Endres, D., Evans, R., Brown, J., Walsh, S., Carle, M., and Podesta, G., Ocean color: availability of the global data set, *Eos*, 70, 634, 1989.

41. Yoder, J. A., McClain, C. R., Feldman, G. C., and Esaias, W. E., Annual cycles of phytoplankton chlorophyll concentrations in the global ocean: a satellite view, *Global Biogeochem. Cycles*, 7, 181, 1993.

42. Sarmiento, J. L. and Toggweiler, R. R., A new model for the role of the oceans in determining atmospheric CO_2, *Nature (London)*, 308, 621, 1984.

43. Sverdrup, H. U., On conditions for the vernal blooming of phytoplankton, *J. Cons. Int. Explor. Mer.*, 18, 287, 1953.

44. Wroblewski, J. S., Sarmiento, J. L., and Flierl, G. R., An ocean basin scale model of plankton dynamics in the Northern Atlantic. I. Solutions for the climatological oceanographic conditions in May, *Global Biogeochem. Cycles*, 2, 199, 1988.

45. Campbell, J. W. and Aarup, T., New production in the North Atlantic derived from seasonal patterns of surface chlorophyll, *Deep-Sea Res.*, 39, 1669, 1992.

46. Charlson, R. J., Lovelock, J. E., Andreae, M. O., and Warren, S. G., Oceanic phytoplankton, atmospheric sulphur, cloud albedo and climate, *Nature (London)*, 326, 655, 1987.

47. Bates, T. S., Charlson, R. J., and Gammon, R. H., Evidence for the climatic role of marine biogenic sulphur, *Nature (London)*, 329, 319, 1987.

48. Deuser, W. G., Muller-Karger, F. E., Evans, R. H., Brown, O. B., Esaias, W. E., and Feldman, G. C., Surface-ocean color carbon flux: how close a connection?, *Deep-Sea Res.*, 37, 1331, 1990.

49. Platt, T., Caverhill, C., and Sathyendranath, S., Basin-scale estimates of oceanic primary production by remote sensing: the North Atlantic, *J. Geophys. Res.*, 96, 15147, 1991.

50. Pettersson, L. H., Johannessen, O. M., Kloster, J., Olaussen, T. I., and Samuel, P., Application of remote sensing to fisheries, Vol. 1 and 2, Nansen Remote Sensing Center, under contract 3348-87-12 ED ISPN.

51. Laurs, R. M., Fiedler, P. C., and Montgomery, D. R., Albacore tuna catch distributions relative to environmental features observed from satellites, *Deep-Sea Res.*, 31, 1085, 1984.

52. Fiedler, P. C. and Bernard, H. J., Tuna aggregation and feeding near fronts observed in satellite imagery, *Continental Shelf Res.*, 7, 871, 1987.

53. Fiedler, P. C., Smith, G. B., and Laurs, R. M., Fisheries applications of satellite data in the eastern North Pacific, *Fisheries Rev.*, 46, 1, 1984.

54. Leming, T. D. and Stuntz, W. E., Zones of coastal hypoxia revealed by satellite scanning have implications for strategic fishing, *Nature (London)*, 310, 136, 1984.

55. Denman, K. L., A time dependent model of the upper ocean, *J. Phys. Oceanogr.*, 3, 173, 1973.

56. Lewis, M. R., Carr, M. E., Feldman, G. C., Esaias, W., and McClain, C., Influence of penetrating solar radiation on the heat budget of the equatorial Pacific Ocean, *Nature (London)*, 347, 543, 1990.

57. Sathyendranath, S. Gouveia, A. D., Shetye, S. R., Ravindran, P., and Platt, T., Biological control of surface temperature in the Arabian Sea, *Nature (London)*, 349, 54, 1991.

Part III

Wind, Wind Waves, and the Marine Boundary Layer

12

Wind Speed from Altimeters

Ella B. Dobson

12.1 Background

Six altimeters have orbited the earth making measurements of ocean parameters since 1973. The satellite altimeter is basically a nadir pointing radar that, in pulse-limited altimeters, transmits a pulse and measures the shape, power, and slope of the returned pulse. From this pulse several physical properties of the ocean can be derived. The general sensor design and specific characteristics of each altimeter are given in Appendix D. The parameter to be examined in this chapter is the surface wind speed as measured by the altimeter. It should be noted that the altimeter makes no direct measurement of wind direction, but as we will demonstrate, wind speed is a useful quantity in itself for many applications.

0-8493-4525-1/95/$0.00+$.50
© 1995 by CRC Press, Inc.

The first altimeter to be flown was S-193 which flew aboard Skylab in 1973 and flew several 1–2-month missions. Compared to altimeters that followed it, Skylab altimeter generated a small database, but proved that an altimeter could measure sea surface height, significant wave height, and wind speed. The next altimeter GEOS-3 flew from 1975 to 1977. Researchers were able to obtain some oceanographic information from GEOS-3, but coverage and accuracy left much to be desired; once again the results that were obtained were encouraging.

The Seasat altimeter was designed for much greater accuracy than GEOS-3 but was short-lived and flew for only 3 months. The wind speed measurements made by Seasat have contributed significantly to the global climatology of wind speed and were helpful in the design of the Geosat and TOPEX altimeters. Both the GEOS-3 and the Seasat altimeters were sponsored by the National Aeronautics and Space Administration (NASA). In 1985 a Navy-sponsored altimeter was launched that flew until 1990. The first altimeter flown by the European community was launched in 1991 and is presently flying and making accurate wind speed measurements. The latest altimeters launched are the TOPEX/Poseidon. These altimeters which fly aborad the same satellite were launched in 1992 and are providing accurate global wind speeds. Having the TOPEX and ERS-1 flying at the same time is providing a wind speed database which in global coverage is the most extensive to date.

In this section we will discuss the reason the altimeter is able to measure wind speed, the way wind speed is extracted from the signal, the wind speed algorithms, the applications and use of the data, and the potential use of altimeter-measured wind speed.

12.2 Electromagnetic Scattering from the Ocean Surface

The sea surface is composed of waves that in simple terms can be divided into two main categories, wind and swell. Wind waves can further by subdivided into capillary and gravity waves. The latter two are driven by the wind and form the basis for our ability to measure wind speed with the altimeter. Theoretical studies of the backscatter from the ocean surface have been performed by researchers for many years. Some of these works can be found in Barrick,[1] Valenzuela,[2] as well as many others. A good review of these theories and the fundamentals of surface backscatter can be found in Ulaby.[3] When a pulse is transmitted from a spaceborne altimeter, the signal is backscattered from the ocean surface to the satellite receiver. This backscattered energy is from waves longer than the incident radiation, and at the frequencies employed by most altimeters this means waves longer than approximately 2 cm. For a radar pointing at or near nadir the backscattered energy is a result of specular reflection, and can be characterized by using geometric optics theory as shown by researchers such as Barrick[4,5] and Jackson et al.[6] When the sea is calm with very few waves, the majority of the energy from the illuminated area is backscattered in the direction of the incident radiation. As the sea becomes rougher, more facets on the surface scatter in different directions, reducing the backscatter received at the altimeter. The waves that are determining this backscatter are driven by the wind so that the higher the wind, the rougher the sea and thus the lower the cross section. Although the sea is very complex and the actual backscatter is a result of the spectrum of the ocean waves, the interaction of the wind and the waves allows the altimeter to measure wind speed. This relationship is independent of wind direction.

12.3 AGC to σ^0 Conversion

Altimeter receiver systems are in general designed to operate within a linear region of response for all parts of the receiver. To accomplish this an automatic gain control (AGC) loop is included in the receiver system. This loop applies attenuation as required to keep the total power received constant. The sigma naught (σ^0) value is actually computed from this AGC. For the Seasat and Geosat altimeters the received signal is gated in bins ranging from 0 to 63 dB and averaged over some fraction of a second. For Seasat and Geosat this time was one twentieth of a second. Using various internal calibrations, loss measurements, biases, etc., the σ^0 is computed from the AGC. A description of the AGC to σ^0 conversion for Seasat can be found in Townsend,[7] and an overall summary can be found in Chelton.[8]

We note here that the power received from the ocean surface is dependent on the area illuminated by the radar. This is why we use σ^0, which is the normalized radar backscattered power per unit area.

12.4 Algorithms for Radar Cross Section to Wind Speed Conversion

To say altimeter wind speed algorithms have been controversial since the launch of the first altimeter is almost an understatement. In this section we will attempt to review most of the wind speed conversion algorithms that have been presented in the literature, describe their origins, and point out the differences. No attempt will be made in this section to determine which algorithm is most accurate or appropriate for any given altimeter, but at this point we wish to summarize each briefly. The wind speed conversions from radar cross section will be presented according to the originator and to some extent in the order they were published in the open literature.

Brown (BR)

Brown et al.[9] derived an expression for σ^0 based on the theory that at normal incidence the power backscattered from the ocean surface is inversely proportional to the mean square slope of a filtered surface. Using this theory Brown expresses the backscatter as:

$$\sigma^0 = \alpha \left| R\left(O^0\right) \right|^2 \Big/ \zeta^2 \qquad (1)$$

where $\left| R\left(O^0\right) \right|^2$ is the Fresnel power reflection coefficient σ^0 backscattered power per unit area and ζ^2 is the mean square slope spectrum. By using the data of Cox and Munk[10] and the wave height spectrum of Phillips,[11] ζ can be expressed as:

$$\zeta^2 = a \ln W + b \qquad (2)$$

By combining Equations 1 and 2, Brown obtains the expression:

$$\sigma^0\left(O^0\right) = -2.1 - 10 \log_{(10)}\left(A \ln W + B\right) \qquad (3)$$

Brown used a set of buoy measurements and the GEOS-3 altimeter radar cross section measurements to perform a least squares fit to Equation 3. When fitting the data, Brown found that they were best fit by applying two sets of A's and B's, where one set was obtained by fitting data ≤9.2 m/sec and the other >9.2 m/sec. Brown acknowledged that the buoy data were not as accurate as required, but the results were positive and altimeter measurement of wind speeds seemed promising. In 1981 Brown et al.[9] used what they considered a better buoy wind speed database to repeat Brown's analysis, and when applied to Equation 3 yielded the following relationship between σ^0 and wind speed at 10 m above the surface. They found that the best fit for this data set required applying Equation 3 to three different wind speed regions. This regression analysis yielded:

$$W = \exp^{\left[(S-B)/A\right]} \qquad (4)$$

Where $S = 10^{\left[-\sigma^0 + 2.1\right]/10}$:

$a = 0.1595$	$b = 0.017215$	$\sigma^0 >= 10.9$
$a = 0.0398934$	$b = -0.031996$	$10.12 <= \sigma^0 <= 10.9$
$a = 0.080074$	$b = -.0124651$	$\sigma^0 = 10.12$

where σ^0 is in decibels.

Although the fit was good in the least squares sense, the distribution of the difference between buoy wind speeds and wind speeds computed from Equation 3, using the coefficients above, resulted in a skewed error in the distribution of the differences between buoy and altimeter wind speeds. To correct this skewness, a fifth-degree polynomial fit was applied to the data, between 0 and 16 m/sec, of the form:

$$W = \sum_{n=1}^{n} a_n W^n \tag{5}$$

where $a_1 = 2.08779$, $a_2 = -0.3649928$, $a_3 = 0.004062421$, $a_4 = 0.00001904962$, and $a_5 = 3.288189 \times 10^{-5}$.

Equation 5 seemed to correct this skewness. As we will discuss later, it did not completely alleviate the problem. We should also note that the maximum wind speed in the buoy database was 18 m/sec, and the total edited buoy-altimeter database was 184 values. With these data Brown obtained a root mean square (rms) difference between the altimeter and buoy wind speeds of 1.74 m/sec.

Chelton and McCabe (CM)

In derivation of this algorithm, Chelton and McCabe[12] used the Seasat scatterometer (SASS)-measured wind speeds to obtain an empirical relationship between σ^0 and wind speed. To derive this, the altimeter σ^0 data were averaged over 2° latitude and 6° longitude and compared to similar areas of wind speed computed from off-nadir Seasat scatterometer data. The global altimeter wind speed algorithm from Chelton and McCabe is given by:

$$\sigma^0 = 10^{\left(\sigma^0/10.-G\right)/H} \tag{6}$$

where $G = 1.502$, $H = -0.468$, and σ^0 is in decibels.

A 96 day data set was used in this analysis, which resulted in 1947 averaged data points. Note also that this algorithm is for wind speeds at 19.5-m altitude above the surface. The range of wind speeds included in the fit are from 4 to 14 m/sec. The CM article was critical of the BR algorithm given in Section 2.1 (Equations 3 and 4) because the Brown data were fit in three sections. They noted that the discontinuities that occur at 10.12 and 10.9 dB caused the wind speed distribution to be trimodal, which is contrary to the knowledge of wind speed distributions over the ocean.

Chelton and Wentz (CW)

In a later publication Chelton and Wentz[13] reevaluated the algorithm and discovered that the SASS database used was flawed, and resulted in several questions as to the validity of the CM algorithm. To correct this, Chelton and Wentz employed a different method of pairing scatterometer and altimeter. In addition, a new SASS scatterometer algorithm was applied. The authors cautioned that since this algorithm was developed by calibrating altimeter wind speeds to SASS wind speeds, they can be only as accurate as the SASS winds. We will discuss later some factors that should be considered when making any assessment of altimeter wind speed

Table 12.1 Chelton-Wentz and Modified Chelton-Wentz Tabular Altimeter Wind Speed vs. Radar Backscattered Cross Section

σ^0	CW $U_{19.5}$ (m/sec)	MCW U_{10} (m/sec)
7.00000	—	20.1540
7.20000	—	19.5970
7.40000	—	19.0380
7.60000	—	18.4630
7.80000	—	17.8770
8.00000	21.0800	17.2770
8.20000	20.3410	16.6550
8.40000	19.5710	16.0110
8.60000	18.767	15.348
8.80000	17.9200	14.6690
9.00000	17.0190	13.9760
9.20000	16.0690	13.2730
9.40000	15.0790	12.5570
9.60000	14.0620	11.8300
9.80000	13.0260	11.0920
10.0000	11.9820	10.3450
10.2000	10.9390	9.59000
10.4000	9.90700	8.82700
10.6000	8.89200	8.05900
10.8000	7.90900	7.29800
11.0000	7.00700	6.57700
11.2000	6.22200	5.92100
11.4000	5.53100	5.32100
11.6000	4.91000	4.76300
11.8000	4.36000	4.25200
12.0000	3.87700	3.79200
12.2000	3.45200	3.37800
12.4000	3.08800	3.01400
12.6000	2.78700	2.70800
12.8000	2.52700	2.44700
13.0000	2.28600	2.20800
13.2000	2.07300	2.20800
13.2000	2.07300	2.20800
13.2000	2.07300	1.99200
13.4000	1.90200	1.81700
13.6000	1.76100	1.67600
13.8000	1.62900	1.54700
14.0000	1.49700	1.41900
14.2000	1.36600	1.29200
14.4000	1.23600	1.16700

accuracy. For completeness, the tabular values are presented in Table 12.1 for σ^0 values ranging from 8.0 to 19.6 m/sec. CW is for 19.5 m and MCW is for 10 m above the surface.

Smoothed Brown (SB)

Prior to the launch of Geosat in 1985, a decision had to be made concerning the algorithm that would be used in onboard processing to compute wind speed from AGC. Since there had been some questions raised concerning the Brown algorithm (Section 1.4.1), JHU/APL was tasked with determining what the onboard algorithm should be. To this end, APL acquired the National Ocean Data Center (NODC) buoy database used to compute the Brown algorithm from Gary Brown.[14] These data were used to compute the smoothed Brown algorithm, by computing the best fit fifth-degree polynomial to the data using regression analysis. This algorithm (Goldhirsh and Dobson,[15] Dobson[16]) is given by a fifth-degree polynomial:

$$W = \sum_{n=0}^{n} a_n \sigma^0 \qquad (7)$$

$8 < \sigma^0 < 15$ dB.
Where $a_0 = 2.087799$, $a_1 = -.3649928$, $a_2 = 0.04062421$, $a_3 = -0.001904952$, and $a_4 = 0.00003288189$

σ^0 is radar backscatter in decibels.

This algorithm is for 10-m height, and was to be used with σ^0 values between 8 and 15 dB or approximately 1.5 and 14 m/sec. However, only four data values above 12 m/sec were used in the fit and the results above that value are not statistically significant. Let us note that although this algorithm was developed to determine which algorithm should be used in the Geosat processing onboard the spacecraft, it was not the algorithm finally used. The decision was made, despite the regions of discontinuity in the Brown algorithm because it was derived from some consideration of backscatter theory (Cox and Munk[10] and Phillips[11]), and had proved fairly accurate in results obtained from Seasat σ^0 data, that the Brown algorithm be used in onboard processing. The smoothed Brown algorithm, when compared with buoy data, had a mean and rms difference of 0.3 and 1.7 m/sec, respectively.

The author notes that in recent years, I have employed a smoothed version of Equation 3. That is, instead of computing a fifth-degree polynomial from the σ^0 vs. buoy wind speed data as was done with the smoothed Brown algorithm, I have fitted a fifth-degree polynomial to tabular data computed from Equation 3 at 0.1 dB steps and smoothed the curve around the discontinuities. This results in a curve that is very close to the Brown algorithm, and eliminates the discontinuities. This polynomial fit has not been published in the open literature, so that the equation will not be given here.

Modified Chelton Wentz (MCW)

Witter and Chelton[17] reported on the results of further modification of the CW algorithm. This algorithm was developed after it was observed that there was a systematic bias between the Seasat and Geosat altimeter σ^0 data. It should be emphasized that this algorithm was developed so that the Seasat altimeter algorithm CW could be applied to the Geosat altimeter. The method used in the MCW development was to compare the histograms from the Seasat and Geosat altimeter σ^0 measurements for July 7–October 10. For Seasat 1978 data were used, and for Geosat 1987 and 1988 data were used. Witter and Chelton found that there was a systematic error between the two histograms that was corrected by utilizing Seasat SASS σ^0 data. To assess the accuracy of the algorithm, Witter and Chelton compared the results of winds computed from the MCW algorithm and 119 National Data Buoy Center wind speed measurements. The rms difference between the buoy and altimeter winds was 1.9 m/sec with a mean difference of 0.45 m/sec. This algorithm is also given in tabular form in Table 12.1 for 10.5-m height.

Two Stick (TS)

Carter[18] assessed the accuracy of altimeter measured wind speeds by using the NODC buoy database made available by Glazman, which consisted of 405 data points. These buoy wind speeds were chosen on the criteria that they be within 1 hour of a Geosat overpass and within one fourth degree of the buoy. They found that the data could be least squares fit to two straight lines, hence the two stick characterization. The straight lines are as follows:

$$u = 44.73 - 3.424 \, \sigma^0 \quad \sigma^0 < 12.2$$
$$u = 5.773 - 0.228 \, \sigma^0 \quad 12.2 < \sigma^0 < 25.3$$

(8)

The rms deviation of the buoy and the altimeter winds was 1.46 m/sec. The discontinuity between the two lines occurs at approximately 3 m/sec. This discontinuity introduces a problem in the wind speed distribution similar to the problem with the Brown algorithm. Carter et al. recognize this to be a problem and suggest some sort of smoother might be used, but make no attempt to eliminate the discontinuity in the article.

Wu

Wu[19] used specular point theory similar to Brown to develop a theoretical algorithm. Using the mean square slope spectrum as measured by Cox and Munk and verified by Wu, and the fact that the Geosat altimeter had a wavelength of 2.16 cm, which is very close to the cutoff wavelength of facets that contribute to radar backscatter, Wu argues that the constants in Equation 3 should be $a = 0.009$ and $b = 0.012$. Using these constants Wu proposes the following algorithm:

$$\sigma^0 = -4.2 - 10 \log(0.009 + 0.012 \ln U) \qquad (9)$$

This equation is valid for 10-m height above the surface.

In a recent article,[20] Wu has refined his algorithm at the lower wind speeds; this change results in using Equation 9 for wind speeds greater than 2.4 m/sec and the following equation for wind speeds less than 2.4 m/sec:

$$\sigma^0 = 15.3 - 6.3 \log U_1 O \qquad (10)$$

We have discussed algorithms that have been published to date. Undoubtedly with ERS-1 and TOPEX altimeters presently flying, there will either be different algorithms or more confidence in some existing ones. Perhaps additional measurements, where high wind speeds are obtained, will serve to sort out the best algorithm for overall accuracy.

12.5 Waves and Wind Speed

There is a school of thought among many researchers that in order for an altimeter wind speed to be more accurate than any presently available, some information on the surface waves must be included.

In several publications Glazman,[21-23] has presented evidence that sea wave maturity has an effect on the relationship between σ^0 and wind speed. Glazman and Pilonz[24] presented theoretical arguments to prove that sea maturity affects the conversion between σ^0 and have developed an empirical approach to quantify the relationship. Spectra from NODC buoys were used in the analysis, and rather strenuous criteria were used to cull the spectra by shape to ensure that only fully developed seas were used. This eliminated two thirds of the data, which leads one to believe that the conditions required in this analysis do not exist in the majority of cases in nature or that the buoys are not located where the conditions exist the most. It is suggested that the articles cited be consulted if one desires more information on the algorithms developed for including fetch. The problem with including fetch is that fetch information is not readily available especially in real time operations.

Monaldo and Dobson[25] added significant wave height as measured by the altimeter to determine whether the rms differences between the buoy and altimeter wind speeds could be reduced. This was done by computing the wind speed using the SB and CW algorithms and then adding $H_{1/3}$ and a parameter called excess $H_{1/3}$ to a linear relationship such as:

$$U_B = a_0 + a_1 U_A + a_2 X_2 \qquad\qquad (11)$$

where U_B is buoy wind speed, U_A is altimeter wind speed from SB or CW, and X_2 can be excess $H_{1/3}$ or $H_{1/3}$. The results indicated some reduction in the rms difference, but was not statistically significant enough to warrant a general conclusion. Carter[18] in his analysis added $H_{1/3}$ to the linear model in an attempt to improve the relationship between σ^0 and wind speed. He found no reduction and observes that σ^0 is related most closely to the sea surface slope variance that does not have a known relationship to $H_{1/3}$.

It is possible that some parameter that describes wave structure should be included in the algorithm for wind speed, but at present that quantity either is not available (i.e., fetch) or is not required.

12.6 Errors in Ground Truth Data

We have discussed three data sources that were used to validate altimeter wind speed measurement. These were buoys, ships, and aircraft radars. Each of these has its own set of errors associated with it. As mentioned earlier, Cardone[26] has looked at trends in ship measurements of wind speeds over more than 150 years, and has found that the winds changed as measuring procedures advanced. This trend is not directly applicable to ship data used in recent years to validate altimeter measurements, but individual instrument calibrations must be considered even with today's more sophisticated anemometers. The other considerations are instrument height, time, and spatial separation of measurements. Monaldo[27] performed a detailed analysis of the errors one can expect when comparing buoy with altimeter wind speeds. All of the factors considered by Monaldo are applicable to ship and airborne radar measurements as well. In his analysis, Monaldo examined instrument errors for altimeter and buoy, time and area averaging, temporal separation, and spatial separation. He found that if both buoy and altimeter were perfect, one can expect a 1.3 m/sec error when spatial measurements are within 50 km. The work of Tournadre and Ezraty[28] indicates that this error could be greater depending on the region of the ocean where the measurements are made.

Wu[29] compared altimeter data to aircraft measurements of wind. All of the above considerations should be applied to an expected difference in these measurements, in addition to the inherent differences in the instruments such as footprint and location (i.e., open ocean measurements vs. bays or gulfs). These errors indicate that no matter how accurate the algorithm, one should never expect perfect agreement between an altimeter and any other instrument.

12.7 General Discussion of Algorithms

We have presented eight altimeter wind speed algorithms in the previous section. Six of these were empirically derived, one pseudo-theoretical, and one theoretical.

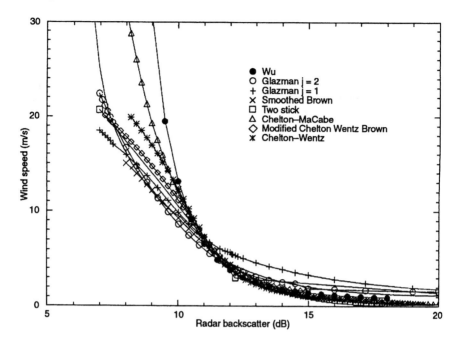

Figure 12.1 Comparison of algorithms for computing wind speed from radar cross section.

Figure 12.1 shows a plot of wind speed in meters per second vs. radar backscattered cross section σ^0 for these algorithms.

It is very obvious that the algorithms give a wide range of wind speeds for the same cross section at most values of σ^0 except in the vicinity of 11 dB. The differences at high wind speeds are particularly large. For example, at $\sigma^0 = 10$ dB the spread is from 8.4 for Glazman j = 2 to 13.01 for Wu. For a radar cross section equal to 9.0 dB, the spread is from 11.9 to 30.8. These differences are there despite the fact that the rms errors quoted are not that different. This author feels that one of the problems with using rms difference as a measure of accuracy for these data is that the largest number of points used in the comparisons, whether with buoys or scatterometer, are for wind speeds between 5 and 12 m/sec; and in these ranges of wind speeds, all algorithms are probably within 2 m/sec rms difference from buoy measurements. The problems occur primarily at the high wind speeds, and a major problem in determining which algorithm is most accurate at high wind speeds is obtaining a sufficiently large database at these high winds to compare with altimeter winds whether that database is from buoys or some other measuring instrument. Comparisons with other microwave instruments to determine accuracy are somewhat suspect because if there are any inherent problems due to the nature of microwave measurements, they may be contained in both data sets.

All algorithms known to the authors have been included here because all have been used by at least one researcher in the recent past. We should, however, point out the current status of these algorithms. The CW and MCW have superseded the CM algorithm, and it appears the CW should be used with Seasat data and the MCM should be applied to the Geosat database. The SB algorithm should be used for wind speeds between 3 and 12 m/sec, because for larger wind speeds, the algorithm underestimates the winds. The Glazman algorithms should be considered only if one has the benefit of fetch or wave history information simultaneous with the altimeter data. One can see from Figure 1 that there is a considerable difference between j = 1 and j = 2, and choosing one over the other without the correct wave age information may result in considerable error even if the algorithms were completely accurate. The criticism of the BR algorithms is still the branches causing discontinuities in the distribution; the MB algorithm eliminates this problem. The Wu algorithm is the only one based on theory solely and would be desirable if accurate for that reason alone. However, the algorithm estimates wind speeds $\sigma^0 > 10$ dB higher than any of the others. It appears to this author that until substantial progress can be made in obtaining accuracy at high wind speeds, none of the algorithms can claim to be more accurate at these higher speeds.

12.8 Research Applications and Validation of Altimeter-Measured Winds

Ultimately any satellite measurements are only as good as the benefit derived from their application to aid in the further understanding and modeling of physical processes. The altimeter-measured winds are no exception to this rule. When we consider the altimeter databases from GEOS-3, Seasat, Geosat, ERS-1, and TOPEX/Poseidon, the spatial and temporal global distribution of the data must be considered and the resolution (i.e., footprint) must be examined to see whether the data are adequate or applicable to the problem to be solved. In some cases, the data will be quite adequate; in others they will serve as an adjunct to other data sets used collectively. Nevertheless, we must concentrate some of our efforts in the research community to determine the best applications for altimeter wind measurements, and to provide the designers of future instruments with those characteristics that will make the data most useful.

In this section we will review some of the research that has utilized altimeter measured winds. This review is not all inclusive and is meant only to familiarize the reader with possible applications.

Ship and Buoy Comparisons

Some of the earliest efforts to validate altimeter wind speeds was reported by Mognard and Lago[30] using GEOS-3 data. The derivation of altimeter wind from AGC and the comparison with buoy wind speeds revealed very early that the altimeter measurements showed promise. Results obtained produced agreement good enough to give confidence that the Seasat altimeter, designed to have better backscatter accuracy, would give wind speeds within the Seasat design specifications that were 2 m/sec at that time.

Dobson et al.[31] compared altimeter geodetic mission winds to NOAA buoy measurements of wind, using the CM, CW, BR, and SB algorithms. In these comparisons the BR and SB algorithms gave rms differences between buoy and altimeter of 1.7 m/sec when data comparisons were limited to 30 min and 50 km between the two measurements. The CM and CW gave values of 2.0 m/sec when only open ocean buoys were used in the comparison. Dobson[16] made further buoy altimeter comparisons with Geosat data during the exact repeat mission and found results similar to the first report, but found that the rms differences dropped slightly when comparisons were limited to less than 25 km. An interesting sidelight of this work was that comparisons were made with individual buoys. The data clearly showed that in some cases the buoys were biased and in need of calibration, while other buoys gave extremely good agreement with no bias. Etcheto and Banege compared Geosat winds to global ship measurements of wind speed and found that of the SB, CW, and MCW algorithms, the CW gave the best comparisons. The authors did not give rms difference values that can be used in comparison with previous values given in this section. The choice of the best algorithm was made based on wind speed bias over the entire range of ship winds and showed that winds computed using CW gave less than a 1 m/sec bias. The authors state that this 1 m/sec bias is all that can be expected from measurements made by ships of opportunity due to their own inherent errors. Potential errors in ship measurements of wind are discussed in detail in Cardone et al.[26]

Ebuchi et al.[32] made comparisons with data from the Japan Meteorological Agency, in a manner similar to Dobson using the BR, SB, and CM algorithms. They found rms differences between buoy and altimeter winds of 2.3, 2.2, and 4.1, respectively. In addition they observed an interesting trend; that is, when the difference between the buoy and altimeter wind speeds was plotted vs. buoy wind speeds, a correlation coefficient of 0.617 was computed. The trend was linear, and the regression relationship was as follows:

$$U = \left(U_A - 2.42 \right) / 0.65 \qquad (12)$$

where U_A is altimeter wind speed before adjustment.

The authors point out that this relationship was developed on a limited set of data and further verification is required. After reading Ebuchi's results, we decided to determine whether this trend was present when some of the buoy data in our possession was used. We used 97 NODC buoy measurements to compare with Geosat wind values computed using the SB and MCW algorithms and found in both cases that the error (i.e., buoy-altimeter wind speed) had a linear relationship with buoy wind speed. The difference data were fit using a linear regression that resulted in a tilt and bias for wind speeds computed using both the SB and CW algorithms. The slope and intercept were different for each of the algorithms and

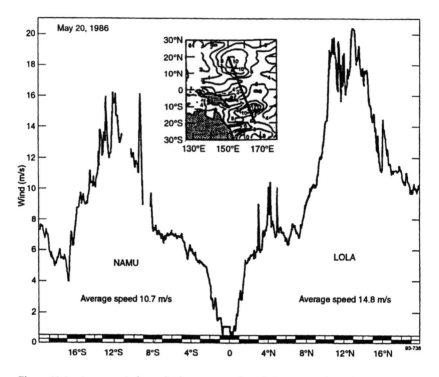

Figure 12.2   Average wind speeds along a ground track that passes through a cyclone pair in the Pacific. Insert shows ground track. (From Porter, D.L., *J. Geophys. Res.*, 95, 3705, 1990. With permission.)

different from Equation 12. This error correlation with buoy wind speed has to be related to the error in the algorithms, but it appears the correction for the tilt and bias cannot be obtained from a small sample of buoy wind speeds.

Mognard et al.[33] combined Geosat altimeter wind speeds with airborne synthetic aperture radar (SAR) data taken during the 1988 Norwegian Continental Shelf Experiment. Mognard analyzed data on a day when there were strong winds and an occluded front parallel to the Norwegian coast. By using the radar cross section from Geosat and the direction of wind streaks in the SAR data, they were able to get both wind speed and direction. Mognard chose the original Brown algorithm for these studies. In addition to the stated results, they show comparisons between SAR cross section at 30° incidence angle and the Geosat nadir backscatter. The data show a linear relationship in decibels and a data scatter that reflects the potential differences between specular and diffuse scattering.

Porter[34] was able to observe a tropical Pacific cyclone pair in 1986 using Geosat winds and verify the storm track by using satellite data and meteorological charts. Figure 12.2 shows the wind speeds measured along a track (see insert) as the altimeter ground track passed through both storms. The SB algorithm was used to compute these winds; thus the estimates are probably low by some amount although at this time that amount is not known exactly. The CW algorithm would probably be around 4 m/sec higher, while the Wu would be higher. Geosat wind and wave measurements were made available in near-real time (i.e., 12 h) to experimenters in the Labrador Sea Extreme Wave Experiment (LEWEX). In addition, the data were used to compare with the LEWEX common winds. These common winds were modeled winds that were developed in hindcast modeling in order to test wave models using the same set of winds for initialization. Although the Geosat sparse ground track coverage in the region did not allow model initialization, the Geosat data did point out those areas where the winds were not modeled well and those regions where there was good agreement. This work is reported in Dobson and Chaykovsky.[35]

The applicability of satellite altimeter in developing climatologies in regions not well covered by altimeters and *in situ* measurements was investigated by Tournadre and Ezraty.[28] The question they addressed was whether the spatial consistency scales of wind speed would allow measurements to be made in one location and then applied to an adjacent or surrounding zone or area. They found while that $H_{1/3}$ measured by an altimeter along a ground track could give the mean and standard deviation within the zone, wind speed measured along a track could not be used to get the climatology of the zone. The wind speed standard deviation computed along a track underestimated the zonal variation by 20%. This is a study that needs to be extended, to determine those areas where this is true and whether there are regions of the oceans where variability in wind speed is sufficiently small to allow development of climatologies.

Altimeter winds have aided recently in helping scientists improve models for the distribution of wave slopes in the ocean. Ultimately the complete modeling of these waves will allow a completely accurate relationship between radar backscatter at nadir and surface wind speeds. The application of altimeter winds used to verify and place bounds on wave slope spectra measurements served as a sort of iterative process to understand the underlying processes involved in scattering by microwaves. Jackson et al.[6] analyzed the results of a near-nadir airborne radar altimeter called ROWS to obtain the parameter that characterizes the ocean wave slopes. The parameter is called mean square slope (mss). Geosat altimeter data were used along with some buoy data to verify the mss measurements and to better understand the differences in interpretation between aircraft and satellite backscatter measurements. Jackson in his article derives a relationship for mss at nadir using a KU-band radar as:

$$m_n^2 \approx \left(1 + 0.0025\, U\right) m_g^2 \qquad (13)$$

Jackson used Geosat exact repeat mission data to refine Equation 13. The relationship that is ultimately derived of course depends on the altimeter algorithm used to compute the wind speeds. However, one can see that the availability of satellite winds in this application can greatly assist researchers who use aircraft data, because the experiments can be designed to take advantage of whatever altimeter is in orbit at the time of the experiment, gives much more spatial coverage than buoys, and provides that coverage more economically.

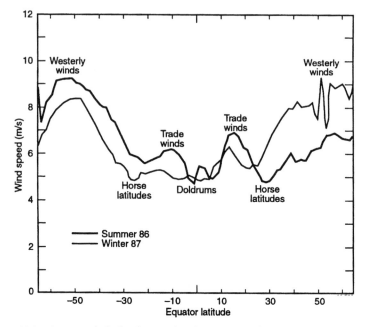

Figure 12.3   Average winds for the North Atlantic Ocean for summer 1986 and winter 1987.[36]

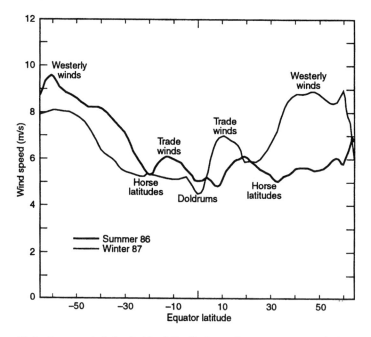

Figure 12.4   Average winds for the North Pacific Ocean for summer 1986 and winter 1987.[36]

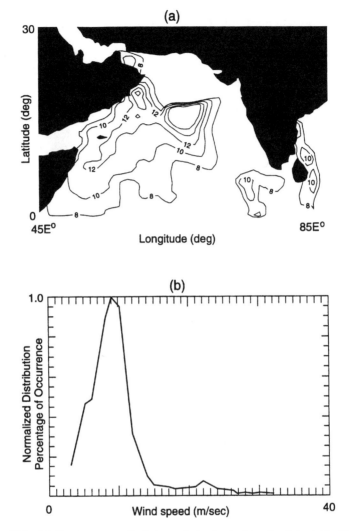

Figure 12.5 (a) Contour map of wind speeds in the Arabian Sea July 6–22, 1988; (b) Histogram of winds in the region.

Another application of altimeter winds is to provide data for an atlas that can be included in other historical databases for use in climatological analysis. As we have mentioned, to date there are over 7 years of global wind speed data from the various altimeters. As an example of some products that may be useful, we present data from the Geosat altimeter. In Figures 12.5a and b we show global mean winds from the North Atlantic and North Pacific oceans for the summer of 1986 and the winter of 1987. These mean winds were computed at 2° latitude bands over all longitudes in these two regions. The plots are from Dobson and Porter.[36] We observe very clearly the westerly winds north and south, the trade winds, the horse latitudes, and the doldrums. In both the Atlantic and the Pacific the shift between summer and winter is rather dramatic (Figures 12.3 and 12.4). The variation in Figure 12.3 around 50° is probably a change in the mean for the summer of 1986 only; Figure 12.5a gives an example of regional maps that can be created, which when generated in a time series can give an indication of the flow in an area. A map of the Arabian Sea area is shown in Figure 12.5a, and the histogram of winds in that region is shown in Figure 12.5b. These winds were computed using the modified Brown algorithm as discussed in Section 1.4. The region of the Somali Jet where the maximum winds around 65° E are 34 m/sec. These data are for one Geosat cycle from July 6 to 22, 1988.

References

1. Barrick, D.E., Rough surface scattering based on specular point theory *IEEE Trans. Antennas Propag.*, AP-16(4) 449, 1968.
2. Valenzuela, G.R., Theories for the interaction of electromagnetic and oceanic waves — A review, *Boundary Layer Meteorol.*, 13, 61, 1978.
3. Ulaby, F.T., Moore, R.K., and Fung, A.D., Microwave remote sensing — active and passive, Vol. 2, Addison-Wesley, 1982.
4. Barrick, D.E., A relationship between the slope probability density function and the physical optics integral in rough surface scattering, *Proc. IEEE*, 36, 1728, 1968.
5. Barrick, D.E., Wind dependence of quasi-specular microwave sea scatter, *IEEE Trans. Antennas Propag.*, AP-22, 1135, 1974.
6. Jackson, F.C., Walton, W.T., Hines, D.E., Walter, B.A., and Peng, C.Y., Sea surface mean square slope from Ku band backscatter data, *J. Geophys. Res.*, 97(C7), 1992.
7. Townsend, W.F., An initial assessment of the performance achieved by the Seasat-1 radar altimeter, *IEEE J. Oceanic Eng.*, OE-5(2), 1980.
8. Chelton, D.B., WOCE/NASA altimeter algorithm workshop, *U.S. WOCE Tech. Rep.*, No. 2, 1988.
9. Brown, G.S., Stanley, H.R. and Roy, N.A., The wind-speed measurement capability of space-borne radar altimeters, *IEEE J. Oceanic Eng.*, OE-6(2), 59, 1981.
10. Cox, C. and Munk, W., Statistics of the sea surface derived from sun glitter, *J. Mar. Res.*, 13, 198227, 1954.
11. Phillips, O.M., *Dynamics of the Upper Oceans,* Cambridge University Press, New York, 1969, 136.
12. Chelton, D.B. and McCabe, P.J., A review of satellite altimeter measurement of sea surface wind speed: with a proposed new algorithm, *J. Geophys. Res.*, 90, 4707, 1985.
13. Chelton, D.B. and Wentz, F.J., Further development of an improved altimeter wind speed algorithm, *J. Geophys. Res.*, 91(C12), 14150–14260, 1986.
14. Brown, G.S., Estimation of surface wind speeds using satellite-borne radar measurements at normal incidence, *J. Geophys. Res.*, 84(B8), 3974, 1979.
15. Goldhirsh, J. and Dobson, E.B., A recommended algorithm for the determination of ocean surface wind speed using satellite-borne radar altimeter, Tech. Rep. S1R85–005, Applied Physics Laboratory, Johns Hopkins University, Laurel, MD, March 1985.
16. Dobson, E.B., Geosat altimeter wind speed and waveheight measure measurements: the ERM mission, Proceedings of the WOCE/NASA Altimeter Algorithm Workshop, Corvallis, Oregon, *U.S. WOCE Tech. Rep.*, No. 2, 1987.
17. Witter, D.L. and Chelton, D.B., A Geosat altimeter wind speed algorithm and a method for altimeter wind speed algorithm development, *J. Geophys. Res.*, 46, 8853, 1991.

18. Carter, D.J.T., Challenor, P.G., and Srokosz, M.A., An assessment of Geosat wave height and wind speed measurements, *J. Geophys. Res.*, 97(C7), 11383, 1992.

19. Wu, J., Near-nadir microwave specular returns from the sea surface-altimeter algorithms for wind and wind stress, *J. Atmos. Oceanic Technol.*, 9(5), 1992.

20. Wu, J., Sea surface slope and equilibrium wind-wave spectra, *Phys. Fluids*, 15, 741, 1972.

21. Glazman, R.E., Near-nadir radar backscatter from a well-developed sea, *Radio Sci.*, 26(1), 1211–1219, 1990.

22. Glazman, R.E., Statistical problems of wind generated gravity waves arising in microwave remote sensing of surface winds, *IEEE Trans. Geosci. Remote Sensing*, 29(1), 135, 1991.

23. Glazman, R.E. and Greysukh, A., Satellite altimeter measurements of surface wind, *J. Geophys. Res.*, 98(C2), 2475, 1993.

24. Glazman, R.E. and Pilorz, S.H., Effects of sea maturity on satellite altimeter measurements, *J. Geophys. Res.*, 95(C3), 2857–2870, 1990.

25. Monaldo, F.M. and Dobson, E.B., On using wave height and radar cross section to improve radar altimeter estimates of wind speed, *J. Geophys. Res.*, 94, 12699, 1989.

26. Cardone, V.J., Greenwood, J.G., and Roy, M.A., On trends in historical marine wind data, *J. Climate*, 3, 113, 127, 1990.

27. Monaldo, F., Expected differences between buoy and radar altimeter estimates of wind speed and significant wave height and their implications on buoy-altimeter comparisons, *J. Geophys. Res.*, 93, 2285, 1988.

28. Tournadre, J. and Ezraty, R., Local climatology of wind and sea state by means of satellite radar altimeter measurements, *J. Geophys. Res.*, 95(C10), 18225, 1990.

29. Wu, J., Altimeter wind and wind-stress algorithms — further refinement and validation, submitted.

30. Mognard, N.M. and Lago, B., The computation of wind speed and have height from Geos 3 data, *J. Geophys. Res.*, 84(B8), 1979.

31. Dobson, E.B., Monaldo, F., Goldhirsh, J., and Wilkerson, J., Validation of Geosat altimeter derived wind speeds and significant wave heights using buoy data, *J. Geophys. Res.*, 92(C10), 1987.

32. Ebuchi, N., Kawamura, H., and Yoshiaki, T., Growth of wind waves with fetch observed by the Geosat altimeter in the Japan sea under winter monsoon, *J. Geophys. Res.*, 97(1), 1992.

33. Mognard, N.M., Johannessen, J.A., Livingstone, C.E., Lyzenga, D., Shuchman, R. and Russell, C., Simultaneous observations of ocean surface winds and waves by Geosat radar altimeter and airborne synthetic aperture radar during the 1988 Norwegian continent shelf experiment, *J. Geophys. Res.*, 96(C6), 1991.

34. Porter, D.L., Geosat observations of the tropical Pacific cyclone pair of May 1986, *J. Geophys. Res.*, 95(D4), 3705, 1990.

35. Dobson, E.B. and Chaykovsky, S.P., Geosat wind and wave measurements during LEWEX, *Directional Ocean Wave Spectra*, Edited by R.C. Beal, Johns Hopkins University Press, 1991.

36. Dobson, E.B. and Porter, D.L., *Global statistics of wind speed and significant waveheight as measured by the Geosat altimeter*, Proceedings of the Twenty-First International Sym. on Remote Sensing of the Environment, Ann Arbor, Michigan, Oct. 1987.

13

Marine Winds from Scatterometers

B. J. Topliss and T. H. Guymer

13.1 Introduction

Surface wind measurements are of high priority to oceanography for a number of reasons. The most obvious manifestation of wind on the ocean is the generation of surface waves. A knowledge of wave conditions is important for a variety of activities including the efficient and safe routing of ships and the design of offshore platforms and coastal defenses. Thus there is a requirement for both real-time dissemination to produce more accurate weather and wave forecasts using numerical models and the acquisition of long time series on which reliable climatological statistics can be based.

Some of the momentum of the wind is transferred into ocean currents. In confined, shallow-water regions storm surges can be set up as a result of these currents; and in combination with reductions in atmospheric pressure, they lead to flooding of low-lying areas. Venice is particularly prone to such events. In some coastal regions, especially on the eastern boundaries of ocean basins, the prevailing

0-8493-4525-1/95/$0.00+$.50

wind induces offshore currents. The resulting upwelling of nutrient-rich waters is often associated with enhanced biological activity.

Over ocean basins persistent spatial variations in the wind vector combine with the effects of earth rotation to drive the major current systems of the world oceans — the gyres. These are responsible for transporting vast quantities of heat and salt and contribute to the balance of climate. The magnitude of the wind is also required in the determination of heat, moisture, and gas fluxes between atmosphere and ocean. At high latitudes the cooling can be sufficient to produce sinking of surface waters that spread out beneath and are replaced by warmer tropical waters as part of a great vertical overturning called the thermohaline circulation.

World War II developments in radar research quickly revealed that when observing targets from ships, the low elevation noise backscattered from the sea surface (clutter) increased with wind speed. Experiments with airborne radar sensors throughout the 1970s quantitatively confirmed these early observations; the Normalized Radar Cross Section (NRCS) was dependent on wind speed. The physical basis for this relationship can be seen as a chain of events: the wind velocity U is related to the sea surface friction velocity U^*, which interacts with the sea surface capillary spectrum; this in turn modifies U^* and the radar scattering cross section, sigma naught (σ^0). There is no adequate theory or mathematical model for understanding or predicting this chain of events, and the whole of present-day wind scatterometer science is based on using an empirical relationship or algorithm between wind vector and scatterometer signal. Confidence in satellite wind speed and direction estimates therefore requires some understanding of the two types of algorithms used: the sensor algorithms derived by the radar engineers and the geophysical algorithms provided by the air-sea interaction scientists.

In July 1991 the European Space Agency (ESA) launched the remote sensing satellite ERS-1 carrying, among other radar devices, an Advanced Microwave Instrument (AMI) with a wind mode, a wave mode, and an image mode. Radars designed to measure a sea surface wind vector are termed scatterometers, and they are at present the only radar sensor capable of giving a wind direction as well as wind speed estimate. The ERS-1 scatterometer (AMI in wind mode) is a C-band (5.3 GHz), vertical-vertical (VV) polarized radar with three antennas; the basis for using this configuration of sensor and the limitations by which radar can measure winds is presented below in terms of both preliminary ERS-1 and previous aircraft and satellite studies.

13.2 Normalized Radar Cross-Section Relationship

Numerous experiments with radar over of the ocean have revealed that NRCS σ^0, increases with wind speed U roughly according to a power law; the exponent b of which is dependent on the radar frequency:

$$\sigma^0 = aU^b \tag{1}$$

Hence a higher radar frequency may give a greater sensitivity to changes in wind speed; and thus a Ku (14.6 GHz) band sensor (such as the Seasat-A Scatterometer System [SASS]) could have a greater potential sensitivity to wind speeds than a C-band sensor. However, the smaller amplitude of the shorter wavelength capillary waves may in some way counterbalance any sensitivity changes. The NRCS is also dependent on the radar beam angle relative to the wind vector with maxima in σ^0 occurring in the upwind and downwind direction and minima in the crosswind directions. This is due to the modulation of small-scale roughness by the longer wind waves. Although the form of the directional dependence is similar for different polarizations, the magnitude of the return changes, vertical polarization showing a stronger return than horizontal polarization. The NRCS also decreases with increasing incidence angle.

These radar relationships are also in part geophysical relationships, describing how radar can be optimized to measure wind, rather than true radar engineering relationships. The engineering aspects of radar are governed by the so-called Radar Equation[1] that relates the NRCS to transmitted power, power received back at the radar, slant range to the target, target area, radar wavelength, atmospheric attenuation, peak antenna gain, and relative antenna gain in the target direction. In estimating the NRCS from the returned power a correction must be made for the attenuation of the signal due to liquid water in the atmosphere (clouds and rain); with SASS data this correction was made by using data from the Seasat/Scanning Multichannel Microwave Radiometer (SMMR), a passive microwave device. However, owing to the viewing geometry this could only be conducted over a portion of the SASS swath. A C-band sensor is less influenced by atmospheric effects than the higher frequency Ku-band sensors and no attenuation correction is made on ERS-1. On the SASS the two antennas were found, after launch, to be imperfectly intercalibrated resulting in a bias. In order to estimate this bias, radar data were needed from a ground target with a known radar return. The suitable target for SASS data was the Amazon rain forest, the leaves being so dense and randomly oriented that they gave an isotropic radar backscatter. Hence sigmas from both antennas should be identical. The Amazon rain forest was also used to calibrate ERS-1 and, together with the use of transponders and European Center for Medium-Range Weather Forecasts (ECMWF) model winds, allowed the sensor to be calibrated to an absolute accuracy of 0.2 dB.

The exact mechanism by which radar views the sea surface is not well understood. It is commonly assumed that the dominant mechanism is Bragg scattering for incident angles, θ, greater than 25° from the vertical. The backscatter is due principally to the in-phase reflections from surface roughness. If a surface is smooth, then an oblique viewing radar receives virtually no return; as the surface becomes rougher, significant backscatter occurs as scattering from periodic structures in the surface roughness constructively interferes. The Bragg scattering equation gives the wavelength, λ_s, of the surface roughness that will give an optimal radar return for a radar wavelength, λ_r:

$$\lambda_s = \frac{n\lambda_r}{2\sin\theta} \qquad (2)$$

For the Ku-band the Bragg wavelength is approximately 3 cm, which is at the short end of the gravity-wave spectrum (capillary effects are restricted to below 1.73 cm). Any increase in σ with wind speed is typically interpreted[2,3] as due to an increased activity at these small-scale wavelengths. The C-band radar, of lower frequency, has a larger Bragg wavelength of approximately 6 cm; thus different radar sensors (including synthetic aperture radar [SAR]) operating at different frequencies will be viewing different scales of sea surface roughness features. However, little is known of the *in situ* (as opposed to laboratory) behavior of these features except by inference from radar measurements.[3] The occurrence of the Bragg wavelets is also ephemeral[1a] and the way this temporary nature affects the radar viewing is largely unexplored.

All the above relationships have been investigated for the C-band AMI on ERS-1 by numerous prelaunch aircraft experiments. Several caveats have been noted from these and previous experiments. The σ^0 values may also be dependent on numerous other nonwind factors such as sea surface temperature, sea state or fetch, surface slicks, etc. (see Section 2.3). These factors may result in anomalous values that may escape standard quality control procedures although new procedures are being developed all the time that improve anomalous data rejection (in particular see later sections on the 3-D σ^0 space).

13.3 Geophysical Relationships

Most of the caveats that effect the radar-capillary spectrum are related to our limited understanding of the *radar-surface roughness-wind speed* relationship. The very short waves associated with radar returns correspond to small-scale stress; however, some percentage of the forcing energy must also go into longer waves. The percentage depends on an equilibrium value of energy transport from capillary to longer waves, and this equilibrium may be situation dependent. In order for the σ-wind relationship to work this percentage must be fairly constant. The short wave spectrum can also be attenuated by the long waves but the mechanisms are not well understood. Not all waves are created by winds; long waves such as swell can travel thousands of kilometers and have no relation to local winds. Surface roughness can also be mechanically created by long waves. Other nonwind processes related to surface roughness are local oceanic and atmospheric mixed layer dynamics that can produce convergence regions; ocean currents can produce a wave-current surface interaction affecting wave shapes independent of wind influences.

Breaking waves and white caps, which will occur at high wind speeds, are expected to interfere with the radar-wind relationship; they also affect flow separation and hence the wind stress. No theory exists so far to cope with these features; yet despite their potential influence, satellite wind speeds have been retrieved up to 30 m/sec and several violent storm wind fields have been reproduced. It is intuitively expected that at high wind speeds there must also be an eventual saturation of the capillary wave field. These wavelengths are very difficult to observe, but measurements by Mitsuyasu and Honda[4] and the mounting evidence from satellite radar do not indicate that saturation is a problem, at least up to 25m/sec. This has caused a revision of theoretical ideas.[5] Part of the problem of estimating the relationship at very high wind speeds is the difficulty of obtaining good *in situ* calibrations in storm conditions; ships will avoid intense storm activity, and buoys and their instrumentation may not function well under conditions of intense wave activity. Also, for C-band wind models a confused (fetch-limited) sea state is more often associated with these higher wind speeds, producing a major problem for correct interpretation of the radar return.

Rain can have the effect either of smoothing the sea surface (intermediate wavelengths) and reducing capillary wave action or of creating greater surface roughness and creating more surface ripples via rain jets. The SASS method of dealing with rain was to identify high rain rate areas with a multifrequency radiometer and then exclude those areas from the wind calculations. The expected tendency is still for satellite-derived winds to underestimate winds in those regions of a hurricane that have a high probability of precipitation. It is thought[5a] that rain effects so far have been less important for ERS-1 wind vector interpretation.

The above two caveats dealt with high wind speeds; other caveats may have a stronger influence at low wind speeds. Sea surface temperature (SST) may affect wind speed estimates; Liu[6] determined a SST-bias for global SASS wind speed estimates less than 6 m/sec. At wind speeds less than 3 m/sec Topliss et al.[7] noted a C-band radar mapped out the thermal gradients of an eddy rather than the wind field, but this radar thermal relationship was lost once a period of higher winds occurred. It is more difficult to verify SST effects from global statistics that cannot take into account directional influences of wind, wave, or satellite orbit. Thus far such global studies[8] have not found an impact of SST on ERS-1 wind vector retrievals.

A useful check on wind speed estimates can be obtained by comparison with other radar sensors such as altimeter or radiometer estimated speeds (which are discussed elsewhere in this book). In any intersatellite comparisons, scientists have to bear in mind that the Seasat winds were calibrated to winds at 19.5 m whereas winds derived from ERS-1 are corrected to 10 m.

13.4 Wind Retrievals

The ERS-1 prelaunch empirical relationship between σ^0 and wind speed V, and direction ϕ as determined from aircraft experiments was termed C-Band Radar Sea Echo Model (CMOD-2).[9] The relationship:

$$\sigma^0 = B_0\left[1 + B_1\cos(\phi) + B_2\cos(2\phi)\right] \qquad (3)$$

has coefficients B_0, B_1, B_2 that depend on the radar beam incidence angle and wind speed; they are expanded as Legendre polynomials to a total of 18 coefficients. The B_0 term has been referred to as the bias term; the B_1 term coefficient, as the upwind/downwind amplitude; and B_2, as the upwind/crosswind amplitude.

Postlaunch comparisons with either *in situ* data or numerical weather prediction (NWP) model data showed discrepancies. The CMOD-2 gave a false bias across the scatterometer swath. Numerous postlaunch studies attempted to fine-tune the 18 coefficients, in particular, the ESA postlaunch validation Haltenbanken campaigns held off Norway in the autumn of 1991. An interim CMOD-3 was adopted by ESA for fast delivery (FD) wind products. Although this model removed the swath dependence, it retained an overall bias of 0.7 m/secec. An intercomparison[10] of several tuned models (also termed transfer functions) derived from field studies or NWP model comparisons showed that although all the new models performed significantly better than the old ones, no one model was best at retrieving both wind speed and wind direction. Hence further improvement in understanding the physical process of extracting both wind speed and direction is still needed, either with a supermodel or via some fundamental improvement in understanding radar-wind processes. A model derived by Stoffelen and Anderson[8] was selected as the CMOD-4 version for implementation by ESA for FD wind products.

The newer models also involve a different approach to interpretation of the empirical fitting functions. With the three antennas of ERS-1, the three parameters have allowed a visualization of the σ-wind vector relationship via a 3-D surface. The three sigmas from the fore, aft, and mid antenna create a cone- or horn-shaped surface. For example, movement along the 3-D σ surface represents changes in wind direction for the same wind speed. The concept that the σ^0 triplets would lie on a surface was first put forward by Cavanier[10a] and was explored fully by Stoffelen and Anderson.[8,11] The concept holds well except at low wind speeds and at the very inner edges of the scatterometer swath (nodes 1–3). The thickness of the cone surface is typically equivalent to a wind speed of 0.3 m/sec. The σ-surface concept has also helped define quality control (QC) steps for rejecting anomalous data; fit non-linearity or second derivative terms to the transfer function; and expand the theoretical insight into backscatter theory.

13.5 Wind Direction Ambiguity

The directional capability of a radar scatterometer is based on having more than one antenna and hence more than one look at the sea surface over time intervals in which it is assumed that the character of the sea surface has not changed. For an empirical equation between σ^0 and wind speed but dependent of angle, it should with multiple looks be possible to resolve for wind speed and direction. The directional solutions, however, are not unique. The sigmas have finite measurement errors, and the radar-wind relationship itself is highly nonlinear, all adding to the inability to provide a unique solution. The SASS had two antennas and aimed to predict winds to a 20° accuracy; this proved possible only if the approximate wind directions were known in order to allow optimal selection. The inability to determine the unique satellite wind direction but to get ambiguous directions (more than one answer) has lead to several schemes for ambiguity removal. These schemes were largely derived to cope with the relatively high ambiguity rates with the SASS dual-beam data.

In order to estimate wind direction it is necessary to extract the answer numerically by minimizing a function related to the sum of squares of the residuals, comparing the measured sigmas to those estimated from the empirical-model function, using an estimate of wind speed and direction. For Seasat the residuals were normalized by the expected noise for each antenna. The instrument noise is related to the system bandwidth, the signal-to-noise ratio, signal plus noise integration time, and noise-only integration time. The estimates are then refined until the sum of squares is minimized. With different initial estimates for wind directions, the analysis converges for a two-beam scatterometer generally on up to four answers or ambiguous wind vectors having similar speeds (differences may be up to 15%) but different directions. With a three-beam scatterometer, the analysis generally yields two different directions that differ by about 180°, with the remaining two solutions being duplicates. The four solutions are ranked in order of increasing residuals (Rank 1 having the smallest residuals and hence being the best estimate).

The directional ambiguity is resolved by strategies or computational schemes that hope to maximize the probability of picking the correct answer. These can be categorized as using (1) nearest-neighbor consistency and smoothing; (2) producing recognizable patterns; and (3) using external data, such as forecast data, for confirmation and feedback. In general, combinations of (1) to (3) have proved most useful. Offiler[1] determined from ERS-1 simulations that taking the Rank 1 wind direction as the true value would be correct 70% of the time. This was later found to be 60% for real ERS-1 data. This would compare with a 25% probability for the two-beam SASS system. By considering an area of data with Rank 1 answers exhibiting some consistency and then changing rank for those nearest-neighbour cells not following the pattern, the probability can be raised to 90%. However, in those areas where the initial, consistent rank was wrong, the ambiguity strategies only reinforce the wrong answer and lower the overall predictive skill. Different wind extraction models also have different levels of reliability, thus potentially adding to the ambiguity problem. External data giving the true direction can help to resolve the ambiguities, but such data are rarely available for any operational use of the satellite wind data and are also biased but in other ways. The use of forecast model data to check for ambiguities can raise prediction probability to 95%, but also had some flaws with SASS wind retrievals. If the forecast incorrectly positions a circulatory feature by 100–200 km, then an area of winds exists between the true and model data where the winds are 180° out; this will result in exactly the wrong satellite data being selected. If in areas of sparse conventional coverage forecast feedback is used to help predict the next satellite pass, this error is perpetuated. In general, it is thought that to date ERS-1 scatterometer errors have not resulted in errors being perpetuated by forecast models.

Prelaunch expectations for the ERS-1 scatterometer were that the fore and aft σ^0 amplitudes would be sufficiently different so as to remove much of the uncertainty in the ambiguity removal. That proved not to be the case, and for reasons that are still not clear there is insufficient difference between the fore and aft σ^0 values to yield an unambiguous direction. Hence external data such as meterologic forecast models have been used to aid in the selection of wind direction for all the ERS-1 FD wind products.

13.6 Wind Studies

The major uses for satellite-derived wind vectors have been for global studies, dictated by the repeat coverage and footprint size. For SASS, the footprint size was 18×70 km with a double-sided 500-km swath coverage but with only one side having SMMR correction data; for the ERS-1 AMI, the single-sided swath width is 500 km with 19 overlapping cells on a 25-km grid with each having a 50×50-km footprint. These swath widths are much smaller than those from the National Oceanic and Atmospheric Administration (NOAA) Advanced Very High-Resolution Radiometer (AVHRR, see Chapter I.A.1) sensor, and as such they cannot provide complete coverage of an area. Even in the ERS-1 3-d repeat cycle there are areas that are never sampled, and the inability to sample any point more frequently than twice in 3 d (using both ascending and descending passes) may be a limitation on the sensor operational impact. One means of overcoming this sampling problem is to assimilate the data into a forecast model.

Accuracy of Scatterometer Winds

Comparing satellite-derived winds with other data can be a problem. The errors and limitations of the *in situ* data also have to be understood.[12] The satellite is providing a wind estimate for a 50×50 km area whereas any ship or buoy measurement is providing a single point value usually averaged over a few minutes. It has been estimated[1] that a point measurement should be averaged for not less than an hour to be comparable with a scatterometer footprint. *In situ* winds are also measured at various heights depending on their origin. For example, winds measured from oil rigs are often at heights greater than 50 m and are of no use as comparisons in a stable atmospheric boundary layer study. Since wind speeds usually increase with height in the boundary layer, any comparisons must be done by correcting wind data to the same height.

For Seasat, in order to assess the accuracy of the wind retrievals from the various proposed model functions, *in situ* ship and buoy data from the Gulf of Alaska Seasat Experiment (GOASEX) were examined. The comparison revealed problems both in the systematic differences between antennas (as already noted) and in the adequacy of the model functions. The latter were modified to bring closer agreement with observed winds; but to validate these satisfactorily, an independent (i.e., withheld from algorithm tuning), high-quality set of surface winds was required. Fortunately the Joint Air Sea Interaction (JASIN) Experiment took place within Seasat operational lifetime. A number of ships and buoys sampled surface meterologic variables in a 200- × 200-km area situated 300 km west of Scotland, U.K. at intervals from 1 min to 6 h. Careful intercalibration between the sensors was made to produce an internally consistent data set in which interplatform differences were less than 0.7 m/sec in speed (0.5 m/sec for buoys) and less than 5° in direction. For convenience one buoy sensor was adopted as a standard to which all other data were corrected, but the absolute accuracy of the JASIN wind speeds is difficult to determine and errors of a few percent are possible.

Different overpasses of the satellite were selected so as to give good spatial coverage of the JASIN area and to sample as wide a range of wind speeds as possible. The range achieved was 0–16 m/sec with 83% of the occasions being distributed between 6 and 14 m/sec. Most points lay within the design specifications of ±2 m/sec and ±20°. However, on some passes, less favorable comparisons were obtained; the most striking was an occasion on which SASS winds exceeded nearby *in situ* winds by up to 15 m/sec. Using JASIN surface and upper-air data in conjunction with IR satellite imagery, it was possible to infer that a midlevel thunderstorm had affected the region at the time of the anomaly.[13] Rain would normally be expected to decrease the backscatter through attenuation, but in this case it was postulated that raindrops falling on a relatively calm surface were generating roughness elements capable of enhancing the radar return. (Anomalous backscatter in rain events is also found in radar altimeter data.[14]) Alternatively, mid-level thunderstorms generate local wind gusts at the surface and strong temporal variability of the wind vector. This means that ship-satellite comparisons may be more susceptible to spatial mismatch, together with the wind and wave conditions not being in equilibrium; all add to the difficulty of obtaining an accurate wind vector under such conditions.

Neither GOASEX nor JASIN provided a validation of the upper wind-speed limit of the SASS (24 m/sec). Using a much larger global data set from the analysis phase of the ECMWF atmospheric model, Anderson et al.[15] concluded that SASS winds were biased high at low winds and vice versa and that there was a possible sea surface temperature dependence to the bias. Coverage was also obtained of several storms, including three hurricanes.[16] Available near-surface data included Air Force reconnaissance winds (through the core on one occasion), ships, and SST surveys. Measurements agreed within design specifications up to about 26 m/sec. In one of the hurricanes the SASS appeared to perform well up to 30 m/sec. However, a clear tendency was noted for SASS to underestimate in those regions of hurricanes where there was a high probability of precipitation. The latter was partly due to an inadequate attenuation correction from the SMMR in intense rain bands due to the SMMRs relatively poor spatial resolution. In the extratropical storm that was studied the high wind-speed comparisons were in closer agreement even for pre-cipitating regions. Rainfall in this system was probably much less intense than in the hurricanes. Analysis of ERS-1 scatterometer data below tropical storms does not support a strong rain effect explanation. Lower than expected wind speeds are retrieved from ERS-1 data in tropical cyclones when examination of cloud imagery has revealed no rain; the current explanation is that the wave field (as examined by SAR imagery) is anomalous when associated with intense fronts, lows, and tropical cyclones.

Evaluations of the accuracy of ERS-1 scatterometer wind retrievals were conducted with several dedicated calibration campaigns. An ESA campaign was organized off the Norwegian coast in the autumn of 1991 (known either as RENE-1991 or as the Haltenbanken field campaign) and involved aircraft, ships, and buoys. The data were used to identify the deficiencies of the prelaunch CMOD-2, provide a means of fine-tuning the required 18 coefficients, and define the accuracy limits of the ERS-1 scatterometer. Within the limits of the application of the final CMOD-4 transfer function the wind retrievals more than met the ERS-1 specifications of speeds within 2 m/sec and directions within 20°. The limitations for application of CMOD-4 are low wind speeds, inner nodes of the swath, and forecast winds being necessary for wind direction selection.

Applications of Scatterometer Data

Despite all the caveats and potential ambiguities concerning wind vectors, there have been numerous examples of satellite-derived wind maps providing realistic and useful information. The potential for such information in near-real time resulted in the ESA making commitments to provide wind data to operational users within 3 h of reception.

One advantage of scatterometer data is that spatial resolution of 50 km often allows more accurate positioning of meteorological features. For Seasat, one case has been presented[15] where after reanalysis of SASS data for an intense extratropical depression, the satellite, not the conventional data gave a more accurate represen-tation of the situation. This storm was on September 9 and 10, 1978 and led to 30-m/sec winds with observed waveheights of 12 m. Because the QEII was caught in this storm, the storm itself has come to be referred to as the QEII storm. The fact that a ship, which would normally move out of the vicinity of such a storm, was present helped to provide evidence for which view of the storm was more correct: the conventional or satellite view. The routine meteorological analysis of the weather

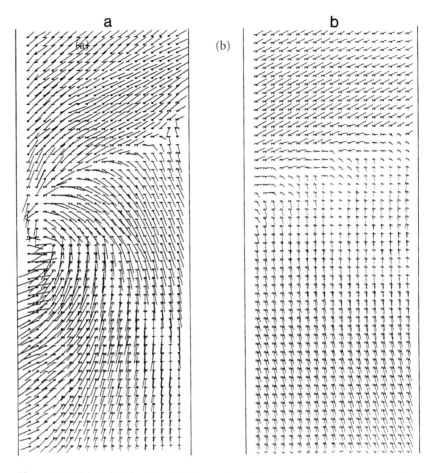

Figure 13.1   (a) A plot of retrieved wind speeds and directions. (b) The forecast; only Rank 1 and 2 directions were used in producing (a). The forecast was used to choose streamline direction locally. Although the forecast was broad, this did not prevent the scatterometer from having considerable structure, which is confirmed by comparison with cloud imagery. (Diagram from Stoffelen, A., and Anderson, D., *Proc. 1st ERS-1 Symp.*, ESA SP-359, Vol. 1, 1992, 41. With permission.)

patterns underestimated the intensity of the storm and its precise position and associated winds. A reanalysis of the storm using both data sets provided more accurate wind estimates and a more accurate positioning of the storm center.

A similar case was reported by Stoffelen and Anderson[11] using ERS-1 data (see Figure 13.1). The scatterometer-derived wind field showed a well-defined low pressure center that was only hinted at in the ECMWF forecast. By comparison with cloud imagery the authors concluded that ERS-1 provided a better representation of reality and that the model was up to 150 km in error in its positioning of the center. Such positional errors could be significant for weather and wave forecasting.

For some purposes, especially when considering the response of the ocean to meteorological forcing, the time-averaged wind is required rather than the detailed wind variations associated with individual weather systems. A 3-month mean wind speed in the tropical Pacific was computed from SASS data by Chelton and O'Brien[18] and showed known climatological features. Strongest winds were found at about 20° N and 15° S, corresponding to the Trades; those in the Southern (winter) Hemisphere were stronger and more extensive in the zonal direction. The light wind region in between corresponded to the Intertropical Convergence Zone (ITCZ). Changes in tropical wind patterns affect currents and SST and are an important part of the El Niño-Southern Oscillation (ENSO) phenomenon.

The directional ambiguity was a major limitation in the application of time-averaged data to large areas. Atlas et al.[19] produced a 3-month data set in which the ambiguities were removed with the aid of an atmospheric forecast model. In an interesting study Chelton and O'Brien[18] used this to construct gridded fields of monthly average wind stress and wind stress curl maps over the global ocean from which the Sverdrup circulation was computed. This is a simple model of ocean circulation in which the transport is obtained solely from the wind stress curl and the application of suitable boundary conditions. An example of the mean fields for the whole 3 months of the Seasat mission is shown in Figure 13.2. The NE Trades of the Northern Hemisphere, the SE Trades of the Southern Hemisphere, and the light winds of the ITCZ can be clearly seen in Figure 13.2(a). The corresponding wind stress curl (computed from the horizontal gradients of the wind components) is plotted in Figure 13.2(b) and reveals alternating bands of positive and negative wind stress curl. In the Sverdrup model these correspond to different ocean currents with the zero contours identifying the boundaries of gyres. The one at 20°N marks the North Equatorial Current, the line at 10°N coincides with the Equatorial Countercurrent, and the one at 10°–20°S is the northern boundary of the subtropical gyre. Results were compared with those using the Hellerman and Rosenstein climatological forcing fields. Except for the region south of 35°S the two estimates were generally very similar. In the Antarctic Circumpolar Current there was a very large discrepancy in the two estimates of the zonal transport. It is not clear whether this was due to the paucity of climatological data; unrepresentativeness of the 3-month SASS period; or errors in the removal of directional ambiguities, which were typically worst at below 40°S.

Although the scatterometer was conceived as an instrument for measuring open-ocean winds on large scales, it has been found to give useful data in coastal and enclosed sea areas. Guymer and Zecchetto[21] considered three main causes for concern with coastal data: (1) potential land contamination giving an incorrect NRCS, (2) potential changes in the wind-radar relationship due to the different characteristics of the wave spectrum in fetch-limited conditions, and (3) incorrect NRCS due to SST effects since shallow coastal areas may experience substantial SST variations in both space and time. Guymer and Zecchetto[21] analyzed a number of scatterometer passes in the North Sea and western Mediterranean. For individual passes realistic spatial distributions of the wind vector were found. Figure 13.3 is an

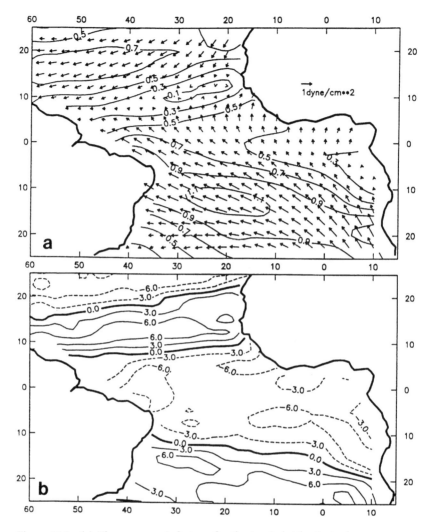

Figure 13.2 (a) The average wind stress for the tropical Atlantic in dynes per square centimeter. (b) The wind stress curl in 10^{-9} dynes/cm³. (From Chelton, D. B. et al., *J. Phys. Oceanogr.*, 20(8), 1175, 1990. With permission.)

example showing the sharp changes in wind observed across a frontal system in much more detail than can be obtained from routine measurements. There is also evidence of the airflow being channeled around Norway. However, when the mean wind speed over the NE Atlantic was calculated from the whole Seasat scatterometer data set, winds in the southern North Sea appeared to be anomalously high. Comparisons with other data did not support these values, and it was suspected that shallow-water effects were playing a role. They speculated that algorithms tuned to

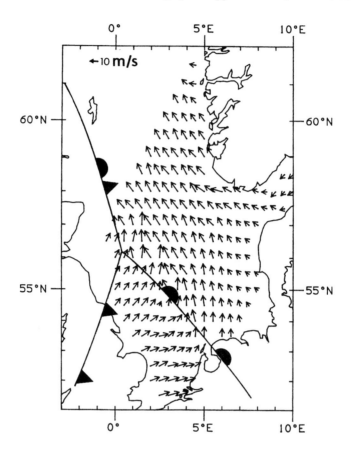

Figure 13.3   SASS scatterometer wind field in the North Sea on September 29, 1978. Interpolation to a 0.5° grid has been conducted. Notice the sharp change in wind direction across the warm front. (From Guymer, T. H. and Zecchetto, S., *Int. J. Remote Sensing*, 12(8), 1699, 1991. With permission.)

open ocean, deepwater conditions will overestimate winds in shallow seas due to the surface drag coefficient being a function of water depth.[8,20]

Scatterometer data for the mistral blowing off southern France over the Mediterranean were found to be realistic and not effected by bathymetry, probably due to the greater water depths in the Mediterranean. SASS-derived sirocco winds in the Adriatic were also considered representative of the local wind regime. Land contamination was not found to be a problem in these studies but the small scale of wind features in coastal areas did produce a greater difficulty in resolving ambiguous wind directions. Guymer and Zecchetto[21] reported on the use of SASS winds for driving a hydrodynamic model of the Adriatic Sea during a storm-surge situation. Despite the considerable simplifications involved, e.g., the compositing of three separate satellite passes into a single wind field that was then used as a fixed forcing function for the model during its integration, the predicted surge was in good agreement with that observed.

Real-Time Wind Estimates from ERS-1

One of the prime objectives of the ERS-1 scatterometer mission was to provide real-time winds for operational purposes. These have been termed fast delivery products (FDPs). The time lag between the signal being received and leaving the processing center in Europe is between 100 and 150 min, the time difference being between the first and the last part of the signal received by the ground station. Typical overall delay times for this product to reach international weather offices, such as regional offices throughout Canada, are between 2 to 3 h. The bulk of this time is still the ground processing time with minimal transmission time losses. It is anticipated that ERS-2 will upgrade the ground processing so as to reduce delays.

The ERS-1 wind speed and direction product is sent out in 500×500-km sections. Due to the greater error associated with determining wind direction at low speeds, wind speeds below 3.5 m/sec are not transmitted but appear as missing values. Because with conventional forms of wind measurements the scatterometer directional measurements are less meaningful for low wind speeds, the speeds are valid (and may be most accurate;) but the directions are not considered to be retrieved with any high degree of accuracy. Although wind data have been received since launch, the early FDP encountered both calibration and computational problems. Improvements were made to the FDP from June 1992 to February 1993 with the aid of the CMOD-3 algorithm. Since that algorithm still retained a bias of –0.7 m/sec and some deterioration of wind direction accuracy, it was again replaced. From February 1993 the CMOD-4 algorithm was implemented with the bias eliminated and an improved wind direction accuracy giving the most reliable data to date. The earlier data will eventually be corrected as archived data are reprocessed and released.

A study to compare ERS-1 FDP winds with model data has been conducted by the Centre Météorologic Canadien. That study found the comparisons to be approximately within the ERS-1 scatterometer design specifications. Speed was compared to a mean absolute deviation of 1.87 m/sec with the corresponding deviation for direction being 28.5° (outside the required specifications). The direction deviations were dependent on wind speed; for light winds from 4 to 5 m/sec the deviation was as high as 54°, whereas for winds between 10 and 25 m/sec the mean absolute deviation met the 20° criterion. For all speeds the deviation value was high because of the number of ambiguous directions, those at 180° to the true value.

13.7 Summary

In general, excellent agreement can be found between the majority of satellite-derived winds and other data sources. The ERS-1 scatterometer data cannot be used within 50 km of land or it will be contaminated by land-based radar return, and it cannot be used to measure winds over ice (although it can be seen from Chapter 23 that the reverse applies; some information about ice may be gained from scatterometer data). The situations where the satellite wind vector is not reliable either are linked to an incorrect ambiguity removal, or are linked to one of the physical processes which interfere with a simple wind-radar empirical relationship. Such anomalous data have to be identified and eliminated. For a three-beam scatterometer this process can be significantly aided by an internal quality control in 3-D σ^0 domain, or otherwise this has to be done or supplemented by human intervention applying experience of meterologic situations or air-sea interaction processes.

Preliminary ERS-1 studies indicate that the wind vector algorithms are now well capable of achieving the desired prelaunch objectives of measuring wind speeds over a range from 4 to 24 m/sec with an accuracy of 2 m/sec and a directional tolerance of 20°. Continual wind coverage cannot be obtained since the ERS-1 scatterometer is forced to compete with the ERS-1 SAR imaging mode and both cannot be operated at the same time (except for the interleaved wave imagette mode; not a full SAR mode). For this reason no scatterometer coverage of Hurricane Andrew was obtained. At least one future scatterometer is planned for full-time operations (as opposed to being switched on by request or as part of the scheduling of other sensors): the NSCAT possibly to be launched around 1996. It is thought that the shape of the NSCAT solution surface in the 3-D/σ^0 space may make wind retrievals more difficult than in the ERS-1 case. The NSCAT sensor will have six antennas for a two-sided look. The midantenna will be offset to operate at 20° from the fore antenna in order to facilitate wind retrievals. Planned future generations of scatterometers should lead both to increased and continual coverage and to a greater understanding and more accurate interpretation of the wind radar process.

References

1. Offiler, D., Wind fields and surface fluxes, in *Microwave Remote Sensing for Oceanographic and Marine Weather-Forecast Models*, Vaughan, R. A., Ed., NATO ASI Series, Kluwer Academic Publishers, 1990, 355.
1a. Dobson, F. and Vachon, P., personal communication.
2. Donelan, M. A. and Pierson, W. J., Radar scattering and equilibrium ranges in wind-generated waves with application to scatterometry, *J. Geophys. Res.*, 92(C5), 4971, 1987.
3. Banner, M. L., The influence of wave breaking on the surface pressure distribution in wind-wave interactions, *J. Fluid Mech.*, 211, 463, 1990.
4. Mitsuyasu, H. and Honda, T., The high frequency spectrum of wind-generated waves, *J. Ocean Soc. Jpn.*, 10, 185, 1974.
5. Philips, O. M., Spectral and statistical properties of the equilibrium range in wind-generated gravity waves, *J. Fluid Mech.*, 156, 505, 1985.
5a. Stoffelen, A., personal communication.
6. Liu, W. T., The effects of the variations in sea surface temperature and atmospheric stability in the estimation of average wind speed by Seasat-SASS, *J. Phys. Oceanogr.*, 14, 392, 1984.

7. Topliss, B. J., Guymer, T. H., and Viola, A., Radar and infrared measurements of a cold eddy in the Tyrrhenian Sea, *Int. J. Remote Sensing*, in press.

8. Stoffelen, A. and Anderson, D. L. T., Wind retrieval and ERS-1 scatterometer radar backscatter measurements, *Adv. Space Res.*, 13(5), 553, 1993.

9. Long, A. E., Towards a C-band Radar Sea Echo Model for the ERS-1 Scatterometer, in Proc. 3rd Int. Colloq. Spectral Signatures of Objects in Remote Sensing, Les Arcs, France, December 1985, *ESA SP-247*, 1986, 29.

10. Offiler, D., Validation and Comparisons of Alternative Wind Scatterometer Models, in Workshop Proc., RENE 1991, April 27–30, 1992, Penhors, Bretagne, France, *ESA-WPP-36*, 1992, 125.

10a. Stoffelen, A., personal communication.

11. Stoffelen, A. and Anderson, D. L. T., ERS-1 Scatterometer Data Characteristics and Wind Retrieval Skill, in Proc. 1st ERS-1 Symp.: Space and the Environment, Cannes, November 4–6, 1992, ESA SP-359, Vol. 1, 1993, 41.

12. Donelan, M. A., The dependence of the aerodynamic drag coefficient on wave parameters, in *Proc. 1st Int. Conf. Meteorol. Air-Sea Interaction of the Coastal Zone*, AMS, Boston, 1982, 381.

13. Guymer, T. H., A Review of Seasat scatterometer data, *Philos. Trans. R. Soc. London*, A309, 399, 1983.

14. Guymer, T. H. and Quartly, G. D., The effect of rain on ERS-1 altimeter data, in Proc. 1st ERS-1 Symp.: Space and the Environment, Cannes, November 4-6, 1992, ESA SP-359, Vol. 1, 1992, 445.

15. Anderson, D., Hollingsworth, A., Uppala, S., and Woiceshyn, P. M., A study of the feasibility of using sea and wind information from the ERS-1 satellite. I. Wind scatterometer data. *ECMWF Contract Report to ESA, ESRIN contract no. 6297/86/HGE-I(SC)*, 1987, 121 pp.

16. Black, P. G., Gentry, R. C., Cardone, V. J., and Hawkins, J. D., SEASAT microwave wind and rain observations in severe tropical and mid-latitude marine storms, *Adv. Geophys.*, 27, 197, 1985.

17. Gyakum, J. R., On the evolution of the QEII storm. I. Synoptic aspects., *Mon. Weather Rev.*, 3, 1137, 1983.

18. Chelton, D. B. and O'Brien, J. J., Satellite microwave measurements of surface wind speed in the tropical Pacific, *Trop. Ocean-Atmos. Newsl.*, 11, 3, 1982.

19. Atlas, R. A., Busalacchi, J., Ghil, M., Bloom, S., and Kalnay, E., Global surface wind and flux fields from model assimilation of Seasat data, *J. Geophys. Res.*, 92, 6477, 1987.

20. Chelton, D. B., Mestas-Nunez, A. M., and Freilich, M. H., Global wind stress and Sverdrup circulation from the Seasat scatterometer, *J. Phys. Oceanogr.*, 10, 1175, 1990.

21. Guymer, T. H. and Zecchetto, S., Winds derived from Seasat's microwave suite, *Int. J. Remote Sensing*, 12(8), 1699, 1991.

22. Geernaert, G. L., Larsen, S. E., and Hansen, F., Measurements of the wind stress heat flux and turbulence intensity during storm conditions in the North Sea, *J. Geophys. Res.*, 92, 13127, 1987.

23. Guymer, T. H. and Zecchetto, S., Applications of scatterometer winds in coastal areas, *Int. J. Remote Sensing*, 14, 1787, 1993.

14

Significant Wave Height from Altimeters

Ella B. Dobson and Frank M. Monaldo

14.1 Physical Principles

A Physical Picture

The concept of a spaceborne radar altimeter is straightforward. Transmit a narrow radar pulse to the surface and precisely measure the time until the return of the pulse from the ocean surface. The two-way return travel time is a measure of the distance to the surface.

In practice, the situation is far more complex. The goal of a satellite altimeter instrument is to measure the geometric distance between the altimeter and the ocean surface to centimeter-level precisions, with return pulses that are sampled in 0.5-m increments. As such, the precise measurement of the return trip travel time requires estimating the mean position of the surface from the shape of the return pulse. In trying to develop the procedures to estimate the distance to the surface from the position of the leading edge of a relatively broad pulse, it became quickly apparent that the return pulse shape contains information not only about the mean position of the surface, but also about ocean wave heights.

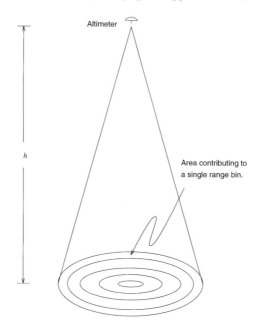

Figure 14.1   Schematic representation of the reflection of a narrow altimeter pulse from the ocean surface. The height of the altimeter is h. Each annular region represents the area contributing to the return in a range gate.

Figure 14.1 is a schematic diagram of the reflection of narrow altimeter pulse from the ocean surface. Since the return pulse is sampled in range gates, it is convenient to think of the pulse during each time increment corresponding to each range gate.

At the first instant of time, in the first range gate, the altimeter pulse illuminates a small circular region on the surface nadir to the altimeter. At subsequent time increments, the altimeter illuminates annular regions surrounding this central spot.

If the ocean surface were perfectly flat and had uniform radar cross section, the return pulse shape would show a quick rise, corresponding to reflection from the initial circular region. The area in each of the subsequent annular regions is constant, resulting in a flat return waveform after the leading edge of the pulse. As the annular regions get farther and farther from the central area, the illumination from the altimeter pulse decreases because of the beam-pattern falloff of the antenna. The last part of the return pulse exhibits a reduction in power associated with the antenna pattern.

If the ocean surface is rough because of the presence of waves, the leading edge of the transmitted pulse intersects the crests of the waves a little sooner than the mean surface and reflects from the wave troughs a little later than it otherwise would. The result is that the pulse shape is broadened. The wider the distribution of wave heights on the surface, the greater the spread in the return pulse. Hence, return pulse shape can be used to infer something about ocean wave heights.

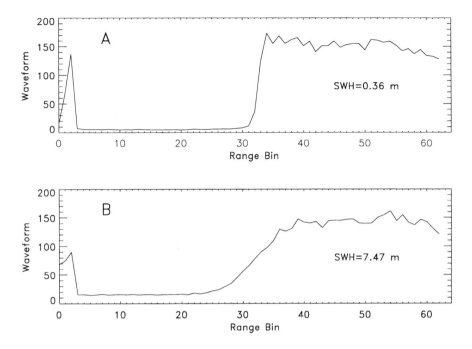

Figure 14.2 Examples of an average of 1000 altimeter pulses acquired over 1 s by the Geosat satellite. Plot A is from a low sea with SWH wave height of 0.36 m, plot B is from a situation with a 7.47 m SWH.

For historical reasons, the distribution of ocean wave heights is often characterized in terms of significant wave height (SWH). Significant wave height is defined for our purposes here as $4\sigma_H$, where σ_H is the root mean square (rms) wave height of the surface.

Figure 14.2 provides illustrative examples of the average of 1000 altimeter pulses acquired over 1 s by the Geosat satellite altimeter. Plot A was obtained from an ocean with an SWH of 0.36, m, essentially a calm sea state. The data were acquired on day 122, 1988 at 1:49:24 GMT, longitude 2.90° W and latitude 9.45° N. Note the sharp leading edge of the pulse and relatively flat tail of the waveform. Plot B represents the waveform from the relatively high sea state of 7.47 m. This waveform was acquired on day 122, 1988 at 4:2:41 GMT, longitude 173.74° W and latitude 56.30° S. The spreading of the pulse at the higher SWH is clearly apparent.

Noise on the return waveform shape is also evident in the return waveform shape. The power received in any one range bin is the result of the returned power from the many ocean surface facets in an annual ring on the ocean surface at a given range from the altimeter. The random phases from each of the facets will add coherently before being squared to produce an estimate of return power. The random addition of phases produces what is known as fading noise or speckle. Averaging subsequent, independent pulses will reduce this noise. Typically, 1000 pulses are averaged over 1 s to produce a waveform.

Fading noise has a χ^2 distribution with $2N$ degrees of freedom, where N is the number of independent pulses averaged to produce a waveform.[1] The average of 1000 return pulses, therefore, has 2000 degrees of freedom and the rms noise in each bin is $1/\sqrt{N}$ of the mean return or 3% of the value in each range bin.

Mathematical Description

As early as 1957, Moore and Williams[2] demonstrated that the return pulse shape for a nadir-looking altimeter is an integral of the transmitted pulse shape, antenna gain, and properties of the reflecting surface.

In 1972, Berger[3] developed a mathematical model for the reflection of a narrow altimeter pulse from the ocean surface. In the model, Berger assumed that the transmitted pulse, the antenna gain pattern, and the surface height distribution are Gaussian-shaped, allowing him to predict analytically the return pulse shape. Berger showed that the pulse shape is clearly a function of the rms height of the surface and spacecraft attitude. Barrick[4] included the effects of the density of surface specular points. See also Brown.[5]

Fedor et al.[13] describe alternate algorithms to extract SWH from the GEOS-3 altimeter pulse shape and provide a good tutorial on the development of models for the return waveform. The reader is directed there for additional information.

If we define $W(x)$ to be the waveform shape as a function of range to the surface x, then we can describe the return waveform with the following convolution equation:

$$W(x) = P_H(x) * R(x) * I(x) * T(x) \tag{1}$$

$P_H(x)$ represents the ocean surface wave height probability density function. If the ocean surface is flat, this function reduces to the Dirac-delta function.

$R(x)$ represents the variation of radar reflectivity as a function of height on the ocean surface. If the reflectivity of the surface is independent on where along the wave that the reflection takes place, this function becomes a constant.

The impulse response function of the altimeter — the return pulse shape from a flat surface — is represented by $I(x)$. This function includes the effects of the finite duration and shape of the transmit pulse as well as the antenna gain pattern. The impulse response function is also a function of spacecraft attitude.

Tracker noise or the misregistration of the waveform will spread the average of many pulses. This spreading is represented by $T(x)$.

Estimation of SWH

The resultant waveform is the convolution of all these functions. Presumably, within the limitations posed by noise, if we can specify the system tracker noise and impulse response function and make the simplifying assumption that reflectivity is independent of height on a wave (i.e., $R(x)$ = a constant), then we can deconvolve the observed waveform shape to estimate the wave height probability density function.

In practice, a full deconvolution is difficult and computationally intensive. As one can see in Figure 2, the difference in pulse shape between a very low and very high sea state SWH is primarily evident in the slope of the leading edge of the return. As a consequence, a number of simpler algorithms based on this observation have been used.[6]

Townsend[8] provides a detailed explanation of the onboard SWH algorithm used for Seasat. Essentially, averages of the returns from range bins before and after the leading edge of the return form early and late gates. These are adaptively differenced to fit the slope of the waveform leading edge. This slope is used to select a SWH from a lookup table. Subsequently, this SWH estimate is corrected for the effects of spacecraft attitude, which itself is estimated from the shape of the trailing edge of the pulse. A similar onboard algorithm is used for both the Geosat and TOPEX altimeters.

14.2 Higher Order Moments

The mean of the ocean wave height distribution gives the distance between the altimeter and the mean ocean surface; the second moment, K_2, is the ocean wave height variance, $K_2 = \sigma_H^2$. The SWH is derived from this moment.

Huang and Long[8] have shown that the third-order moment of the wave height probability height distribution, skewness, is related to a parameter they call significant slope, §, by the equation:

$$K_3 = 8\pi\S \tag{2}$$

Significant slope is defined as the ratio σ_H/λ_0, where λ_0 is the dominant wavelength on the surface. For simple, unimodal seas, it is in principle possible to retrieve the dominant wavelength of the ocean surface from an altimeter pulse.

Fading noise makes the extraction of higher order moments of the wave height distribution difficult. In addition, the return waveform can be subtly distorted by the attenuation caused by rain and clouds within the altimeter footprint.[9] Though not significantly affecting the retrievals of height and SWH from the waveform shape, the effect of variable cloud and rain attenuation obscures the manifestation of higher order moments of the wave height distribution in the return pulse. We are unaware of the routine, successful extraction of more than the first and second moments of the ocean wave height distribution from spaceborne altimetry.

14.3 Altimeter Significant Wave Height as an Operational and Research Tool

Validation

Altimeter measured significant wave height (SWH) has been applied to various problems by various researchers for over a decade. The advantage of spaceborne instruments is the global coverage obtained in a relatively short period of time and in the case of SWH the ease with which the data may be retrieved and used. That is, there is very little computation required in order to apply the data. In this section we will review the various methods of validating the altimeter data for accuracy and discuss some of the problems where SWH has been applied. Each altimeter flown has been designed to operate at a particular specification. These specifications for five altimeters are given in Table 14.1. In all cases listed in the table, the specification accuracies were met or surpassed.

Methods of validation have been similar to those used to validate altimeter wind speed. Some of the first such validations were performed by Parsons.[10] Parsons compared GEOS-3 SWH data to aircraft, buoy, and ship measurements and to National Weather Service (NWS) model hindcasts. The overall mean and standard deviation of differences between satellite measurements and the aircraft and buoy data were 0.34 and 0.61 m, respectively. Comparisons with the model SWH estimates indicated that, in general, the models did not track the magnitude and location of many of the major wave fields measured by the altimeter. Comparison of shipboard measurements with GEOS-3 altimeter SWH measurements gave varying results. When the separation between the ship and satellite ground track were 200 km or greater, Parsons found that the GEOS-3 values underestimated sea state when the SWH was greater than 8 m and overestimated SWH at very low sea states. He attributes this to the limiting factor of along-track measurements. The satellite sometimes misses the highest or lowest waves.

McClain[11] reported on GEOS-3 SWH and aircraft laser profilometer comparisons. Apparently at the time of the Parsons report there were some problems with the laser measurements due to inappropriate correction for aircraft motion. After applying certain techniques for correcting the laser data, good SWH comparisons were found with the altimeter. For three out of four data points, the altimeter-lidar comparisons were as good as buoy, hindcast, and aircraft altimeter comparisons. Other investigators such as Queffeulou et al.,[12] Fedor and Brown,[13] and Mognard and Lago[21] compared GEOS-3 and Seasat satellite-measured SWH with buoys and found errors within specification values.

National Ocean Data Center (NODC) buoy data were used by Dobson et al.[14] to validate Geosat wave heights. Their results, computed for 116 comparisons over a 7-month period, produced in a mean and standard deviation of 0.4 and 0.49 m, respectively. Their results indicate that the Geosat SWH measurements were about 0.4 m smaller than the buoy measurements. Figure 14.3 shows a comparison of

Table 14.1 Altimeter Specifications

GEOS-3	25% of SWH for $2 \leq SWH \geq 10$
Seasat	±0.5 m or 10% of SWH
Geosat	±0.5 m or 10% of SWH
TOPEX	±0.5 m or 10% of SWH
ERS-1	±0.5 m or 10% of SWH

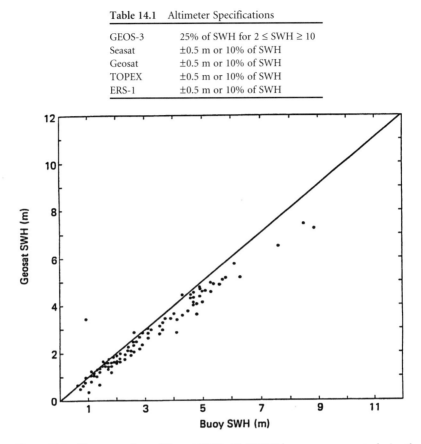

Figure 14.3 The comparison of Geosat SWH with NODC buoy measurements during the Exact Repeat Mission. (From Dobson, E.B. et al., *J. Geophys. Res.*, 92, 719, 1987. With permission.)

buoy and Geosat SWH for the Exact Repeat Mission where a mean and rms difference of 0.5 and 0.4, respectively, were found. Carter et al.[16] compared buoy and Geosat SWH for 164 data points. They found that the Geosat measurements were 13% lower than buoy measurements. Glazman and Pilorz[17] found that the Geosat measurements were on the average 0.4 m lower than buoy measurements in comparisons of over 400 data pairs.

Using 1 week of data, Guillaume and Mognard[18] compared Geosat SWH data to the METRO FRANCE VAG wave model. Because a model was used, a large number (2260) of comparisons were possible over SWH ranges from less than 1 to 9 m. They also found that there is an underestimation of wave height by Geosat when compared to the VAG model.

Figure 14.4 shows this error for a range of significant wave heights. Note that for SWH greater than 5 m there is a 24% error. Guillaume and Mognard's[18]

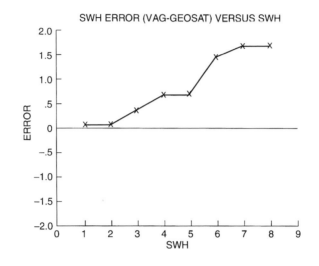

Figure 14.4 Mean SWH differences between the VAG model and Geosat, calculated for each class of SWH. Results shown are in meters. (From Guillaume, A. and Mognard, N., *J. Geophys. Res.*, 97, 9705, 1992. With permission.)

comparisons had 84 data points in the range greater than 5 m. None of the buoy comparisons cited previously had anywhere near this number of high wave conditions.

It should be noted that in all of the Geosat altimeter comparisons discussed above, the NOAA-distributed Geophysical Data Records (GDR) Geosat tape values were used. Some of the error in Geosat measurements can be attributed to the waveform fit used onboard the satellite to estimate SWH. Hayne and Hancock[19] have suggested that while a quadratic fit was generally used for Seasat and Geosat onboard processing, at times a higher order fit may be necessary. As a result, they propose a correction that can be applied to the data taken from the NOAA GDR. Anyone using these data should take all of these findings into consideration. Care should be exercised when obtaining data from other sources, to determine the corrections that have already been applied.

Carter[16] also compared global mean averages from Geosat and Seasat for the period from July 7 to October 10. The year was 1987 for Geosat and 1978 for Seasat. They concluded that there are differences between the two data sets that cannot be attributed to the different time periods, and they suggest that the Seasat SWH data may be biased high and Geosat data biased low. The authors point out that before these data are used for climatological studies, discrepancies between the different altimeter measurements of SWH should be resolved. This will probably also be true when attempting to incorporate TOPEX/Poseidon and ERS-1 data into climatological databases.

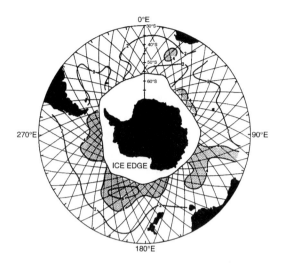

Figure 14.5 Southern Ocean map of the maximum wind waves (meters) deduced from SWH. (From Mognard, N.M. et al., *Wave Dynamics and Radio Probing of the Ocean Surface,* Plenum Press, New York, 1986, 479. With permission.)

Global and Regional Applications

Tournadre and Ezraty[20] report that altimeter-measured SWH along a ground track allows determination of the wave climatology in a zonal region represented by an ellipsoid of major axes ranging from 50 to 300 km. These findings are very important in light of the fact that altimeter SWH can be used more extensively in regions where one can safely assume that measurement along a ground track is representative of the SWH in a larger region. With two altimeters currently flying it may be possible to identify all regions and time of the year when zonal application is appropriate. Mognard et al.[22] studied waves in the southern oceans that result from cyclones using Geosat wind and wave measurements. Using the premise that any SWH values greater than those derived from the spectrum of a fully developed sea are due to swell, Mognard et al.[22] were able to generate maps of wind waves and swell. Figures 14.5 and 14.6 show these maps. Note the high wind wave areas in Figure 14.5 and the correspondence to high minimum swell in Figure 14.6.

Expected maximum wave heights were computed from the expression:

$$H_{ww} = 0.022(1.08U_{10})^2 \tag{3}$$

where U_{10} is the wind speed at 10 m height and was derived from the Seasat altimeter wind speeds. When H_{ww} is less than the measured SWH, the assumption is made that the excess is due to swell. Thus the swell energy can be computed by calculating the total energy, which is proportional to the square of SWH, and

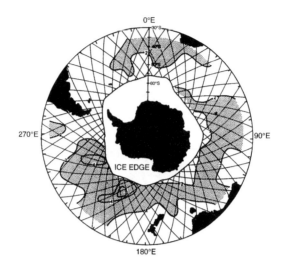

Figure 14.6 Southern Ocean map of the minimum swell (meters) deduced from SWH. (From Mognard, N.M. et al., *Wave Dynamics and Radio Probing of the Ocean Surface,* Plenum Press, New York, 1986, 479. With permission.)

subtracting the energy due to wind waves. When there was no swell present, a fully developed sea is assumed, allowing Mognard et al.[22] to estimate the wave spectrum associated with the cyclones in the region. For these wind waves, they employed the JONSWAP spectrum (Hasselmann et al.[25]) given by:

$$E(f) = \alpha g^2 (2\pi)^{-4} f^{-5} \exp\left\{ -\frac{5}{4}\left(\frac{f_m}{f}\right)^4 + \ln \gamma \exp\left[-\frac{(f - f_m)^2}{2\sigma^2 f_m^2} \right] \right\} \quad (4)$$

where α is a constant, g is gravitational acceleration, f is wave frequency, and f_m is the peak frequency of the spectrum.

In another analysis,[23] significant swell is derived from the SWH by assuming that a fully developed sea is present and any SWH in excess of that predicted by the Pierson-Moskowicz spectrum is associate with swell. The SWH from this spectrum is given by:

$$H_{1/3} = 2\alpha U_{19.5}^4 / \beta g^2 \quad (5)$$

where $H_{1/3}$ is SWH and β is a constant.

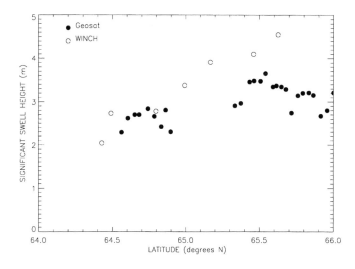

Figure 14.7   Geosat-derived significant swell heights and significant swell height hindcasts from the WINCH model. (From Mognard, N.M. et al., *J. Geophys. Res.*, 96, 467, 1991. With permission.)

Significant swell was compared to the Norwegian wave numerical model WINCH. The results showed that the swell could be extracted accurately in those regions where a fully developed sea existed, but could not be in areas where the sea is less than fully developed.

Figure 14.7 shows the comparison with the WINCH hindcast. Altimeter- and model-derived swell compare well below 65° N, but the altimeter-derived values are lower than the model north of 65°.

The estimation of wave growth from swell is also derived from Geosat SWH data as reported by Ebuchi et al.[24] Fetch is derived from weather maps and weather stations, while the nondimensional significant wave height is derived from Geosat. The nondimensional fetch and wave height are given by:

$$\hat{F} = gF\big/U_{10}^{2} \tag{6}$$

$$\hat{H} = gH_{1/3}\big/U_{10}^{2} \tag{7}$$

U is the wind speed at 10 m (in this case determined from Geosat wind speed data). Figure 14.8 shows the relationship between $\hat{F}$ and $\hat{H}$. The curves marked J and W are the empirical formulas for the Joint North Sea Wave Project (JONSWAP)[25] and Wilson,[26] respectively.

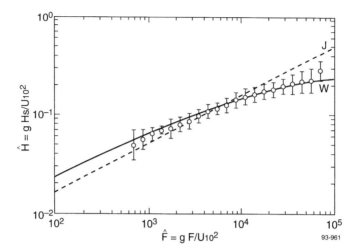

Figure 14.8   The relationship between the nondimensional fetch $\hat{F}$ and the nondimensional wave height $\hat{H}$. The circles are average values in a section and the bars represent the standard deviation. (From Ebuchi, N.H. et al., *J. Geophys. Res.,* 97, 809, 1992. With permission.)

14.4 Potential of Future Measurements

In the previous sections, we have examined the significant wave height as measured by satellite-borne altimeters. Although some problems exist in some of the data sets, there are no inherent problems in making accurate estimates of SWH with a satellite altimeter. When we consider the historical data sets that now exist from GEOS-3, Seasat, and Geosat; and the data sets that are presently being collected by TOPEX/Poseidon and ERS-1, altimeters are providing a wealth of data that can be potentially used in ocean science research. In addition, a Geosat follow-on altimeter (GFO) is planned for launch in 1995 along with ERS-2; and a TOPEX follow-on is being discussed also. The challenge to researchers should be how to best utilize these data sets both as stand-alone data and in conjunction with other satellite sensors. An example of this synergistic use of multiple data sets will be the real-time reception of GFO data onboard U.S. Navy ships. An integrated system is intended to use altimeter data along with other satellite data such as infrared and perhaps SAR data to give the ship an accurate assessment of the operating environment.

References

1. Ulaby, F.T., Moore, R.K., and Fung, A.K., *Microwave Remote Sensing: Active and Passive, II,* Addison-Wesley, Reading, MA, 463, 1982.
2. Moore, R.K. and Williams, C.S., Radar terrain at near vertical incidence, *Proc. IRE,* 45, 228, 1957.
3. Berger, T., Satellite altimetry using ocean backscatter, *IEEE Trans. Antennas Propag.,* AP-20, 295, 1972.

4. Barrick, D.E., Determination of mean surface position and sea state from the radar return of a short-pulse satellite altimeter, sea surface topography from space, *Tech. Rep. TR ERL 288-AOML* 7, 1, 16–1–16–19, National Oceanic and Atmospheric Administration, Rockville, MD, 1972.

5. Brown, G.S., The average impulse response of a rough surface and its applications, *IEEE J. Oceanic Eng.,* OE-2, 67, 1977.

6. Fedor, L.S., Godbey, T.W., Gower, J.F., Guptill, R., Hayne, G.S., Rufenach, C.L., and Walsh, E.J., Satellite altimeter measurements of sea state — an algorithm comparison, *J. Geophys. Res.,* 84, 3991, 1979.

7. Townsend, W.F., An initial assessment of the performance achieved by the Seasat-1 radar altimeter, *IEEE J. Oceanic Eng.,* OE-5, 80, 1980.

8. Huang, N.E. and Long, S.R., An experimental study of the surface elevation probability distribution and statistics of wind generated waves, *J. Fluid Mech.,* 101, 179, 1980.

9. Walsh, E.J., Monaldo, F.M., and Goldhirsh, J., Rain and cloud effects on a satellite dual-frequency radar altimeter system operating at 13.5 and 35 GHz, *IEEE Trans. Geosci. Remote Sensing,* GE-22, 615, 1984.

10. Parsons, C.L., Geos-3 wave height measurements: an assessment during high sea state conditions in the North Atlantic, *J. Geophys. Res.,* 84, 4011, 1979.

11. McClain, C.R., Chen, D.T., and Hammond, D.L., Comment on GEOS 3 wave height measurements: an assessment during high sea state conditions in the North Atlantic by C.L. Parsons, *J. Geophys. Res.,* 84, 4027, 4028, 1979.

12. Queffeulou, P., Braun, A., and Brossier, C., A comparison of Seasat-derived wave height with surface data, *Oceanography from Space,* Gower, J.F.R., Ed., Plenum Press, 1986, 637.

13. Fedor, L.S. and Brown, G.S., Wave height wind speed measurements from the Seasat radar altimeter, *J. Geophys. Res.,* 87, 3254, 1982.

14. Dobson, E. B., Monaldo, F., Goldhirsh, J., and Wilkerson, J., Validation of Geosat altimeter-derived wind speeds and significant wave heights using buoy data, *J. Geophys. Res.,* 92, 719, 1987.

15. Dobson, E.B., Geosat altimeter wind speed and wave height measurements: the ERM mission, Proc. WOCE/NASA Altimeter Algorithm Workshop, Corvallis, OR, U.S. WOCE Tech. Rep. No. 2, 1987.

16. Carter, D.J.T., Challenor, P.G., and Srokosz, M.A., An assessment of Geosat wave height and wind speed measurements, *J. Geophys. Res.,* 97, 11,383, 1992.

17. Glazman, R.E. and Pilorz, S.H., Effects of sea maturity on satellite altimeter measurements, *J. Geophys. Res.,* 95, 2857, 1990.

18. Guillaume, A. and Mognard, N., New method for the validation of altimeter-derived sea state parameters with results from wind and wave models, *J. Geophys. Res.,* 97, 9705, 1992.

19. Hayne, G.S. and Hancock, D.W., Corrections for the effects of significant wave height and attitude on Geoscat radar altimeter data, *J. Geophys. Res.,* 95, 2837, 1990.

20. Tournadre, J. and Ezraty, R., Local climatology of wind and sea state by means of satellite radar altimeter measurements, *J. Geophys. Res.,* 95, 8255, 1990.

21. Mognard, N.M. and Lago, B., The computation of wind speed and wave height from Geos-3 data, *J. Geophys. Res.,* 84(B8), 3979, 1979.

22. Mognard, N.M., Campbell, W.J., Cheney, R.C., Marsh, J.G., and Ross, D., Southern ocean waves and winds derived from Seasat altimeter measurements, *Wave Dynamics and Radio Probing of the Ocean Surface,* Phillips, O.M. and Hasselmann, K., Eds., Plenum Press, 1986, 479.

23. Mognard, N.M., Johannessen, J.A., Livingstone, C.E., Lyzenga, D., Shuchman, R., and Russel, C., Simultaneous observation of ocean surface winds and waves by Geosat radar altimeter and airborne synthetic aperture radar during the 1988 Norwegian continental shelf experiment, *J. Geophys. Res.,* 96, 10,467, 1991.

24. Ebuchi, N., Kawamura, H., and Toba, Y., Growth of wind waves with fetch observed by the Geosat altimeter in the Japan sea under winter monsoon, *J. Geophys. Res.*, 97, 809, 1992.

25. Hasselmann, K., Barnett, T.P., Bouws, E., Carlson, H., Cartwright, D.E., Enke, K., Ewing, J.A., Gienapp, H., Hasselmann, D.E., Krusemann, P., Meerburg, A., Muller, P., Olbers, D.J., Richter, K., Sell, W., and Walden, H., Measurements of wind-wave growth and swell during the Joint North Sea Wave Project (JONSWAP), *Dtsch. Hydrogr. Z.*, Suppl. 8(12), 1973.

26. Wilson, B.W., Numerical prediction of ocean waves in the North Atlantic for December, 1959, *Dtsch. Hydrogr. Z.*, 18, 114, 1965.

15

Synthetic Aperture Radar Imagery of Ocean Waves

Paris W. Vachon and Harald E. Krogstad

Synthetic Aperture Radar (SAR) imaging of moving targets is reviewed along with the analytic theory for forward mapping an ocean wave spectrum into a SAR image spectrum. The inversion of a SAR image spectrum into an ocean wave spectrum is discussed with examples. Topics of incomplete and ongoing SAR/ocean wave research are summarized.

15.1 Introduction

The capability and utility of synthetic aperture radar for measuring ocean waves has been vigorously debated since well before the launch of Seasat in 1978. Today, based on spaceborne (References 1–5, for example) and airborne (References 6–9, for example) field programs and most recently results from the ERS-1 SAR (Reference 10, for example), there is little doubt that SAR imagery does contain useful information about the ocean directional wave field. However, extraction of that

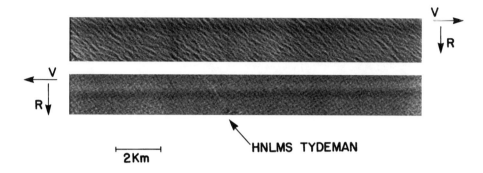

Figure 15.1   SAR images of the same ocean conditions recorded during the Labrador Extreme Waves Experiment (LEWEX) in March 1987 by the CCRS CV-580 C-band SAR.[8] The main difference is the relative flight direction, as indicated.

information is still a difficult and somewhat controversial process due to the nonlinear nature of the mapping of the moving ocean surface into a SAR image (References 6, 11–17, for example).

Figure 15.1 shows a pair of SAR images acquired a short time apart by the same airborne radar. For these images, the sea state was nominally the same and the radar parameters were identical; the only significant difference was the platform flight direction with respect to the waves. The substantial difference in the appearance of these images well illustrates the complexity of obtaining meaningful ocean wave information from SAR images.

In Section 15.2 the main issues in SAR imaging of static and dynamic scenes are reviewed. In Section 15.3 SAR ocean wave imaging is discussed and a description of the nonlinear ocean to SAR spectral transform of Hasselmann and Hasselmann[18] is given. This transform is an analytic representation of the SAR ocean wave imaging process and is a significant theoretical accomplishment.

The Hasselmann transform solves the forward problem, that is, the mapping from an ocean wave spectrum into the corresponding SAR image spectrum. The inverse problem consists of finding an ocean wave spectrum from an observed SAR image spectrum. SAR spectral inversion is discussed and demonstrated in Section 15.4. Some current ongoing research topics are discussed in Section 15.5. The current understanding of the capability of SAR for imaging ocean waves is summarized in Section 15.6.

15.2 SAR Imaging

The basic theory of SAR operation is described and discussed in many sources (Reference 19, for example). Briefly, the SAR creates a high-resolution image by recording the phase and amplitude of reflected electromagnetic radiation from the collection of scatterers that constitute the scene, and subsequently processing this information into a high-resolution image in an operation called compression. The

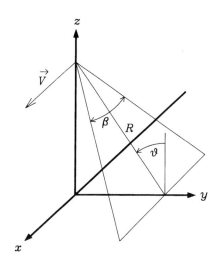

Figure 15.2 SAR geometry: R is the scene range; $\vec{V}$ is the platform velocity, ϑ is the local incidence angle; β is the azimuth beamwidth; and the coordinate system (x,y,z) represents azimuth, ground range, and height, respectively.

compression process includes the application of a filter that greatly enhances the image resolution when compared to that of a conventional real aperture radar.

For a radar operating with wavelength λ and a side-looking antenna aperture of dimension D, the azimuth (along-track) beamwidth is $\beta \sim \lambda/D$. The case of flat earth geometry is shown in Figure 15.2. If the platform is moving with velocity V, a target at range R remains in the antenna beam for the observation time:

$$T = \frac{R\beta}{V} \tag{1}$$

which is typically on the order of 1–10 s. The radar image azimuth resolution is $\rho = R\beta$ in the absence of SAR processing. During this integration time, the received signal from each scatterer undergoes a linear Doppler shift (often referred to as linear frequency modulation [FM]) due to the relative motion between the scatterer and the radar platform. The total change in frequency, or Doppler bandwidth, is

$$B = \frac{2V\beta}{\lambda} \tag{2}$$

The azimuth compression filter matches the expected phase of the received signal, thus improving the resolution by a factor equal to the time-bandwidth product $T \cdot B \gg 1$ compared to the case of no SAR processing at all. The SAR resolution turns out to be

$$\rho_1 = \frac{V}{B} \sim D \tag{3}$$

which is independent of the distance (range) to the scene and the radar wavelength.

Most SAR processors produce N looks, or independent images of the scene, by partitioning (bandpass filtering) the available Doppler bandwidth. The individual looks are usually non-coherently summed. This procedure trades spatial resolution for reduced effects of speckle, the grainy noiselike phenomenon that is characteristic of coherent imaging systems. The N-look resolution is

$$\rho_N = N\rho_1 \tag{4}$$

and the standard deviation of the speckle fluctuations is correspondingly reduced by the factor $\sqrt{N}$.

The range processing is analogous to the azimuth processing and is usually based on transmitting and compressing linear FM (chirp) pulses. The range location of the target is determined from the travel time of the radar pulse and the range resolution from the width of the pulse after compression.

The theoretical SAR azimuth resolution is achieved only if the received signal is coherent: that is, if the magnitude and phase of the signal received from each target in the scene remains self-consistent and predictable throughout the operation period T. For the dynamic ocean surface, this is almost never the case. Then, it is useful to define the scene coherence time τ as the timescale over which the scene actually does retain the necessary coherence. In this case, it can be shown that the azimuth resolution is degraded to:[20]

$$\rho = \rho_1 \sqrt{N^2 + \left(\frac{T}{\tau}\right)^2} \tag{5}$$

Further signal processing cannot recover the lost resolution (or information).

The Effect of Sensor Resolution

The ability of a SAR to measure scene detail is reduced as the detail of interest becomes comparable in scale to the SAR resolution. This loss in imaged detail may be partially compensated in the image spectral domain by a deconvolution process involving the Fourier transform of the system impulse response, commonly called the system transfer function (STF).[1] The correction, which is illustrated in Figure

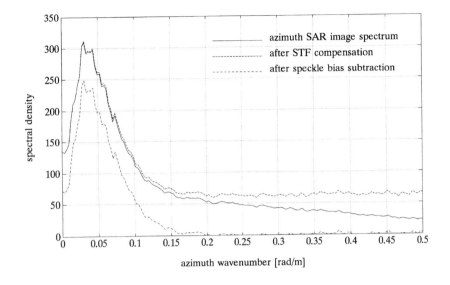

Figure 15.3 A SAR image azimuth wave number spectrum (the result of integrating over all range wave numbers) from the CCRS CV-580 C-band SAR, at various stages of STF and speckle bias correction.

15.3, is critical for spacecraft SARs for which the resolution is on the order of 30 m, especially if the properties of waves having a 50-m wavelength are of interest, for example. The correction is less important for airborne SARs for which the resolution may be on the order of 6 m or less.

The STF may be estimated from point target measurements and a model form for the impulse response, such as Gaussian pulse. Alternately, the STF may be measured from the autocorrelation function of a white noise input to the radar (under the assumption of Gaussian statistics).[20] One source of such a scene is an enclosed, wind-roughened body of water.[1]

The Effect of Speckle Noise

The relevant effect of speckle noise is the introduction of an approximately constant bias level in the image spectral domain.[21] This speckle bias must be estimated and removed from the SAR image spectrum. Analytic expressions for the bias level do exist. However, these expressions depend on exact compensation for the STF and details of the SAR processor implementation. In practice, a straightforward bias measurement at wave numbers that are not expected to contain ocean wave information is conducted. An example of the effects of the speckle noise in the spectral domain, and its compensation, is illustrated in Figure 15.3.

Coherent Imaging Effects

Coherent imaging effects are any that alter the phase structure of the received signal data. One coherent effect is caused by Doppler frequency shifts introduced by differential scene motion and is called velocity bunching. A target with a radial velocity component u_r (as observed along the radar line of sight) has its received signal Doppler shifted by the frequency $f_D = 2u_r/\lambda$, where λ is the radar wavelength. This Doppler shift results in an azimuth position shift of the target image by the amount:[22]

$$d = \frac{R}{V} u_r \tag{6}$$

Thus, differential motion, as occurs when scatterers on the ocean surface are advected by ocean wave orbital velocities, produces azimuth position shifts and wave-dependent patterns in SAR imagery.[23–25] Velocity bunching is a nonlinear imaging phenomenon and has been shown to be the dominant SAR imaging mechanism for ocean waves that are not traveling in the range direction.[6,26] Velocity bunching is discussed further in Section 2.4.

Another coherent effect is caused by the scene coherence time. The dynamic ocean surface generally lacks the necessary coherence time relative to the SAR integration time to allow the SAR to produce an image that meets the theoretical azimuth resolution. The scene coherence time is affected by decorrelation between scatterers and the intrinsic lifetime of the scatterers themselves.[6,16,27,28] This absence of coherence for the ocean surface limits the achievable resolution in the azimuth dimension for ocean wave imagery and is responsible for the azimuth cutoff that is characteristic of SAR ocean wave image spectra.

The degree of velocity bunching nonlinearity and the azimuth cutoff caused by coherence time limitations are proportional to the scene range-to-platform velocity ratio (R/V), as well as depending on the actual wind and wave conditions. For polar orbiting spaceborne SARs, such as Seasat and ERS-1, $R/V \sim 115$ s; while for airborne SARs, this parameter could be an order of magnitude smaller. The ERS-1 SAR is subject to severe azimuth cutoff for some wave conditions, as illustrated in Figure 15.4.

Noncoherent Imaging Effects

Noncoherent imaging effects are those that occur independent of the phase structure of the received signal data. These effects primarily concern the translation of the ocean wave reflectivity pattern at the wave phase velocity.[29] One such effect is scanning distortion. Since a SAR is a range-line scanning device, it builds up an image of the scene on a range-line by range-line basis. Therefore, if the wave phase velocity is an appreciable fraction of the platform velocity, the imaged wave pattern

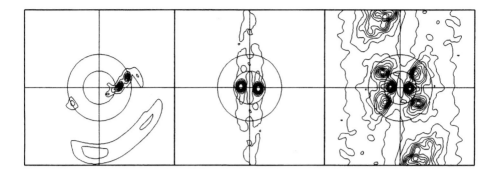

Figure 15.4 A comparison of a directional wave buoy slope spectrum (left), a spaceborne SAR image spectrum (ERS-1, © ESA 1991 — center), and an airborne SAR image spectrum (CCRS CV-580 C-band SAR — right) of the same ocean wave field acquired during the ERS-1 SAR Wave Spectra Validation Experiment on the Grand Banks in November 1991.[10] Azimuth is left to right and the circles represent 100-m (outer) and 200-m (inner) wavelengths.

becomes distorted. For waves with wave number $\mathbf{k} = (k_x, k_y)$, the imaged waves have a modified azimuth wave number:[30]

$$k'_x = k_x - \frac{\omega}{V} \tag{7}$$

where $\omega^2 = g|\mathbf{k}|\tanh(|\mathbf{k}|h)$ is the ocean wave dispersion relation, g is the local acceleration of gravity, and h is the local water depth. Scanning distortion should be compensated in airborne SAR-derived ocean wave spectra.[8] An example of the effect of scanning distortion correction is illustrated in Figure 15.5.

Another noncoherent effect is look misregistration. Since the pattern of reflectivity created by the wave field is translating at the wave phase speed $C = \sqrt{g/|\mathbf{k}|}$ and since the SAR requires the time T to collect the data to form a synthetic aperture, the wave pattern image is smeared over a distance CT, which is observable in the imagery if $CT \geq \rho_N$. However, due to the large time-bandwidth product characteristic of all practical SAR systems, the smearing may be largely compensated for by properly processing the individual look data. For example, a pair of looks based on signal data separated by Doppler frequency ΔB essentially correspond to images acquired at discrete time steps of:

$$\Delta t = \frac{T}{B}\Delta B \tag{8}$$

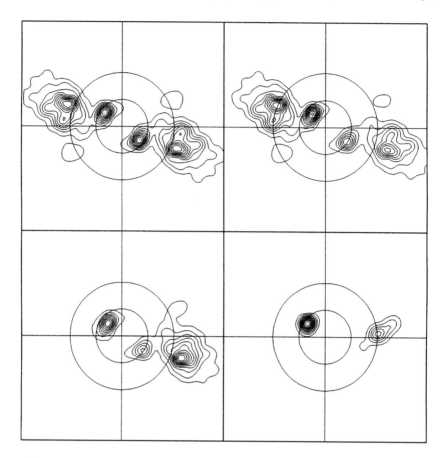

Figure 15.5   A comparison of SAR image spectra derived from data acquired during the Labrador Ice Margin Experiment (LIMEX) in March 1989 by the CCRS CV-580 C-band SAR.[27] The spectra represent the effects of various spectral corrections: raw SAR image spectrum (upper left); after scanning correction (upper right); and directionally resolved (the dominant peak of the ambiguous pair) based on processing the individual looks (lower left). A directional slope spectrum derived from an ice motion package is also included (lower right). Azimuth is left to right and the circles represent 100-m (outer) and 200-m (inner) wavelengths.

Ignoring scanning distortion, the individual look images of the wave field are misregistered by $D = C\Delta t$. Look misregistration with its consequences for the SAR processor focus setting has been an area of controversy for many years. Plant[15] has recently reviewed the focus controversy.

   To recover the expected distribution of wave spectral intensities in the SAR image spectrum, look misregistration may be compensated for on a look-by-look basis.[31] This may be accomplished in the image domain for a single wave mode, or in the spectral domain in the event of multiple wave modes. In the spectral domain the look misregistration becomes a deterministic phase term. However, the individual looks must be available for this analysis.

Look misregistration may be tracked and the direction of wave propagation established for all wave modes actually present in the scene.[31,32] An example of a directionally resolved SAR image spectrum based on individual look analysis is shown in Figure 15.5.

Another coherent effect is layover. For this wave-imaging mechanism, the apparent displacement of the scatterers in the range direction is modulated by their relative heights.[14,33] Layover is in this respect similar to velocity bunching, but is not an important imaging mechanism for ocean waves and is usually neglected.

15.3 SAR Ocean Wave Imaging

The basic physical principles underlying SAR imaging of ocean waves have been known and accepted for over 10 years.[6] The backscattering of electromagnetic waves from the ocean surface to the radar is dominated by Bragg scattering.[34,35] The Bragg-scale waves are assumed to be modulated by longer waves in a two-scale model. The Bragg-scale waves are defined by the resonance condition:

$$\Lambda_B = \frac{\lambda}{2\sin(\vartheta)} \tag{9}$$

where ϑ is the local incidence angle and λ is the electromagnetic wavelength. Usually $\Lambda_B \sim 2$ to 25 cm and the Bragg-scale waves are riding on long-scale waves ($\Lambda \sim 10$–1000 m), which in turn affect the local conditions at the Bragg scale.

It is helpful to consider the dynamic ocean surface as a collection of Bragg-scattering elements, with each element being composed of an ensemble of co-phase, small-scale waves satisfying the Bragg wavelength condition. Local tilting of the scatterers induced by the long waves changes the subset of waves fulfilling the Bragg condition, hence, the local backscatter, in a process called tilt modulation. Also, the local surface current induced by the long waves modulates the Bragg-scale waves in a process called hydrodynamic modulation. A hypothetical image resulting from tilt and hydrodynamic modulations would be the image obtained by a real aperture radar (RAR) with the same resolution as the SAR. This image is termed the RAR image.

Velocity bunching, caused by the wave orbital motion, prevents the RAR image from ever being observed. The SAR image may deviate significantly from the RAR image. However, velocity bunching only redistributes the scatterers in the final image and does not affect the average backscatter level.

The ocean scene may be represented as an ensemble of Bragg scatterers distributed randomly over the ocean surface. There may be many such scattering elements within one SAR resolution cell. Computer programs that simulate the ocean-to-SAR mapping have been in use for some time[36] and have been shown to give reasonable results when compared to actual imagery.[37,38] These computer programs first simulate the ocean surface amplitude, slope, and velocity using Monte Carlo techniques. Then, each patch of the surface is mapped into the SAR image by means of the mechanisms discussed above. The SAR image spectrum is obtained by averaging the modulus squared of the Fourier transforms of an ensemble of these simulated surfaces.

The Hasselmann Ocean-SAR Transform

An analytical theory for the ocean-to-SAR mapping for a general sea state has only recently been developed. Hasselmann's nonlinear transform[18] is the spectral counterpart of the simulation programs mentioned above.

Ocean waves are accurately described in a stochastic sense by Gaussian linear wave theory (LWT).[39] In this theory, the spatial and temporal wave field amplitude is represented as a random summation of sinusoidal components that obey the wave dispersion relation. The wave field is characterized by the directional wave number spectrum $\Psi(\mathbf{k})$ that is essentially the wave energy density as a function of wave number. The wave spectrum may also be written as a function of frequency and the direction of the wave number. Properties related to the wave field, such as the surface slope and surface orbital velocity, may be derived from the amplitude representation using linear (that is, wave field-independent) filters. The filters are defined by wave number-dependent transfer functions, and all stochastic properties such as spectra and correlation functions are given in terms of the wave spectrum and the transfer functions.

Although some uncertainties persist in this theory, it is generally believed that the effects of tilt and hydrodynamic modulations in a SAR image of the ocean may be obtained by linear filters applied to the wave number spectrum. Thus, the RAR image $I^R(\mathbf{x})$ may be obtained by LWT. The same applies to the relative motion of any point on the ocean surface; hence the field of azimuth target shifts in the image $d(\mathbf{x})$, in accordance with Equation 6. Neglecting scene coherence time limitations, the mapping from the RAR image to the SAR image $I^S(\mathbf{x})$ may be expressed in a conservation of scatterers equation of the form:[18]

$$I^S(\mathbf{x}) = \sum_{\mathbf{x}'} I^R(\mathbf{x}') \left| \frac{d\mathbf{x}'}{d\mathbf{x}} \right| \qquad (10)$$

where the summation extends over all solutions of $\mathbf{x} = \mathbf{x}' - d(\mathbf{x}')$. The Hasselmann transform expresses the spectrum $S(\mathbf{k})$ of $I^S(\mathbf{x})$ in terms of the correlation functions of d and I^R. A straightforward derivation of the transform was found by Krogstad.[40] The Hasselmann transform for $\mathbf{k} \neq 0$ may be written:

$$S(\mathbf{k}) = \frac{1}{(2\pi)^2} \int_{\mathbf{x}} e^{-i\mathbf{k}\mathbf{x}} G(\mathbf{x}, \mathbf{k}) d^2\mathbf{x} \qquad (11)$$

where the *G*-function is given by:

$$G(\mathbf{x}, \mathbf{k}) = I_0^2 e^{-k_x^2 [\rho_{dd}(0) - \rho_{dd}(\mathbf{x})]} \left\{ 1 + \rho_{II}(\mathbf{x}) + ik_x \left[\rho_{Id}(\mathbf{x}) - \rho_{Id}(-\mathbf{x}) \right] \right.$$
$$\left. + k_x^2 \left[\left(\rho_{Id}(0) - \rho_{Id}(\mathbf{x}) \right) \left(\rho_{Id}(0) - \rho_{Id}(-\mathbf{x}) \right) \right] \right\} \qquad (12)$$

Here, I_0 is the mean value of the RAR image, and the ρ's are spatial correlation functions between the shift field (d) and the RAR image (I), as indicated by the subscripts. The correlation functions are obtained from the wave spectrum $\Psi(\mathbf{k})$ and the corresponding transfer functions T_d and T_I by forming appropriate auto and cross spectra using an inverse Fourier transform.

The Hasselmann transform resembles a Fourier transform, but the *G*-function depends on both **x** and **k**, making it genuinely nonlinear. A discussion of the numerical evaluation of the transform was given by Krogstad.[40]

It is helpful to discuss the Hasselmann transform in terms of the dimensionless azimuth wave number $\kappa = k_x \rho_{dd}(0)^{1/2}$.[40] When $|\kappa| \ll 1$, G is approximately equal to $I_0^2 (1 + \rho_{II}(\mathbf{x}))$, and $S(\mathbf{k}) = S_{II}(\mathbf{k})$, which is the RAR image spectrum. Expanding the *G*-function to second order in κ leads to a linear transform, which includes both the RAR and the velocity bunching modulations and which is valid for somewhat larger values of $|\kappa|$. This linear transform was known before the nonlinear Hasselmann transform had been derived.[6] As $|\kappa|$ increases, the nonlinear character of the Hasselmann transform becomes dominant. The strongest nonlinearity results from the exponential factor in G. When $|\kappa|$ is larger than unity, this exponential factor suppresses all other contributions in G and the details of the RAR modulation are then only of minor importance. Figure 15.6 shows the result of applying the Hasselmann transform to an *in situ* measured wave spectrum.

Spectra of SAR ocean wave images have a characteristic azimuth cutoff. There have been numerous studies that attempt to relate the observed cutoff to ocean and SAR parameters. The spectrum computed from the nonlinear Hasselmann transform also has an intrinsic azimuth cutoff that in many cases fits very well with actual observations.[18,41,42] The intrinsic azimuth cutoff of the Hasselmann transform occurs around $|\kappa| = 2$; and this relates the cutoff directly to the standard deviation of the azimuth shift, which may be compactly related to fundamental sea state parameters via:[43]

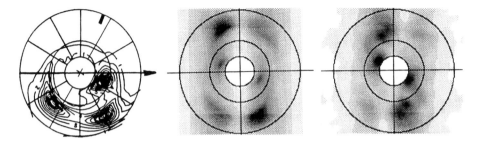

Figure 15.6 A wave buoy measured ocean spectrum (left), airborne SAR image spectrum computed using the buoy spectrum and the nonlinear Hasselmann transform (center), and the observed SAR image spectrum with $R/V = 70s$ (CCRS CV-580 C-band SAR — right) acquired during the Norwegian Continental Shelf Experiment (NORCSEX) in March 1988.[41] In each case, azimuth is left to right and the circles represent 100-m (outer), 200-m, and 400-m (inner) wavelengths.

$$\rho_{dd}(0)^{1/2} = \left(\frac{R}{V}\right)\sqrt{\frac{1}{2}\left(1-\bar{a}_2\right)\sin^2(\vartheta) + \cos^2(\vartheta)}\,\frac{H_s}{T_0}\frac{\pi}{2} \qquad (13)$$

where H_s is the significant wave height, T_0 is the mean wave period, ϑ the radar incidence angle, and $\bar{a}_2$ is the average value of the second cosine Fourier coefficient of the directional wave distribution. This equation represents the most direct connection between a fundamental quantity of the SAR spectrum and bulk sea state parameters.

However, as shown by Vachon et al.,[27] there are cases in which the azimuth cutoff observed in SAR data is stronger than is predicted by the Hasselmann transform. The additional cutoff can be attributed to a shortened coherence time, perhaps as a result of forcing of the Bragg-scale waves by the local wind field. This effect may be completely included in the Hasselmann transform by using the generalization derived by Bao et al.[44]

Before the derivation of the fully nonlinear theory, ocean-to-SAR transforms typically used a linear transfer function multiplied by a wave field-dependent exponential azimuth cutoff factor.[1,6,16] These quasi-linear transforms are fully adequate in many cases[28] and may be derived as an approximation to the fully nonlinear transform.[18] Neglecting scanning distortion, the quasi-linear transform takes the form:

$$S_Q(\mathbf{k}) = A\big(k_{az}, \rho_{dd}(0)\big)\Bigg\{ \frac{\big|T_I(\mathbf{k}) + ik_{az}T_d(\mathbf{k})\big|^2}{2}\Psi(\mathbf{k})$$

(14)

$$+ \frac{\big|T_I(-\mathbf{k}) + ik_{az}T_d(-\mathbf{k})\big|^2}{2}\Psi(-\mathbf{k})\Bigg\}$$

where k_{az} is the azimuth component of the wave number. The function A represents the azimuth cutoff, e.g., $A(k_{az}, \rho_{dd}(0)) = \exp(-k_{az}^2 \rho_{dd}(0))$.[18]

15.4 SAR Spectral Inversion

Inversion is the process of deriving an ocean wave spectrum from an observed SAR image spectrum. The inversion problem is formulated as the minimization of a function of the form:[18]

$$\big\|\Psi - \Psi_0\big\|_{W_\Psi}^2 = \int_k \big|\Psi(k) - \Psi_0(k)\big|^2 W_\Psi(k)\,d^2k$$

(15)

Here, Ψ_0 is the best guess *a priori* ocean wave spectrum, which would usually come from a wave model; $T(\Psi)$ is the ocean-to-SAR forward mapping transform; and S_0 is the observed SAR image spectrum. The norms use nonnegative weight functions W_Ψ and W_S, respectively:

$$\big\|\Psi - \Psi_0\big\|_{W_\Psi}^2 = \int_k \big|\Psi(\mathbf{k}) - \Psi_0(\mathbf{k})\big|^2 W_\Psi(\mathbf{k})\,d^2\mathbf{k}$$

(16)

and similarly for the second norm.

Various choices of the weight functions and other inversion constraints have been considered.[18,42,43] For example, the minimization should be conducted subject to the constraint $\Psi \geq 0$ since there is no restriction in the function that prevents the solution from being negative. The *a priori* spectrum may be used as a guide in resolving the wave propagation direction (though not reliably for multimodal wave systems) and as a source of high-frequency information for waves that would otherwise be subject to azimuth cutoff in the SAR imaging process.

The iterative inversion procedure may be conveniently combined with a quasi-linear approximation to the nonlinear Hasselmann transform to produce a fast and simple inversion algorithm.[43] The fully nonlinear Hasselmann transform may be used for fine-tuning the solution near the end of the iterative procedure. An example of SAR spectral inversion is given in Figure 15.7.

15.5 Ongoing Research

There has been significant progress in the last few years in the general understanding of the SAR ocean wave imaging process and the utilization of SAR-derived wave information. However, there are still areas of active research in SAR ocean wave imaging and SAR spectral inversion theory.

Scene Coherence Time

The width of the SAR azimuth passband is generally assumed to be dependent only on the distribution of orbital velocities associated with the wave field. However, any phenomenon that modulates the surface roughness could also contribute to the effective scene coherence time and hence to the width of the azimuth passband when imaging ocean waves.

As one example, the local surface roughness depends on the wind speed. The local scene coherence time also depends on the wind speed, in the sense that an increased wind speed results in a shorter scatterer lifetime and correspondingly a narrowing of the SAR azimuth passband. This effect has been considered for waves in ice as well as open ocean waves.[27,28] The wind speed, possibly from meteorological surface analyses or based directly on the SAR-derived radar cross section, may be necessary to augment the parameterization of the SAR azimuth spectral width.

As another example, the surface roughness could also depend on the surface current field.[45] Thus, for some regions, ocean current information may also be a helpful input to the SAR inversion procedure.

Inversion Constraints

It is possible to introduce additional constraints into the iterative SAR inversion process. One possible constraint is to use directionally resolved SAR image spectra based on individual look techniques. Such information would allow independence in the inversion process from the *a priori* spectrum, at least for resolving the wave propagation direction. The *a priori* spectrum would then only be required to supply information outside of the SAR azimuth passband.

The calculation of directionally resolved image spectra depends on the integration time being long enough to observe wave motion relative to the SAR resolution scale. For a given SAR platform geometry, this may be accomplished by using a longer wavelength (hence, a broader azimuth antenna pattern) or by steering the antenna in the azimuth direction (that is, spotlight mode operation).

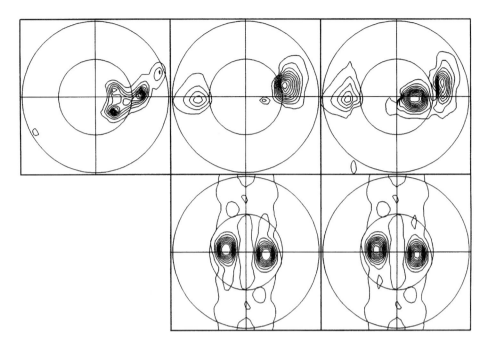

Figure 15.7   An example of SAR spectral inversion based on the quasi-linear transform. The data are from the ERS-1 Grand Banks Wave Spectra Validation Experiment of November 1991:[28] directional wave buoy spectrum (upper left); first guess model spectrum (upper center); inverted spectrum (upper right); measured ERS-1 SAR image spectrum (© ESA 1991) (lower center); and forward mapped inverted spectrum (lower right). Note how the ERS-1 data have modified the strength of the swell system in the inverted spectrum. Azimuth is left to right and the circles represent 200-m (inner) and 100-m (outer) wavelengths.

Another possible constraint is to force the inverted spectrum to have a parameterized spectral shape. For example, each wave mode found could be forced to conform to a Joint North Sea Wave Project (JONSWAP) spectrum with a predefined directional spread function. While this may not force the inversion to be more correct, the inverted result would at least be realistic in its spectral form and presumably would be easier to assimilate into a wave model that uses the same spectral form.

RAR MTF

The mapping of the initial pattern of radar cross section into a SAR image is now very well understood in terms of the fully nonlinear problem and possible linearizations. However, a weakness still appears to be the understanding of the initial RAR image. To date, the RAR modulation transfer function (MTF) is assumed to be due to tilt and hydrodynamic modulation of the short-scale roughness induced by the long waves of interest. Hydrodynamic modulation is often neglected relative to tilt modulation for steep incidence angle SARs.[13] Recent experience with inversion models shows anomalous growth of the inverted wave spectrum in the range direction, indicating that the magnitude of the RAR MTF has been underestimated.[42] Complete and consistent quantification of the magnitude and phase of the RAR MTF remains to be demonstrated. Empirical techniques have been developed to estimate the RAR MTF directly from the SAR image spectrum.[46]

Wave Model Assimilation

One reason for deriving ocean wave data from SAR imagery is as a source of information to improve operational wave modeling. As noted, the inversion process depends on the existence of an *a priori* wave spectrum. In practice, the *a priori* spectrum would come from a wave model. Thus, a tight coupling between SAR inversion and wave modeling is essential. The SAR is best equipped to provide low-frequency information about the wave field due to its inherent azimuth cutoff. The SAR information is necessarily restricted to those portions of the wave number plane that are not subject to azimuth cutoff. Under some conditions, the inverted SAR spectra can supply improved initial conditions for the wave model. However, a robust assimilation should update both the wave and the wind fields. If the wind field is not updated, the winds may become inconsistent with the blended wave field. The effects of the assimilation could then be wiped out in the next model step. The inversion and assimilation of SAR-derived ocean wave data into operational wave models is the subject of ongoing research at several national wave forecasting facilities.

15.6 Summary

SAR imaging of ocean waves has been reviewed. Apart from some contributions from variations of surface slope and hydrodynamic straining, a SAR is above all dependent on scene motion to create the ocean wave image through the velocity bunching imaging mechanism. However, the scene motion also acts to degrade the image by introducing look misregistration and a scene-dependent azimuth smearing. The smearing leads in turn to the azimuth cutoff in the SAR image spectrum.

Look misregistration results from scene motion at the wave phase velocity over the time period required to form the SAR image. It may be used to advantage to resolve the wave propagation direction for SARs with long coherent integration times.

On the other hand, the azimuth smearing and cutoff is an artifact of the imaging mechanism and can only be regarded as a hindrance. The azimuth cutoff results in a narrow azimuth passband that only allows wave spectral energy with a small azimuth wave number component (that is, in the platform flight direction) to pass through to the SAR image. A rough estimate of the width of the azimuth passband follows from Equation 13:

$$|k_{az}| \leq \frac{2}{\rho_{dd}(0)^{1/2}} \approx 1.2 \frac{1 + \vartheta^2/4}{R/V} \frac{T_o}{H_s} \tag{17}$$

The azimuth cutoff becomes more severe with a larger range-to-velocity ratio (R/V) and for higher sea states. In addition, a shorter scene coherence time will enhance the azimuth cutoff. Useful wave spectral information may not be recovered from outside of the azimuth passband by using the SAR image alone.

A spaceborne SAR, such as ERS-1, has a rather large range-to-velocity ratio of $R/V = 115$ s. Thus, only long wavelength swell (greater than 150-m wavelength as a rule of thumb) may be imaged by the ERS-1 SAR without significant distortion. Of course, since a SAR is the only spaceborne sensor that can provide information on ocean wavelength and direction, even distorted ocean wave spectra such as those that may be derived from the ERS-1 SAR are a unique and important information source.

An aircraft SAR may be operated with R/V an order of magnitude smaller than that of the ERS-1 SAR. Thus, airborne SAR imagery of ocean waves may be much less susceptible to imaging nonlinearity.

Unfortunately, the relationship between the ocean wave spectrum and the SAR image spectrum is genuinely nonlinear. However, there has recently been an important theoretical advance, the development of the Hasselmann transform, which represents velocity bunching analytically in the spectral domain, complete with nonlinearity. This forward transform has proved to be of central importance in completing our understanding of the velocity bunching imaging mechanism, and is a key part of many SAR spectral inversion algorithms. Quasi-linear transforms have been developed as approximations of the fully nonlinear transform. Both the Hasselmann transform and the quasi-linear transform may be extended to include processes that reduce the scene coherence time and further narrow the azimuth passband. Effects of the local wind speed on the scene coherence time, for example, may then be included in the forward transform in a straightforward manner.

SAR spectral inversion is the process of deriving a wave spectrum from a SAR image spectrum and has been demonstrated successfully in a number of measurement campaigns. In fact, SAR inversion is now often regarded as a routine procedure. SAR inversion is usually based on constrained iterative techniques and a first guess or *a priori* ocean wave spectrum that would usually come from an operational wave model. However, useful results depend on the nature of the wave field and the selection of weighting functions for the SAR and the *a priori* spectra. SAR inversion is usually conducted with the objective of assimilating the SAR-derived spectral wave information into the wave model as a way of updating the initial conditions of the model. In the case of ERS-1, the biggest effect from the SAR-derived information will be on the swell components of the wave spectrum.

Further research is required to complete our understanding of some details of the SAR ocean wave imaging process and to further constrain and improve the SAR inversion and assimilation process. The main weakness in our understanding of SAR imaging of ocean waves appears to be the RAR modulation transfer function. Meanwhile, long-term monitoring, assimilation, and validation projects will ultimately establish the operational utility of SAR data for ocean wave measurement.

References

1. Beal, R.C., Tilley, D.G., and Monaldo, F.M., Large- and small-scale spatial evolution of digitally processed ocean wave spectra from SEASAT synthetic aperture radar, *J. Geophys. Res.*, 88(C3), 1761, 1983.
2. Beal, R.C., Monaldo, F.M., Tilley, D.G., Irvine, D.E., Walsh, E.J., Jackson, F.C., Hancock, D.W., III, Hines, D.E., Swift, R.N., Gonzalez, F.I., Lyzenga, D.R., and Zambresky, L.F., A comparison of SIR-B directional ocean wave spectra with aircraft scanning radar spectra, *Science*, 232, 1531, 1986.
3. Holt, B., Introduction: studies of ocean wave spectra from the shuttle imaging radar-B experiment, *J. Geophys. Res.*, 93(C12), 15,365, 1988.
4. Macklin, J.T. and Cordey, R.A., SEASAT SAR observations of ocean waves, *Int. J. Remote Sensing*, 12(8), 1723, 1991.
5. Vesecky, J.F. and Stewart, R.H., The observation of ocean surface phenomena using imagery from the Seasat synthetic aperture radar: an assessment, *J. Geophys. Res.*, 87(C3), 3397, 1982.
6. Hasselmann, K., Raney, R.K., Plant, W.J., Alpers, W., Shuchman, R.A., Lyzenga, D.R., Rufenach, C.L., and Tucker, M.J., Theory of synthetic aperture radar ocean imaging: a MARSEN view, *J. Geophys. Res.*, 90(C3), 4659, 1985.
7. Shemdin, O.H., Tower ocean wave and radar dependence experiment: an overview, *J. Geophys. Res.*, 93(C11), 13,829, 1988.
8. Vachon, P.W., Olsen, R.B., Livingstone, C.E., and Freeman, N.G., Airborne SAR imagery of ocean surface waves obtained during LEWEX: some initial results, *IEEE Trans. Geosci. Remote Sensing*, 26(5), 548, 1988.
9. Rufenach, C.L., Olsen, R.B., Shuchman, R.A., and Russel, C.A., Comparison of airborne synthetic aperture radar and buoy spectra from the Norwegian Continental Shelf Experiment of 1988, *J. Geophys. Res.*, 96(C6), 10,423, 1991.
10. Dobson, F.W. and Vachon, P.W., The Grand Banks ERS-1 SAR wave spectra validation project: experiment overview and data summary, *Atmos.-Ocean*, 32(1), 7, 1994.
11. Burridge, D.A., Synthetic aperture radar imaging of ocean surface waves: multi-look and single-look systems, *Int. J. Remote Sensing*, 13(12), 2199, 1992.
12. Lyzenga, D.R., An analytic representation of the synthetic aperture radar image spectrum for ocean waves, *J. Geophys. Res.*, 93(C11), 13,859, 1988.
13. Monaldo, F.M. and Lyzenga, D.R., On the estimation of wave-slope and wave height variance spectra from SAR imagery, *IEEE Trans. Geoscience Remote Sensing*, 24(4), 543, 1986.
14. Ouchi, K., Synthetic aperture radar imagery of range travelling ocean waves, *IEEE Trans. Geosci. Remote Sensing*, 26(1), 30, 1988.
15. Plant, W.J., Reconciliation of theories of synthetic aperture radar imagery of ocean waves, *J. Geophys. Res.*, 97(C5), 7493, 1992.
16. Tucker, M.J., The imaging of waves by satelliteborne synthetic aperture radar: the effects of sea-surface motion, *Int. J. Remote Sensing*, 6(7), 1059, 1985.
17. West, J.C., Two-dimensional modelling of synthetic-aperture-radar imaging of ocean waves using the subwindowing technique, *Int. J. Remote Sensing*, 13(4), 615, 1992.
18. Hasselmann, K. and Hasselmann, S., On the nonlinear mapping of an ocean wave spectrum into a SAR image spectrum and its inversion, *J. Geophys. Res.*, 96(C6), 10,713, 1991.
19. Raney, R.K., SAR on SEASAT, ERS-1 and RADARSAT, in *Oceanographic Application of Remote Sensing*, Ikeda, M. and Dobson, F.W., Eds., CRC Press, Boca Raton, FL, 1995.
20. Raney, R.K., Wave orbital velocity, fade, and SAR response to azimuth waves, *IEEE J. Oceanic Eng.*, OE-6(4), 140, 1981.
21. Goldfinger, A.D., Estimation of spectra from speckled images, *IEEE Trans. Aerospace Electron. System*, AES-18(5), 675, 1982.
22. Raney, R.K., Synthetic aperture imaging radar and moving targets, *IEEE Trans. Aerospace Electron. Syst.*, AES-7(3), 499, 1971.
23. Larson, T.R., Moskowitz, L.I., and Wright, J.W., A note on SAR imagery of the ocean, *IEEE Trans. Antennas Propag.*, AP-24, 393, 1976.

24. Alpers, W. and Rufenach, C.L., The effect or orbital motions on synthetic aperture radar imagery of ocean waves, *IEEE Trans. Antennas Propag.*, AP-27(5), 685, 1979.

25. Swift, C.T. and Wilson, L.R., Synthetic aperture radar imaging of moving ocean waves, *IEEE Trans. Antennas Propag.*, AP-27(6), 725, 1979.

26. Alpers, W., Ross, D.B., and Rufenach, C.L., On the detectability of ocean surface waves by real and synthetic aperture radar, *J. Geophys. Res.*, 86(C7), 6481, 1981.

27. Vachon, P.W., Olsen, R.B., Krogstad, H.E., and Liu, A.K., Airborne synthetic aperture radar observations and simulations for waves-in-ice, *J. Geophys. Res.*, 98(C9), 16,411, 1993.

28. Vachon, P.W., Krogstad, H.E., and Paterson, J.S., Airborne and spaceborne synthetic aperture radar observations of ocean waves, *Atmos.-Ocean*, 32(1), 83, 1994.

29. Raney, R.K. and Vachon, P.W., Synthetic aperture radar imaging of ocean waves from an airborne platform: focus and tracking issues, *J. Geophys. Res.*, 93(C10), 12,475, 1988.

30. Raney, R.K. and Lowry, R.T., Ocean wave imagery and wave spectra distortions by synthetic aperture radar, in Proc. 12th Int. Symp. Remote Sensing of the Environment, Manila, Philippines, April 20–29, 1978, 1978, 683.

31. Vachon, P.W. and West, J.C., Spectral estimation techniques for multilook SAR images of ocean waves, *IEEE Trans. Geosci. Remote Sensing*, 30(3), 568, 1992.

32. Vachon, P.W. and Raney, R.K., Resolution of the ocean wave propagation direction in SAR imagery, *IEEE Trans. Geosci. Remote Sensing*, 29(1), 105, 1991.

33. Gower, J.F.R., Layover in satellite SAR imagery of ocean waves, *J. Geophys. Res.*, 88(C12), 7719, 1983.

34. Wright, J.W., Backscattering from capillary waves with application to sea clutter, *IEEE Trans. Antennas Propag.*, AP-14(6), 749, 1966.

35. van Zyl, J.J., Zebker, H.A., and Elachi, C., Imaging radar polarization signatures: theory and observation, *Radio Sci.*, 22(4), 529, 1987.

36. Alpers, W., Monte Carlo simulation for studying the relationship between ocean wave and synthetic aperture radar image spectra, *J. Geophys. Res.*, 88(C3), 1745, 1983.

37. Brüning, C., Alpers, W., and Hasselmann, K., Monte-Carlo simulation studies of the non-linear imaging of a two dimensional surface wave field by a synthetic aperture radar, *Int. J. Remote Sensing*, 11, 1695, 1990.

38. Lyzenga, D.R., Numerical simulation of synthetic aperture radar image spectra for ocean waves, *IEEE Trans. Geosci. Remote Sensing*, GE-24(6), 863, 1986.

39. Borgman, L.E., Irregular ocean waves: kinematics and forces, in *The Sea: Ocean Engineering Science, Volume 9A*, LeMehaute, B. and Hanes, D.M., Eds., Wiley Interscience, New York, 1990, 2121.

40. Krogstad, H.E., A simple derivation of Hasselmann's nonlinear ocean-SAR transform, *J. Geophys. Res.*, 97(C2), 2421, 1992.

41. Krogstad, H.E., Reliability and resolution of directional wave spectra from heave, pitch, and roll data buoy, in *Ocean Wave Spectra: Measuring, Modeling, Predicting, and Applying*, Beal, R.C., Ed., Johns Hopkins University Press, Baltimore, MD, 1991, 66.

42. Engen, G., Johnsen, H., and Krogstad, H.E., Inversion of synthetic aperture radar ocean image spectra, in Proc. 1992 Int. Geosci. Remote Sensing Symp. (IGARSS'92), May 1992, Houston, TX, 1992, 1331.

43. Krogstad, H.E., Samset, O., and Vachon, P.W., Generalizations of the nonlinear ocean-SAR transform and a simplified SAR inversion algorithm, *Atmos.-Ocean*, 32(1), 61, 1994.

44. Bao, M., Brüning, C., and Alpers, W., A generalized nonlinear ocean wave-SAR spectral integral transform and its application to ERS-1 SAR ocean wave imaging, in Proc. 2nd ERS-1 Symp., Space at the Service of Our Environment, October 11–14, 1993, Hamburg, Germany, ESA SP-361, 1994, 219.

45. Carande, R.E., Dual baseline and frequency along-track interferometry, in Proc. 1992 Int. Geosci. Remote Sensing Symp. (IGARSS'92), May 1992, Houston, TX, 1992, 1585.

46. Høgda, K.A. and Jacobsen, S., Estimating the RAR modulation transfer function directly from SAR ocean wave imagery, in Proc. 1993 Int. Geosci. Remote Sensing Symp. (IGARSS'93), Better Understanding of Earth Environment, August 18–21, 1993, Tokyo, Japan, 1993, 19.

16

Microwave Radiometers for Studies of the Ocean and the Marine Atmosphere

Kristina B. Katsaros and W. Timothy Liu

16.1 Introduction

Measurements of oceanographic and marine meteorological parameters by microwave instruments on satellites started in the 1960s on several Russian satellites. In this chapter we discuss microwave radiometers only, leaving out discussion of sounders and active microwave sensors (radars). The list of microwave radiometers on satellites is quite long, see Appendix C. We give a short recapitulation of the experience gained with the early microwave radiometers in space, and then concentrate on the characteristics of the current operational system, the Special Sensor Microwave/Imager (SSM/I), on satellites in the Defense Meteorological Satellite Program (DMSP). Some characteristics of the atmosphere with respect to microwave radiation and emission characteristics of the sea surface are recalled. The instrument, its sampling characteristics, and a few applications are then discussed, followed by a selected bibliography containing review articles that pursue the subject in greater depth.

The earliest satellites carrying microwave instruments, being Russian, were particularly aimed at observing the sea ice of the Arctic Ocean. Therefore, an early start was made in this area of research that has advanced possibly to the point of its technical limits. The Snow and Ice Data Center of the U.S. National Oceanic and Atmospheric Administration (NOAA) relies heavily on microwave radiometer data from polar orbiting satellites for its analyses. In the 1970s the U.S. flew the microwave radiometers on Skylab, a manned mission of short duration, and flew several instruments on polar orbiting satellites such as the Electrically Scanning Radiometers and the Scanning Multichannel Microwave Radiometers (SMMRs) on the Nimbus series experimental satellites and on Seasat.[1] Currently, a series of operational microwave radiometers are being launched on a routine basis (see Appendix C).

An advantage of microwave measurements is that the sea surface can be observed at many microwave frequencies. At other frequencies the atmospheric emission, absorption, or scattering provide information on integrated atmospheric properties throughout the whole column due to the relatively high transparency of the atmosphere. In regions of intense precipitation, however, the emission or scattering may be so strong that the sea surface and the lower atmosphere are obscured.

Parameters of interest for the sea surface include:

Wind speed
Sea surface temperature
Sea ice cover and type (1-year or multiyear ice)

and for the marine atmosphere:

Column integrated water vapor
Column integrated cloud liquid water
Precipitation rate (or precipitation index)
Column integrated ice scattering by atmospheric ice particles

Figure 16.1[2] shows the calculated transmittance in the frequency band 0–120 GHz (0.2–10 cm) of clear atmospheres at an incidence angle of 53° (typical value of incidence angle) for five representative climatic conditions. The weak absorption band of water vapor at 22 GHz allows measurements even through the humid tropical

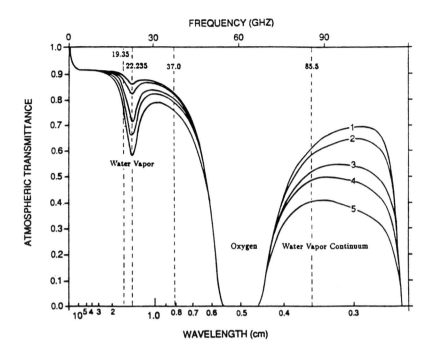

Figure 16.1   Total atmospheric microwave transmission for a zenith angle of 53°. Model atmospheres used are (1) wintertime polar, (2) summertime polar, (3) wintertime midlatitude, (4) summertime midlatitude, and (5) tropical. Plots were generated using the FASCOD3 line-by-line package. (After Petty, G. W., Ph.D. Thesis, University of Washington, Seattle, WA, 1990. With permission.)

atmosphere without saturation of the microwave signal. The oxygen band near 60 GHz is employed by microwave sounders to measure atmospheric temperature profiles. Radiometers operate in the windows above and below this band. The atmosphere is viewed against the mirrorlike background of the sea surface, which has low emissivity (hence high reflectivity). Direct emission by the atmosphere and its effects on transmission of the sea surface signal must therefore be considered. At the microwave wavelengths Planck's radiation law becomes linearly dependent on the thermometric temperature, a relationship referred to as the Rayleigh-Jeans law. It has become customary to speak of radiance in this wavelength region in terms of brightness temperature, TB. The electrically conducting sea surface emits polarized radiation, and the state of polarization of the signal received at the satellite also provides information about the source of the radiation. The signals emitted by the ocean are polarized such that the vertically polarized component has a brightness temperature about 20° K higher than the horizontally polarized one. Interference by absorption in rain clouds, nonpolarized emission by the rain, and scattering by the large graupel particles in deep convective clouds modify the polarization. Figure 16.2 illustrates these properties in a schematic manner.[3]

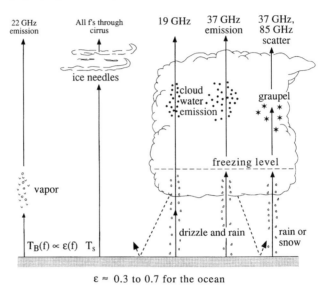

PROPERTIES OF ATMOSPHERIC WATER
(vapor, liquid and ice) WITH RESPECT TO MICROWAVES

$\varepsilon \approx 0.3$ to 0.7 for the ocean

Figure 16.2　Schematic of atmospheric transmission and scattering emission at SSM/I frequencies.[3]

Sea ice has higher emissivities than seawater, an effect due to the loss of sea salt in the freezing process, and the emissivity is frequency dependent. Whether the sea ice is of the smooth, first-year or of the rough, multi-year type also affects the polarization of the emitted radiation. The frequency dependence and the polarization are employed in deducing the extent of sea ice cover and the ice type.[4]

16.2 Microwave Satellite Radiometers

Microwave radiometers have operated exclusively on polar orbiting satellites. Low orbit satellites are necessary so that ground resolution will be of reasonable size. They still have very low resolution, of about 10–50 km (100 km in the worst case), compared to visible or infrared instruments (resolution of a few kilometers). The major satellite systems flown or scheduled to be launched in the near future are listed in Appendix C, Table 1. They are the Electrically Scanning Microwave Radiometer (ESMR), flown on the Nimbus-5 and -6 satellites, followed by the Scanning Multichannel Microwave Radiometer (SMMR).[5,6] Two of the SMMR instruments have been flown: one in 1978 on the Seasat satellite that survived only a few months, and one on the Nimbus-7 satellite that operated for 9 healthy years. The series of the Special Sensor Microwave/Imager (SSM/I), which is launched operationally on satellites in the Defense Meteorological Satellite Program, is expected to extend this long continuous series into the next millenium.[7] Between July 1987 and the end of 1993 3 have been launched, and a total of 16 are scheduled. These satellites are in a dusk-dawn sun-synchronous orbit. Appendix C, provides the main characteristics of the SMMRs and SSM/Is.

16.3 Algorithms

Discussions of the derivation of algorithms for obtaining atmospheric and sea surface properties from microwave sensors are found in the articles by Wilheit and Chang,[8] Wentz,[9] Petty and Katsaros,[10-13] Petty,[2,14,15] and Katsaros.[3] Radiative transfer calculations are usually involved, with empirical fits to the functions derived from these calculations providing the ultimate, often rather simple, formulas.

Care must be taken when using any satellite data source to obtain information about specific corrections required due to slight orbit variations and resulting incidence angle changes. The operational data centers take care of these variations when providing geophysical parameters; however, when starting from brightness temperatures, each user may have to apply his own corrections.

Algorithms for calculating the wind speed and various atmospheric water parameters are found in the following publications.[10,11,16-20]

As long as there is no precipitation in the field of view and the integrated cloud liquid water content is low, comparisons of wind speed estimates from SSM/I with buoy measured wind speeds show standard deviations of about 2 m/sec. Integrated water vapor values have the same standard deviation as the values obtained by integrating values obtained by radiosondes. Cloud liquid water is a less tested quantity for lack of comparison data. Petty and Katsaros[12,13] and Petty[14,15] discuss rain algorithms. Katsaros[3] summarizes these algorithms. Future improvements may include additional information derived from visible and infrared data from either the Operational Linescan System (an instrument also on the DMSP satellites) or from geostationary satellites observing the same clouds. Improved accuracy is expected from information on footprint filling by clouds or rain cells.[21,22]

Sea ice algorithms have been discussed by Gloersen et al.[4] and Cavalieri et al.[23] for the Nimbus-7 SMMR, and by Comiso[24] for the SSM/I.

16.4 Applications

For more than 25 years microwave radiometry has been used to study the oceans, sea ice, and marine atmosphere. The fundamental problem of footprint filling by clouds and rain cells and the lack of comparison data over the sea for cloud liquid water and rain resulted in tentative algorithms for these quantities. However, qualitative data on distributions and general structure of weather systems can often be useful and can help guide theoretical and model developments, especially when other types of data are not available. The following examples and the references should give the reader access to most of the relevant literature, although the list is by no means exhaustive.

16.5 Sea Surface Temperature

This parameter has limited use for oceanographic studies, because of the very low resolution obtainable with the 6.6 GHz frequency used for this purpose on the SMMRs. However, the SMMR ability to see through clouds allows for better climatological sampling of some regions of the global ocean than is available with infrared data (see References 25–28).

16.6 Sea Surface Wind Speed

Climatological studies of the surface wind speed obtained from microwave radiometers are described by Wentz[9] and Wentz et al.[9,29] The wind speed from the SMMRs or SSM/Is have been used to estimate turbulent surface fluxes by bulk methods. For instance, Liu and Niiler[30] and Liu[31,32] calculate evaporation rate in the tropical Pacific Ocean using SMMR estimates of wind speed and the column integrated water vapor as a stand-in for surface humidity. Similarly, Etcheto and Merlivat[33] and Boutin and Etcheto[34] calculate global climatologies of carbon dioxide flux from or to the sea using surface wind speeds estimated from SMMR or SSM/I.

16.7 Integrated Atmospheric Water Vapor

Since the atmospheric water vapor is one of the most robust parameters provided by microwave radiometers with accuracies equivalent to values obtained by integrating atmospheric soundings,[17,35,36] this parameter has been employed in numerous studies. Using data from the Seasat and Nimbus-7 SMMRs, McMurdie and Katsaros[37] showed that the numerical model of the European Center for Medium-Range Weather Forecasts (ECMWF) tends to load the atmosphere with far too much water vapor. Liu et al.[38] also found an overestimate of water vapor by ECMWF and TIROS Operational Vertical Sounder (TOVS) in the dry air masses over the eastern tropical-subtropical ocean, particularly in the southern hemisphere, when compared with SSMI and radiosound data. Claud et al.[39] analyze the water vapor structure in high latitude regions during cold air outbreaks and discuss the compatibility of these data with water vapor information from the TOVS that has long been launched on NOAA polar orbiting operational satellites.

Atmospheric water vapor affects electromagnetic propagation in the atmosphere, and coincident measurements of atmospheric water vapor is crucial to the accurate determination of sea level by microwave altimeters.[39,40] The delay of radar propagation in the atmosphere caused by atmospheric water vapor is measured by specifically designed nadir-looking microwave radiometers onboard both ERS-1 and TOPEX/Poseidon.

Atmospheric water vapor is more transparent to the shortwave radiation from the sun than to the longwave radiation from the earth and is an important greenhouse gas. The relation between greenhouse warming and water vapor suggested by Raval and Ramanathan[42] has been examined using data from the Earth Radiation Budget Experiment (ERBE) and coincident measurement by SSM/I by Liu.[43,44]

From the conservation principle, the surface hydrologic flux (the difference between evaporation and precipitation) has to be balanced by the divergence of water vapor transport integrated over the depth of the atmosphere under steady-state assumption. The atmospheric water measured by spaceborne microwave radiometers could be used in combination with cloud-drift wind or surface level wind to determine hydrologic forcing over the ocean.[43,45]

16.8 Evaporation Rate

Evaporation is a major component of both the thermal and hydrologic forcings on the ocean. For any practical large-scale application, evaporation is estimated by bulk parameterization models that require the measurement of surface wind speed, sea surface temperature, and surface level humidity. Sea surface temperature, surface level wind speed, and integrated water vapor can be estimated from observations of the spaceborne microwave radiometer, but not surface level humidity. Liu et al.[46] found that the dominant mode of humidity profile variability in the atmosphere, which is coherent through the entire depth, is decoupled from other modes coherent only through limited depth. For periods longer than a week, the integrated water vapor is a good estimator of the humidity profiles and surface level humidity. A global relation between integrated water vapor and surface level humidity derived by Liu[31] has been used to estimate ocean surface latent heat flux in a number of studies.

At present, the relative accuracy of latent heat flux estimated from satellite data is sufficient to resolve the seasonal cycle over most of the ocean and interannual variability of El Niño and the Southern Oscillation (ENSO) in the tropical ocean. Surface latent heat flux in the tropical Pacific has been estimated using SMMR data to study the evolution of major climate signals, such as ENSO.[32] It has been estimated over global oceans, using observations of SSM/I and Advanced High-Resolution Radiometer (AVHRR) to examine the feedback on sea surface temperature changes.[47]

The surface heat flux could be integrated to give the mean meridional heat transport by the ocean. In the past, only meteorological reports from volunteer ships were used, but satellite data have the potential of providing better coverage. To adequately resolve the meridional heat transport, an absolute accuracy better than $10°W\ m^{-2}$ in the total heat flux is required.[48] The estimation of most components of surface heat flux, including latent heat, needs further improvement to meet such stringent requirements. Improvements are expected in the next decade with the launching of advanced sensors for surface wind and humidity soundings. Field experiments, such as Tropical Ocean Global Atmosphere, Coupled Ocean Atmosphere Response Experiment (TOGA COARE) and Humidity Exchange Over the Sea (HEXOS), have provided much needed data to improve the bulk parameterization technique at high and low wind conditions.

Miller and Katsaros[49] used a similar method to that of Liu and Niiler[30] to estimate the latent heat flux in the environment of a rapidly deepening cyclone in midlatitudes, while Claud et al.[50] estimated surface latent heat flux in a polar low.

16.9 Midlatitude Cyclones

The distribution of integrated atmospheric water and rain in midlatitude cyclones over the oceans has been studied extensively by McMurdie and Katsaros.[36,37] They found characteristic patterns and variability with region and season. Katsaros et al.[51,52] and Katsaros[3] suggest that the gradients in integrated water vapor associated with frontal zones could be used to identify such regions, especially in rapidly developing situations and far from radiosonde stations. The gradient in water vapor is characteristically strong in the rear of a cold front, and marine warm fronts also appear to be identifiable with this parameter. The exact location of the surface front with respect to this gradient may vary from case to case, but the gradient calculation can provide an objective method for locating frontal zones.[51,52] McMurdie et al.[53] also showed that convergence patterns in the surface wind obtained by the scatterometer on the Seasat satellite were related to the integrated water vapor, cloudiness, and rain in such storms as seen by the SMMR.

16.10 Integrated Cloud Liquid Water

There has been very little verification of this parameter from *in situ* data, but Young et al.[22] discuss the use of several satellite systems. Point comparison to upward looking radiometers have been performed (see Reference 18) for the SSM/I, giving good results, but the statistical base is weak. Prabhakara et al.[27] took an interesting approach for deducing climatological distributions using the 10-GHz frequency band on SMMR that has a more linear response to cloud liquid water. Raustein et al.[54] used the SSM/I qualitative estimates of cloud liquid water to test a mesoscale numerical cloud model.

16.11 Rain Rate

There is great interest in obtaining estimates of rain rate or accumulated precipitation over the sea, since this is one of the often missing parameters in the oceanic salt budget and in climate characterizations of oceanic regions. The difficulty with microwave radiometer estimates of rain is due to the large footprints of the microwave radiometer and the essentially nonlinear dependence of the emission on the rain intensity, making footprint averages nonrepresentative. Petty and Katsaros[55] illustrate this effect by comparing coastal radar data for the Taiwan area. Two algorithm comparison tests have been conducted with several research groups participating.

16.12 Ice Scattering in Convective Clouds

This parameter can be used to detect thunderstorms and intense convective activity. It was first suggested by Spencer[56] and used in statistical studies of thunderstorms. Petty[2] shows how this parameter illustrates the convective activity in a low pressure system and along a cold front. Claud et al.[38,57] use this parameter to identify polar lows over the Norwegian Sea. Carleton et al.[58] discusses similar observations of so-called mesocyclones in the Southern Hemisphere, and Mognard and Katsaros[59,60] also show the relationship of this parameter to atmospheric frontal zones in midlatitude depressions.

16.13 Tropical Cyclones

The development of tropical cyclones depends to a large extent on the release of latent heat energy in the clouds and in precipitating regions. Katsaros et al.,[61] following a suggestion by Glass and Felde,[62] showed that microwave parameters (based on 19-GHz polarization difference or 85-GHz brightness temperature) allow estimates of tropical storm intensity, defined in terms of the surface maximum wind speed. The microwave measurement gives an index of the increasing organization of the rain clouds as a tropical cyclone intensifies, invoking a physical relationship to provide a quantitative relationship between the microwave index and storm intensity. However, as with most of the microwave-derived parameters applied to oceanic regions, there are few *in situ* or other direct measurements allowing empirical calibrations.

16.14 Sea Ice Studies

Sea ice studies based on microwave active or passive data are discussed in other chapters of this book. For completeness, we mention here the comprehensive reference works on the use of SMMR data for sea ice analysis by Gloersen et al.,[4] and an article on sea ice analysis with SSM/I data by Comiso.[24]

16.15 Conclusion

Microwave radiometry has a bright future for studies of atmospheric weather systems over the sea, for sea surface wind measurements supplementing observations made by active systems (scatterometers or altimeters), and for climatological studies where the low resolution of these instruments can be compensated for by temporal averages (as long as nonlinear averaging effects are not present or can be accounted for). Microwave radiometers are already used routinely to obtain the atmospheric correction for the travel time delay of altimeter signals. As microwave radiometers are relatively simple and inexpensive instruments, numerous radiometers will be in use over the coming decades. The 1400-km wide swaths from two satellites operating simultaneously provide adequate sampling for traveling weather systems, especially at mid- to high latitudes. Their data could be made available via local receiving antennas, as are AVHRR measurements of infrared sea surface temperatures and cloud data. Meteorologists would have the microwave data available in real time and could provide early warning of severe weather due to polar lows, midlatitude depressions, and tropical cyclones.

References

1. Njoku, E. G., Passive microwave remote sensing from space — a review, *Proc. IEEE*, 70, 728, 1982.
2. Petty, G. W., On the Response of the Special Sensor Microwave/Imager to the Marine Environment — Implications for Atmospheric Parameter Retrievals, Ph.D. thesis, Department of Atmospheric Sciences, University of Washington, Seattle, WA, 1990, 291. (Requests for copies, copying or reproduction of dissertation: University Microfilms, 300 North Zeeb Road, Ann Arbor, MI 48106.)
3. Katsaros, K. B., Measurements of atmospheric water parameters and wind speed with the Special Sensor Microwave/Imager and the ERS-1 scatterometer, in Proc. ECMWF/EUMETSAT Semin. Developments in the Use of Satellite Data in Numerical Weather Prediction, September 6–10, 1993, ECMWF, Reading Park, U.K., 1994.
4. Gloersen, P., Campbell, W. J., Cavalieri, D. J., Comiso, J. C., Parkinson, C. L., and Zwally, H. J., Arctic and Antarctic Sea Ice, 1978–1987: Satellite Passive Microwave Observations and Analysis, Science and Technical Information Program, National Aeronautics and Space Administration, Washington, D.C., 1992, 290.
5. Gloersen, P. D. and Barath, F. T., A Scanning Multichannel Microwave Radiometer for Nimbus-G and Seasat-A, *IEEE J. Oceanic Eng.*, OE-2, 172-178, 1977.
6. Njoku, E. G., Stacey, J. M., and Barath, F. T., The Seasat Scanning Multichannel Microwave Radiometer (SMMR): instrument description and performance, *IEEE J. Oceanic Eng.*, OE-5, 100, 1980.
7. Hollinger, J. P., Lo, R., Poe, G., Savage, R., and Pierce, J., *Special Sensor Microwave/Imager User's Guide*, Naval Research Laboratory, Washington, D.C., 1987, 177.
8. Wilheit, T. T. and Chang, A. T. C., An algorithm for retrieval of ocean surface and atmospheric parameters from the observations of the Scanning Multichannel Microwave Radiometer (SMMR), *Radio Sci.*, 15, 525, 1980.
9. Wentz, F. J., A model function for ocean microwave brightness temperatures, *J. Geophys. Res.*, 88, 1892, 1983.
10. Petty, G. W. and Katsaros, K. B., New geophysical algorithms for the Special Sensor Microwave/Imager, *Prepr. Vol., 5th Int. Conf. Satellite Meteorol. Oceanogr.*, September 3–7, 1990, London, U.K., 1990.
11. Petty, G. W. and Katsaros, K. B., Precipitation observed over the South China Sea by the Nimbus 7 Scanning Multichannel Microwave Radiometer during Winter MONEX, *J. Appl. Meteorol.*, 29, 273, 1990.
12. Petty, G. W. and Katsaros, K. B., The response of the SSM/I to the marine environment. I. An analytic model for the atmospheric component of observed brightness temperatures, *J. Atmos. Oceanic Technol.*, 9, 746, 1992.

13. Petty, G. W. and Katsaros, K. B., The response of the SSM/I to the marine environment. II. A parameterization of roughness effects on sea surface emission and reflection, *J. Atmos. Oceanic Technol.,* 11, 617, 1994..

14. Petty, G. W., Physical retrievals of over-ocean rain rate from multichannel microwave imagery. I. Theoretical characteristics of normalized polarization and scattering indices, *Meteorol. Atmos. Phys.,* 54, 79, 1994.

15. Petty, G. W., Physical retrievals of over-ocean rain rate from multichannel microwave imagery. II. Algorithm implementation, *Meteorol. and Atmos. Phys.,* 54, 101, 1994.

16. Goodberlet, M. A., Swift, C. T., and Wilkerson, J. C., Remote sensing of ocean surface winds with the Special Sensor Microwave/Imager, *J. Geophys. Res.,* 94, 14,547, 1989.

17. Alishouse, J. C., Snyder, S., Vongsathorn, J., and Ferraro, R. R., Determination of oceanic total precipitable water from the SSM/I, *IEEE Trans. Geosci. Remote Sensing,* 28, 811, 1990.

18. Alishouse, J. C., Snider, J. B., Westwater, E. R., Swift, C., Ruf, C., Snyder, S., Vongsathorn, J., and Rerraro, R. R., Determination of cloud liquid water content using the SSM/I, *IEEE Trans. Geosci. Remote Sensing,* 28, 817, 1990.

19. Wentz, F. J., Mattox, L. A., and Peteherych, S., New algorithms for microwave measurements of ocean winds: applications to SEASAT and the Special Sensor Microwave/Imager, *J. Geophys. Res.,* 91, 2289, 1986.

20. Wentz, F. J., User's Manual, SSM/I Antenna Temperature Tapes, RRS Tech. Rep. 032588, Remote Sensing Systems, Santa Rosa, CA, 1988.

21. Miletta, J., Meteorological Applications of Coincident Visible, Infrared, and Microwave Observations from the Defense Meteorological Satellite Program. M.Sc. thesis, Department of Atmospheric Sciences, University of Washington, Seattle, WA, June 7, 1993, 86 pp.

22. Young, D. F., Minnis, P., Katsaros, K., Dybbroe, A., and Miletta, J., Comparison of Techniques for Deriving Water-Cloud Microphysical Properties from Multiple Satellite Data, *Prepr. Vol. 11th Int. Conf. Clouds Precipitation,* Montreal, Quebec, Canada, August 17–21, 1992.

23. Cavalieri, D. J., Gloersen, P., and Campbell, W. J., Determination of sea ice parameters with the Nimbus 7 SMMR, *J. Geophys. Res.,* 89, 5355, 1984.

24. Comiso, J. C., Arctic multiyear ice classification and summer ice cover using passive-microwave satellite data, *J. Geophys. Res.,* 95, 13,411, 1990.

25. Lipes, R. G., Bernstein, R. L., Cardone, V. J., Katsaros, K. B., Njoku, E. J., Riley, A. L., Ross, D. B., Swift, C. T., and Wentz, F. J., Seasat Scanning Multichannel Microwave Radiometer: results of the Gulf of Alaska workshop, *Science,* 204, 1415, 1979.

26. Njoku, E. G. and Swanson, L., Global measurements of sea surface temperature, wind speed, and atmospheric water content from satellite microwave radiometry, *Mon. Weather Rev.,* 111, 1977, 1983.

27. Prabhakara, C., Wang, I., Chang, A. T. C., and Gloersen, P., A statistical examination of Nimbus-7 SMMR data and remote sensing of sea surface temperature, liquid water content in the atmosphere and surface wind speed, *Bull. Am. Meteorol. Soc.,* 22, 2023, 1983.

28. Gloersen, P. D., Cavalieri, D. J., Chang, A. T. C., Wilheit, T. T., Campbell, W. J., Johannessen, O. M., Katsaros, K. B., Kunzi, K. F., Ross, D. B., Staelin, D., Windsor, E. P. L., Barath, F. T., Gudmandsen, P., Langham, E., and Ramseier, R. O., A Summary of Results from the First Nimbus-7 SMMR Observations, *J. Geophys. Res.,* 89, 5335, 1984.

29. Wentz, F. J., Cardone, V. J., and Fedor, L. S., Intercomparison of wind speeds inferred by the SASS, altimeter and SMMR, *J. Geophys. Res.,* 87, 3378, 1982.

30. Liu, W. T. and Niiler, P. P., Determination of monthly mean humidity in the atmospheric surface layer over oceans from satellite data, *J. Phys. Oceanogr.,* 14, 1451, 1984.

31. Liu, W. T., Statistical relation between monthly mean precipitable water and surface-level humidity over global oceans, *Mon. Weather Rev.,* 114, 1591, 1986.

32. Liu, W. T., Moisture and latent heat flux variabilities in the tropical Pacific derived from satellite data, *J. Geophys. Res.,* 93, 6749, and 6965, 1988.

33. Etcheto, J. and Merlivat, L., Satellite determination of the carbon dioxide exchange coefficient at the ocean-atmosphere interface: a first step, *J. Geophys. Res.,* 93, 669, 1988.

34. Boutin, J. and Etcheto, J., Seasat scatterometer versus Scanning Multichannel Microwave Radiometer wind speeds: a comparison on a global scale, *J. Geophys. Res.*, 95, 22,275, 1990.

35. Alishouse, J. C., Total precipitable water and rainfall determinations from the Seasat Scanning Multichannel Microwave Radiometer (SMMR), *J. Geophys. Res.*, 88, 1929, 1983.

36. McMurdie, L. A. and Katsaros, K. B., Atmospheric water distribution in a mid-latitude cyclone observed by the Seasat Scanning Multichannel Microwave Radiometer, *Mon. Weather Rev.*, 113, 584, 1985.

37. McMurdie, L. A. and Katsaros, K. B., Satellite derived integrated water vapor distribution in oceans midlatitude storms: variation with region and season, *Mon. Weather Rev.*, 119, 589, 1991.

38. Liu , W.T., Tang, W., and Wentz, F.J., Precipitable water and surface humidity over global oceans from Special Sensor Microwave Imager and European Center for Medium Range Weather Forecasts, *J. Geophys. Res.*, 97, 2251, 1992.

39. Claud, C., Katsaros, K. B., Petty, G. W., Chedin, A., and Scott, N. A., A cold air outbreak over the Norwegian Sea observed with the Tiros-N Operational Vertical Sounder (TOVS) and the Special Sensor Microwave/Imager (SSM/I), *Tellus*, 44A, 100, 1992.

40. Tapley, B. D., Lundberg, J. B., and Born, G. H., The Seasat altimeter wet tropospheric range correction, *J. Geophys. Res.*, 87, 3213, 1982.

41. Liu, W. T. and Mock, D., The variability of atmospheric equivalent temperature for radar altimeter range correction, *J. Geophys. Res.*, 95, 2933, 1990.

42. Raval, A. and Ramanathan, V., Observational determination of the greenhouse effect, *Nature (London)*, 342, 758, 1989.

43. Liu, W. T., Evaporation from the ocean, in *Atlas of Satellite Observations Related to Global Change*, Gurney, R. J., Foster, J. L., Parkinson, C. L., Eds., Cambridge University Press, 1993, 265.

44. Liu, W. T., Water vapor and greenhouse warming over ocean, in *Satellite Remote Sensing of the Ocean Environment*, Jones, A., Sugimori, Y., and Steward, R., Eds., Seibutsu Kenkyusha Co., Tokyo, Japan, 1993, 214.

45. Liu, W. T., Tang, W., and Niiler, P. P., Humidity profiles over ocean, *J. Climate.*, 4, 1023, 1991.

46. Liu, W. T., Tang, A., and Bishop, J. K. B., Evaporation and solar irradiance as regulators of sea surface temperature in annual and interannual changes, *J. Geophys. Res.*, in press.

47. Heta, Y. and Mitsuta, Y., An evaluation of evaporation over the tropical Pacific Ocean as observed from satellites, *J. Appl. Meteorol.*, 32, 1242, 1993.

48. Dobson, F. W., Bretherton, F. P., Burridge, D. M., Crease, J., and Kraus, E. B., vondar Haar, T. H., The CAGE (Can the Atlantic Gain Energy) Experiment: A feasibility study, *World Climate Program*, 22, WMO, Geneva, 1982, 95 pp.

49. Miller, D. K. and Katsaros, K. B., Satellite-derived surface latent heat fluxes in a rapidly intensifying marine cyclone, *Mon. Weather Rev.*, 120, 1093, 1992.

50. Claud, C., Mognard, N. M., Katsaros, K. B., and Scott, N. A., Synergistic satellite study (TOVS, SSM/I and GEOSAT) of a rapidly-deepening cyclone over the Norwegian Sea, *The Global Atmosphere Ocean System*, 3(1), in press.

51. Katsaros, K. B., Bhatti, I., McMurdie, L., and Petty, G., Identification of atmospheric fronts over the ocean with microwave measurements of water vapor and rain, *Weather Forecasting*, 4, 449, 1989.

52. Katsaros, K. B., Petty, G. W., Bhatti, I., and Miller, D., Application of Special Sensor Microwave/Imager Data for Analysis Of Cyclonic Storms in Midlatitudes over the Sea, 2nd Annual Report on Contract N00014-86-K-0453, Department of Atmospheric Science, University of Washington, Seattle, WA, 1989.

53. McMurdie, L. A., Levy, G., and Katsaros, K. B., On the relationship between scatterometer derived convergences and atmospheric moisture, *Mon. Weather Rev.*, 115, 1281, 1987.

54. Raustein, E., Sundqvist, H., and Katsaros, K. B., Quantitative comparison between simulated cloudiness and clouds objectively derived from satellite data, *Tellus*, 43A, 306, 1991.

55. Petty, G. W. and Katsaros, K. B., Nimbus-7 SMMR precipitation observations calibrated against surface radar during TAMEX, *J. Appl. Meteorol.,* 31, 489, 1992.

56. Spencer, R. W., A satellite passive 37 GHz scattering-based method for measuring oceanic rain rates, *J. Climate Appl. Meteorol.,* 25, 754, 1986.

57. Claud, C., Mognard, N. M., Katsaros, K. B., Chedin, A., and Scott, N. A., Satellite observations of a polar low over the Norwegian Sea by Special Sensor Microwave Imager, Geosat, and TIROS-N Operational Vertical Sounder, *J. Geophys. Res.,* 98, 14,487, 1993.

58. Carleton, A. M., McMurdie, L. A., Katsaros, K. B., Zhao, H., Mognard, N. M., and Claud, C., Satellite-derived features and associated atmospheric environments of Southern Ocean mesocyclone events, *The Global Atmosphere Ocean System,* in press.

59. Mognard, N. M. and Katsaros, K. B., Statistical comparison of the Special Sensor Microwave/Imager and the Geosat altimeter wind speed measurements over the ocean, *The Global Atmosphere Ocean System,* 2, 291, 1995.

60. Mognard, N. M. and Katsaros, K. B., Weather patterns over the ocean observed with the Special Sensor Microwave/Imager and the Geosat altimeter, *The Global Atmosphere Ocean System,* 2, 301, 1995.

61. Katsaros, K. B., Zhao, H., Miletta, J., and Quilfen, Y., Monitoring severe storms by passive and active satellite sensors in the microwave frequency range, in *Tropical Cyclone Disasters,* Proc. ICSU/WMO Int. Symp., October 12–16, 1992, Beijing, China, Lighthill, J., Zhemin, Z., Holland, G., Emanuel, K., Eds., Peking University Press, Beijing, China, 1993, 69.

62. Glass, M. and Felde, G. W., Intensity Estimation of Tropical Cyclones Using SSM/I Brightness Temperatures, Prepr., 6th Conf. Satellite Meteorol. Oceanogr. American Meteorological Society, Atlanta, GA, 1992, J8-J10.

17

Remote Sensing of Surface Solar Radiation Flux and PAR over the Ocean from Satellite Observations

Catherine Gautier

17.1 Introduction

An accurate knowledge of sun influence on earth biological, physical, and chemical processes and radiative response of the surface-atmosphere system is vital for a complete understanding of the state and evolution of the earth. While it has long been recognized that sun energy drives the earth system, only recently have we begun to produce accurate, large-scale surface radiation flux data sets, and we are

271

just learning how to effectively employ them in scientific studies. While climatologies of surface radiation fluxes have been available,[1] their accuracy is limited and they do not offer the time and space resolution now required in numerical climate models (e.g., 1 d and 2.5° × 2.5° grids).

The most clearly stated need for large-scale surface radiation flux estimates is for input to or validation of numerical atmosphere and ocean circulation (or coupled) models in order to quantitatively characterize physical and biological interaction processes between radiation and the surface of the earth.[2] For applications investigating physical processes, the important parameter to estimate is the net (solar or shortwave and longwave) surface radiation flux. It allows one to infer the surface heating or cooling by radiation absorption and emission. For applications related to biological processes, the critical parameter is the photosynthetically available radiation (PAR), which covers the part of the solar spectrum active in the photosynthesis of plants over both land and ocean.

Accurate, large-scale ocean surface solar radiation flux data sets are recent developments for several reasons. First, making surface radiation flux measurements over large oceanic expanses and long periods is complex and expensive. Surface radiation measurement systems are only now being envisioned, but still over a limited region of the oceans (e.g., the western tropical Pacific as part of the Atmospheric Radiation Measurement (ARM) program). Second, only with the advent of satellite-based instruments and the concurrent development of remote sensing retrieval techniques have the first accurate large-scale surface radiation fluxes been produced and their low frequency (interseasonal and interannual) variability analyzed. While problems remain, satellite remote sensing represents the most suitable means for monitoring and conducting global studies of surface radiation fluxes.[3] The largest obstacle to producing fields of surface radiation flux is that they require voluminous data sets and enormous amounts of computing power, which have only recently become available for this type of application.

The development of satellite remote sensing methods for estimating surface radiation fluxes has benefited from the rapid progress in radiative transfer computations in the atmosphere for both solar and longwave spectral ranges and in clear and cloudy conditions. The validation of a number of such radiative transfer models has been performed under the auspices of the International Comparison of Radiation Codes for Climate Models (ICRCCM)[4] project. These intercomparisons have provided additional confidence in the overall quality of the models, and have also uncovered some of their deficiencies.

The shortwave, or solar, radiation flux was the first component to be estimated because observational studies demonstrated high correlations between visible planetary satellite observations and surface solar radiation flux measurements.[5] As a consequence, simplified radiative transfer models provided relatively accurate surface solar radiation flux estimates.[6] The longwave component, however, has proved more difficult to estimate due to the decoupling between surface and planetary longwave radiation fluxes.[7] This component will not be discussed here.

Today a number of investigators are routinely producing accurate satellite-based estimates of surface solar radiation flux and PAR. The first phase of this activity has culminated in intercomparisons of the various method results over local and global scales.[8] This has been made possible by the availability of global and continuous (starting in July 1983) data sets from the International Satellite Cloud Climatology Project[9] (ISCCP) and from the Earth Radiation Budget Experiment[10] (ERBE) but for a shorter time period. Multiyear global fields are thus now available for conducting diagnostic studies, verifying atmospheric circulation models, and forcing ocean circulation models. In this chapter, Section 17.2 reviews the advances in radiative transfer theory that have been central to computing surface solar radiation flux and PAR. Section 17.6 describes and discusses the methods to compute the net solar surface radiation flux and PAR. Section 17.7 presents some global surface radiation flux fields obtained with one method and compares them to those obtained using other similar approaches. Section 17.8 discusses the perspectives for the next decade, including new sensors and scientific advances to be expected.

17.2 Radiative Transfer Computations of Solar and PAR Fluxes

Several approaches, encompassing varying levels of sophistication and precision, have been developed for estimating surface fluxes from satellite observations. Central to most of these approaches has been the development of radiative transfer models to solve the forward problem, that is, to compute radiation field within, at the top and at the bottom of the atmosphere from known surface and atmospheric properties. These models have allowed us to understand all the physical processes involved in the transfer of radiation and its interaction with atmospheric components and the surface, and to study its sensitivity to a variety of parameters and conditions. Such sensitivity studies have guided the selection of the minimum set of parameters required in simplified satellite retrievals of surface solar flux. The existence of such accurate radiative transfer models has also provided the necessary confidence in the results obtained by the simplified approaches.

17.3 Definitions

Since the terminology associated with radiation quantities varies with the disciplines in which it is used, the terms used below are now defined. For further definitions the reader is referred to Reference 11, Chap. 2.

Irradiance or flux (F) refers to the ratio of the flux of radiation incident on an infinitesimal element of surface to the area of the element. It thus involves radiation coming from an entire hemisphere (or 2π steradians). This is, for instance, the type of measurement made at the surface by a pyranometer.

The radiant intensity or radiance (I) measures the radiation in a given direction and thus corresponds to the radiation within an infinitesimal cone containing the direction.

The relationship between irradiance and radiance is given by:

$$F = d\phi I \cos\theta \sin d\theta$$

for a radiation falling on a surface at an angle θ from the normal to the surface. The angle ϕ represents the azimuth angle of the incident radiation. For radiation that has the same properties in all directions (isotropic radiation), it can be shown that $F = \pi I$.

The radiation from the sun is emitted at all wavelengths (broadband) and when impinging at the top of the atmosphere (TOA) it is referred to as extraterrestrial irradiance. When radiation is limited to a spectral interval such as measurements made by a satellite sensor, it is called narrowband radiation. The radiation is monochromatic if it is at a particular wavelength.

Radiation impinging on the top of the atmosphere can be scattered and absorbed by air molecules and reflected and absorbed by cloud liquid water droplets. The transmitted radiation reaches the surface where it is reflected and absorbed. Part of the reflected component is scattered back to the surface by the atmosphere or reflected back to the surface by the cloud. An infinity of mutliple reflections can take place between the surface and the atmosphere below cloud base.

The two important radiative parameters to know about a surface with regards to net surface solar flux are the direction in which it reflects the incoming radiation (bidirectional reflection properties) and also the spectral impact the reflection has on the incoming radiation (spectral reflectance). When the reflected radiation is isotropic, the surface is said to be a Lambertian reflector.

17.4 Cloud-Free Atmospheres

In the solar spectral interval (0.25–4.0 μm, also referred to as shortwave region) in cloud-free conditions, solar radiation is scattered by air molecules and aerosols; and it is absorbed primarily by water vapor, ozone, carbon dioxide, oxygen, and aerosols. The spectral radiance depends on solar zenith angle, zenith and relative azimuth viewing angles, extraterrestrial irradiance, and atmospheric and surface properties. The ocean surface affects solar radiation through its reflection, which varies spectrally and as a function of both incident radiation and viewing directions. The most complex radiative transfer models compute spectral radiances by accounting for all orders of scattering (scattering with a single particle but also multiple scattering). For air molecules, Rayleigh scattering approximations are introduced. For larger particles with more complex shapes, such as aerosols, sophisticated (Mie) computations are used. While scattering and absorption processes interact in a complex manner, gaseous absorption processes can be accurately handled separately and thus the computations can be simplified.[12] In general, then, the absorption of ozone, water vapor, carbon dioxide, oxygen, and aerosols is computed over each spectral interval and is usually accounted for by way of absorption coefficients.

Simplifications can be introduced in the radiative transfer computations by representing aerosol effects by means of a visibility parameter and also assuming a Lambertian and uniform surface, in which case the solar irradiance reaching the surface can be expressed as a simple function of the atmospheric radiative properties. Another alternative is to simplify the radiative transfer computations by employing analytical formulas for small spectral intervals.[13]

The accuracy of the most sophisticated (highly spectrally resolved) radiative transfer computations in the solar region is probably around 1–2%, although no validation has yet been performed to confirm this very high level of accuracy. Ongoing field experiments, such as the Atmospheric Radiation Measurement (ARM) Program are expected to provide the data for such validation in the future. The main sources of uncertainty are the Mie scattering computations for aerosols and aerosol optical properties. The results of sophisticated methods usually serve as a reference against which the accuracy of more simplified approaches are evaluated. The ICRCCM program has initiated a validation of the radiation codes used in climate models[14] and has uncovered a number of limitations in the shortwave radiation models, illustrated by differences among them. In clear conditions, agreement among all the models is the best and on the order of 1–2%.

Several studies have quantified the sensitivity of the surface solar irradiance to the various atmospheric scatterers and absorbers.[15,16] Relatively large perturbations (e.g., 30%) in ozone and water vapor only minimally affect the magnitude of the atmospheric absorption (0.003 and 0.007%, respectively) and surface solar irradiance by a few watts per square meter. The high sensitivity is definitely due to aerosols. Aerosol optical depth changes such as those observed in nature over the ocean can induce changes of about 1% in absorption, while changes in aerosol type from maritime to continental can modify atmospheric absorption by up to a few percent, due to the increase in absorption of continental aerosols.

17.5 Cloudy Atmospheres

In cloudy atmospheres, radiation transfer modeling is much more complex because of absorption by water vapor and scattering and absorption by ice crystals water droplets (and possibly aerosols) within clouds. Solutions to this radiative transfer problem are usually obtained by assuming that clouds are homogeneous and plane parallel; a good summary of cloud radiative properties (i.e., albedo, transmission, and absorption) under these conditions is provided in a series of articles by Stephens[17,18] and Stephens collaborators.[19] Stephens[18] proposed parameterization schemes for cloud albedo and absorption as a function of optical thickness, single scattering albedo, scattering phase function, and solar zenith angle. Theoretical and parameterized variations of albedo, transmission, and absorption with integrated cloud liquid water, which is obtained from optical thickness, show that cloud absorption is relatively small. They also imply that total cloud liquid water is the main parameter determining cloud radiative properties and that cloud transmission can be expressed as a function of cloud albedo, a parameter measurable from space. The existence and validity of such a simple relationship between cloud albedo and transmission is at the heart of the success of satellite-based retrieval methods.

In the case of PAR, however, the situation is further simplified by the fact that cloud liquid water does not absorb radiation in the PAR wavelengths. Once the cloud albedo transmission relationship is established, several issues remain that concern total atmospheric absorption (e.g., water vapor vs. liquid water and cloud vs. clear water vapor absorption) and second-order corrections to this relationship for broken cloud conditions. For total atmospheric absorption (liquid and water) within clouds and outside clouds, theoretical results from radiative transfer models have been used to demonstrate that it was reasonable to decouple both: (1) clear and cloudy conditions, because total absorption in the atmosphere is small; and (2) liquid water and water vapor absorption, because water vapor absorption dominates above clouds, while water droplet absorption prevails within clouds.[20] Although a second-order effect, cloud droplet size distribution affects both reflected and transmitted shortwave radiation, with larger reflection resulting from the presence of smaller size droplets in clouds. This effect is clearly observed in satellite reflectance observations over the ocean in the vicinity of ships. Ship tracks, corresponding to regions of enhanced reflection, are due to an increase in small particle concentration resulting from the emission of smoke by ships.[21,22]

The issue of cloud absorption by liquid water and its parameterization still remains problematic and central to an accurate determination of surface solar irradiance from satellite observations. Measurements have shown significantly larger absorption in clouds than those predicted by simple energy budget considerations.[23-28] The discrepancies can be explained, in part, by measurement uncertainties, which in some cases are almost as large as the magnitude of the absorption (absorption is derived as a small difference between two large terms). Other hypotheses have been proposed to explain the abnormally high cloud absorption and have been reviewed by Stephens and Tsay.[29]

The impact of finite cloud fields and broken cloudiness on reflected radiation and surface solar radiation flux (and PAR) has been examined by many researchers.[30-39] This has been principally achieved by means of Monte Carlo calculations, which are applied to photon packets in order to compute radiation scattering by finite size clouds and assess radiation diffusion through cloud sides and cloud-cloud interactions. Different, but simple, cloud shapes and liquid water distributions were tested, all demonstrating markedly different directional reflectance variations with sun zenith angle between plane parallel and finite clouds. This issue remains unsettled[39-41] and is the one that will require the most attention in the near future for accurate determination of surface shortwave radiation flux over the ocean.

17.6 Net Surface Solar Flux and PAR Computations

Rationale

This section addresses the use of satellite measurements to derive the surface solar flux and PAR over the oceans. None of the physically based methods presented below has been specifically designed for oceanic conditions because the behavior of radiation is the same over land and over the ocean insofar as surface properties are correctly specified. There is no evidence that clouds behave differently even though their water content may be different.

All the methods for estimating the surface solar flux rely on the high correlation between planetary albedo and surface solar irradiance, which is relatively independent of cloud amount or properties[41-43] for plane parallel clouds. Similarly, but to a lesser extent, a correlation exists between top of the atmosphere (TOA) spectral radiance (as measured by satellite) and downwelling surface solar irradiance. These correlations result from the fact that absorption of shortwave radiation in the atmosphere is relatively small, as we have seen earlier, and negligible in the visible part of the spectrum (0.4–0.8 µm). Cloud geometry and brokenness (discussed in an earlier section) may have some effect on this relationship, but they are expected to be of smaller magnitude in comparison with the dominant solar zenith angle effects.

Downwelling Surface Solar Radiation Flux

Both statistical and physically based approaches can be applied to estimate the downwelling surface solar radiation flux, but since statistical methods require high-quality and spatially representative surface measurements, these methods are rarely applied over the ocean. Simple physically based methods replace the statistical relationship between planetary albedo and surface solar irradiance by a quantitative expression that involves bulk atmospheric radiative properties. As previously discussed, cloud transmittance can be reasonably well expressed as a function of planetary albedo (corrected for scattering and absorption occurring above clouds and multiple reflections occurring between cloud base and the surface) and cloud absorption can be parameterized as a function of cloud albedo. Either the cloud transmittance is estimated from narrowband visible radiance measurements or spectral transformation is performed to transform narrowband to broadband radiances.

The existing physical methods vary in complexity and procedure, with some methods handling the clear and cloudy atmospheric conditions in different ways. Some of the simplest methods[6,44,45] are based on energy conservation principles that account for the significant contributions to atmospheric column radiative energy budget, atmospheric scattering and absorption, cloud scattering and absorption, and surface reflection.

In the case of Gautier et al.,[6] once overcast regions have been delineated based on a cloud threshold approach, clear and overcast conditions are handled separately. The radiative transfer treatment is highly developed for clear conditions only. The input to these models are climatological values of water vapor and aerosol concentrations; their effects are parameterized as a function of surface dew point temperature and visibility, respectively. Cloudy downwelling irradiance is simply the product of computed clear irradiance by cloud transmission, which is estimated from satellite radiance. Cloud transmission is obtained as the difference between cloud albedo and absorption. Cloud albedo is computed from radiance at the top of the atmosphere over clouds. Absorption is estimated from cloud albedo following a parameterization similar to Stephens.[18] A small correction term is usually added to account for multiple reflections between the ocean surface and the atmosphere (including clouds) above. A constant ocean surface albedo is used for both threshold and surface solar irradiance computations.

A similarly simple and accurate method has been developed by Moeser and Raschke[46] for application to Meteosat data. For these simple methods, neither narrowband to broadband nor radiance field anisotropy corrections are included. Despite their simplicity and limitations, the computed surface solar irradiance is surprisingly accurate with reported uncertainties on the order of 15% (or better) on an instantaneous basis, and 7–8% on a daily basis when using at least three observations per day and appropriately calibrated satellite data.[47]

Another method, by Pinker and Laszlo,[48] is based on more complete radiative transfer computations using a three-layer model and the delta-Eddington approximation. This approximation simplifies scattering computations. The model is run to construct lookup tables relating cloud optical thickness to TOA spectral reflectance for a variety of observation and sun zenith angles, and surface albedo conditions. Together with this lookup table, the satellite data are subsequently used to estimate cloud optical depth, which is then employed by the radiative transfer model to compute the surface solar irradiance. The accuracy on a daily timescale is similar to that of the simpler models.

While the methods just discussed were initially based on geostationary satellite observations, Darnell et al.[49] proposed a simple method that initially used sun-synchronous TOVS and Advanced High-Resolution Radiometer (AVHRR) data sets. It estimates broadband daily surface solar irradiance as the product of the TOA solar irradiance, clear sky and cloud transmittances. Clear sky conditions are expressed as a function of broadband atmospheric optical depth estimated from TOVS water vapor concentration and ozone burden data. Cloud transmittance is estimated from TOA albedo derived from AVHRR data. No narrowband to broadband or bidirectional transformation is performed. The reported accuracy is on the order of 19% for daily averages based on one observation per day and 2.7% for monthly averages. This method has recently been modified to handle ISCCP data.

Since PAR covers part of the solar spectrum, it can be estimated in a similar, though more direct, fashion than solar irradiance. The only difference is in the extraterrestrial solar irradiance for clear conditions, which is modified to account for only the PAR portion of the solar spectrum (e.g., Frouin et al.[13]). In cloudy conditions, the cloud transmission is directly estimated from measured narrowband cloud albedo because standard satellite visible spectral measurements (e.g., Visible Infrared Spin-Scan Radiometer [VISSR], AVHRR channel 1) approximately cover the PAR spectral range for which cloud absorption is negligible.[50] Other investigators have developed PAR models, which are in general simplifications of their original shortwave radiation flux models.[51]

All the methods described above are based on plane-parallel cloud assumptions and clear/overcast conditions. None of these methods model the complexities of the three-dimensional solar radiation field in cloud conditions, in part because observations do not provide the necessary information to support such modeling. Nevertheless, they are remarkably equivalent and justify, in large part, their extreme simplicity. These models presently represent the best available tools for generating global climatologies of the surface solar irradiance over the ocean. Most of these methods rely on some estimations of cloud cover (explicitly or implicitly). Difficulties exist in retrieving cloud cover, however, in the case of large viewing angles and vertically extending clouds (such as deep convection), conditions relevant for geostationary satellite observations at the edges of the satellite coverage in tropical regions. Cloud cover overestimation can result in these conditions due to the reflected radiation on cloud sides measured by the satellite instrument, which is mistakenly interpreted for radiation coming from the cloud top. Such an overestimation has been found for instance in ISCCP cloud cover data.[52] This overestimation leads to a bias in monthly cloud cover data, since with a geostationary satellite the same scene is always observed under the same viewing angle. Corrections need to be developed and applied to eliminate such a bias.

Net Surface Solar Radiation Flux

For applications involving interactions between solar radiation and earth surface, the parameter to estimate is the net surface solar radiation flux. The net flux is the difference between the downwelling surface solar radiation and the surface reflected component. To obtain the reflected component, the surface albedo is usually estimated separately. Some approaches, however, bypass the surface albedo estimation by directly relating the TOA net solar radiation flux to that at the surface,[43,53-55] following the suggestion of Ramanathan.[42]

Upwelling Surface Solar Radiation Flux and Surface Albedo

Over the ocean, albedo determination is rather simple, particularly outside sunglint regions (i.e., the region of sun specular reflection) since the spectral dependence of seawater is well known for pure water. Water contamination effects, natural (e.g., phytoplankton) or artificial (e.g., oil slicks), are, however, minimal on the reflected solar radiation because this term is small. Some complexity is added by the fact that the surface albedo depends on surface conditions (roughness and foam coverage, in particular) and is also a function of cloudiness,[56] as a result of the increase in diffuse radiation. The cloud dependence is rather small and therefore is rarely included in methods deriving the net surface solar radiation flux. The effect of surface roughness and foam on surface albedo can be large locally, but are small (on the order of a few percent only) on the scale of the satellite measurements.

Direct Methods

It is possible to compute directly the net surface solar radiation flux when broadband measurements are available such as those from ERBE data. Using radiative transfer computations, Cess and Vulis[43] demonstrated that there is a linear relationship between surface and surface-atmosphere solar absorption in clear conditions. This relationship appears mildly dependent on scene type and only requires moderate correction for atmospheric absorption processes. In cloudy conditions for plane parallel cloud assumptions, the slope of the linear relationship only slightly varies with cloud optical thickness; and thus the broadband flux measurements can be used directly to infer the absorbed (or net) surface solar radiation flux. This has been taken advantage of in modeling by Li and Leighton[54] and Bréon et al.[55] As with other approaches, for accurate results, however, the effects of the three-dimensional properties of broken cloud fields discussed above should still be taken into consideration when deriving the relationship between TOA and surface absorption. This remains to be achieved with models capable of simulating radiation through three-dimensional clouds.

Satellite Radiometer Calibration

In the discussion above, it is implicitly assumed that the radiances employed in the various retrieval methods are absolutely calibrated. The surface solar irradiance sensitivity to the visible calibration has been discussed by Gautier and Frouin.[57] They showed that a 10% uncertainty in the calibration could lead to 70 W m^{-2} uncertainty (i.e., bias) in instantaneous surface solar irradiance estimations in overcast conditions. Although this uncertainty lessens over monthly averaged values, it remains significant and on the order of 15 W m^{-2}. Operational visible radiometers (e.g., VISSR, Meteosat, AVHRR), with data that are employed in the determination of the surface solar irradiance, are insufficiently characterized before launch. Furthermore, they have either no internal calibration or unreliable internal calibration that often deteriorates after launch. Research instruments such as ERBE,[10] on the other hand, are usually adequately characterized before launch, and suitable in-flight calibration procedures are continually performed to ensure homogeneity of the data. For most visible radiometers used in surface solar irradiance computations, however, an indirect in-flight calibration must be performed after launch in order to obtain reasonably calibrated radiances. The accuracy of most in-flight calibrations, nevertheless, remains limited and is hard to assess. It depends not only on the method employed, but also on the instrument deterioration. Several calibration methods have been proposed[58-60] and their results have been recently intercompared.[61] They displayed agreement to within about 8%, while distinctly illustrating a slow differential deterioration of the two AVHRR visible channels on National Oceanic and Atmospheric Administration (NOAA)-7. In the case of the VISSR, the calibration coefficients computed for the sensors on most of the recent satellites show a more erratic variability that results, in part, from irregular adjustments introduced at the ground station[62] (see Gautier and Frouin[57] for more details). Additional slow seasonal changes have been observed but remain unexplained.

Finally, vicarious in-flight calibration cannot be performed when the atmosphere is loaded with unusual aerosol amounts such as was the case after the eruption of Mt. Pinatubo in June 1991. This had the adverse effect of eliminating the ability to perform the calibration of operational instruments for several months.

In Situ Validation

Validation of satellite estimations is extremely important since most climate research programs now require accurate surface solar irradiance estimates on a variety of time- and space scales (e.g., 10 W m^{-2} over a month for the Tropical Ocean and Global Atmosphere [TOGA] program). The absolute accuracy of individual estimates and the nature of their error (random or bias) must be carefully assessed and examined. For climate studies, the mean values estimated from satellite data must have no bias (introduced as a result of tuning or incomplete process parameterization), so that small changes can be detected. Such a careful validation requires high-quality and long-term surface measurements not only to assess the absolute accuracy of individual measurements, but also to characterize the space-time variability of the surface radiation flux so that the discrepancies between surface measurements (which are local and time averaged) and satellite estimates (which are instantaneous and spatially averaged) can be elucidated. Such space-time statistics are not available from surface measurement networks at the resolution and the accuracy required. The satellite and the surface information must therefore be combined in some way to infer these statistics over some time and space scale.

Satellite-derived estimates of downwelling surface solar radiation flux are most frequently validated by comparison with pyranometer measurements. Pyranometers are relatively simple broadband flux radiometers that measure the global (direct plus diffuse) radiant flux. The precision and accuracy of currently available pyranometers vary measurably, with the best calibrated and characterized instruments, achieving a total measurement uncertainty of about 3%. At present, the best possible network measurements cannot be expected to reach an accuracy better than 5%. Other types of standard instruments, such as pyheliometers, can be used for limited comparisons since they only measure the direct beam and use a shaded pyranometer to measure the diffuse component. The uncertainty in the direct beam measurement can be very small (0.3%), and thus these measurements are very accurate for relatively clear days when the diffuse component is small.

Accuracy

Many of the methods discussed above report similar root mean square (rms) differences for hourly (about 20% or 50 W m^{-2}) and daily estimations (10% or 15 W m^{-2}) of the downwelling surface solar irradiance over land.[41] These findings were recently supported by a preliminary intercomparison of methods.[8] Only limited validations have been made for oceanic conditions[6,23,65] because of the difficulties of performing long-term measurements at sea. Their reported accuracy is similar to that obtained over land, however. This is to be expected since the models are physically based (thus accounting for differences in surface types), and the uncertainties associated with surface reflection are small because of the dark ocean surface.

17.7 Global Surface Solar Radiation Flux over the Oceans

Data Availability

One important component in the production of global surface radiation flux fields is the availability of global satellite data sets. During the last 10 years, the International Satellite Cloud Climatology Project (ISCCP) has been collecting satellite data from several geostationary (Geostationary Operational Environmental Satellite [GOES], Meteosat, GMS) and polar-orbiting (NOAA series) satellites in order to produce a global climatology of cloud cover and properties. These global data sets, which are available in radiance form at high (1-km) and medium (32-km) resolution (B format) or in a cloud property form at a lower resolution (250 km; C format), serve as the basis for the production of global surface radiation fluxes.

Another set of data that is now very useful for surface radiation flux studies is the well-calibrated Earth Radiation Budget Experiment (ERBE) data set.[10] Since algorithms have recently been developed to employ these data for surface solar irradiance computations, global fields are now being generated routinely.

Surface Solar Radiation Flux: Example

The Gautier et al.[6] method has recently been implemented so that global shortwave radiation flux can now be routinely computed over the oceans from ISCCP C1 data. The computations require cloud cover and cloudy radiances from which the shortwave irradiance is computed according to:

$$SW = N \cdot SW_{\text{cloudy}} + (1 - N) \cdot SW_{\text{clear}} \qquad (1)$$

$$SW_{\text{cloudy}} = f(R_{\text{cloudy}}) \qquad (2)$$

where N and R_{cloudy} are ISCCP parameters provided every 3 h. The function f represents a simple radiative transfer model we have developed that uses the cloudy radiance R_{cloudy} to estimate cloud albedo and in which cloud absorption is parameterized as a function of cloud albedo. SW_{clear} is computed using an accurate radiative transfer model[11] (5 S) run for every half hour, in order to adequately simulate the effects of sun zenith angle variations on the clear surface solar radiation flux. The surface albedo necessary for computing both SW_{clear} and SW_{cloudy} is taken to be a constant equal to 0.06.

Global monthly mean fields have been produced with this algorithm for the period 1983–1990, for which ISCCP C1 data are presently available, and have been extended to include earlier periods (1976–1983) with additional cloud data from other satellites.[51]

Comparison with Results from Other Methods

Other methods have also been applied to a subset of the same ISCCP C1 data (i.e., N and R_{cloudy}) as part of an effort of the World Climate Program to produce global shortwave fluxes for the scientific community. Two other methods have been applied, that of Pinker and Laszlo[48] and that of Darnell et al.[49] modified by Staylor to be used with ISCCP C1 data. We have begun to intercompare results from these two methods with those from our method, for 4 months in 1986. Results from these intercomparisons are presented in Figures 17.1 and 17.2, where monthly mean fields for January for all the methods and a scatter plot of their comparison are presented. Figure 17.1 shows that very similar patterns are obtained by the three methods, with high values of shortwave radiation flux in the descending branch of the Hadley cell and low values in convective regions of the tropics and in the storm track regions in the Northern Hemisphere. The quantitative comparisons of Figure 2 indicate a high correlation between the individual estimates (>0.96) and a small bias between each of them. With such similar results but containing a non-negligible bias between methods, the difficulty now becomes deciding which one is best, if possible. This is a rather difficult task since no absolute validation exists. An intercomparison with available climatologies (e.g., Oberhuber[67]) cannot settle the differences because it indicates discrepancies much larger than those found between the various satellite-based methods. Only comparisons with surface measurements can offer a possible validation, but these (as discussed earlier) contain their own uncertainties and are not representative of spatial averages as are satellite observations.

One of the deciding factors may be the ease with which the method can be applied. In the case of Gautier et al.[6] method it does not require any ancillary data but only calibrated radiances. It can therefore be applied to any region of the oceanic globe.

17.8 New Data Sets

During the 1990s new instruments important to the remote sensing of surface radiation fluxes are expected to be launched on Earth Observing System (EOS) polar platforms. Two facility instruments, the Moderate Resolution Imaging Spectrometer (MODIS), and the Atmospheric Infrared Sounder (AIRS), together with

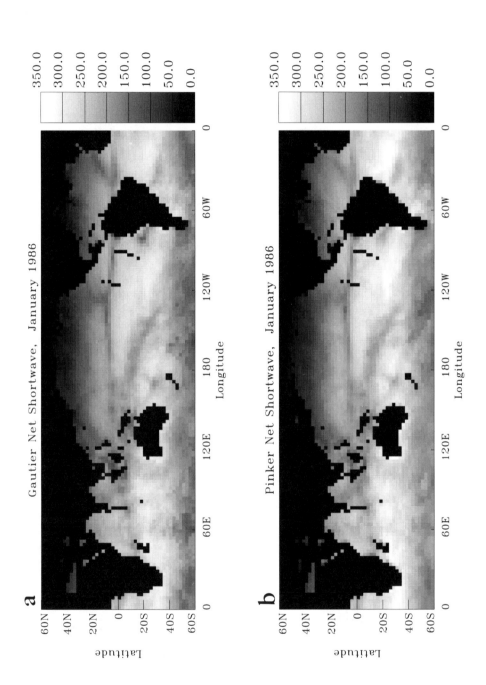

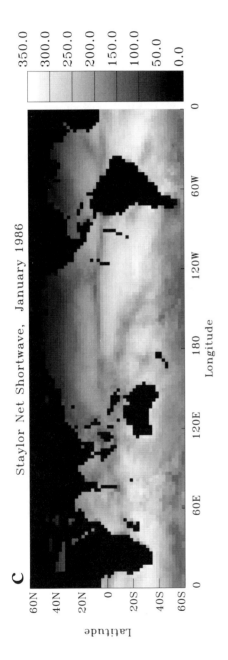

Figure 17.1 Monthly surface solar radiation fluxes over the global oceans computed with ISCCP C1 data by applying different methods: (a) Gautier et al.,[6] (b) Pinker and Laszlo,[48] (c) Staylor.[60]

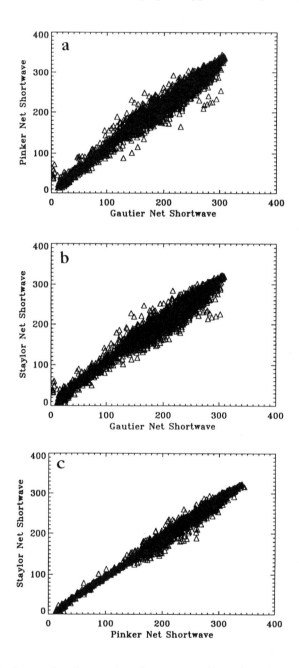

Figure 17.2 Scatter plot of comparisons between monthly surface solar radiation fluxes computed with different methods: (a) Gautier vs. Pinker, (b) Gautier vs. Staylor, (c) Pinker vs. Staylor.

the Cloud and Earth's Radiant Energy System (CERES), will be on the first U.S. EOS platform. These new sensors will provide an exceptional array of new sensors.

The MODIS is composed of a sensor that provides images in 40 spectral bands in the spectral range of 0.4–14.2 μm at surface resolution varying from 500 to 1000 m. Surface and cloud spectral properties can be derived; and because of this sensor resolution, studies of broken cloud field effects on the surface solar flux will be possible.

An improved version of the ERBE, CERES has been selected to fly on several U.S. EOS platforms, in addition to smaller (explorer-type) satellites. This instrument is expected to provide broadband planetary albedo measurements in two different directions, which will be used to estimate the net surface solar flux as has been done by a number of investigators with ERBE data.

The AIRS instrument is an array-grating spectrometer that will provide measurements in about 2000 infrared spectral regions and over 5 visible spectral regions, including a broadband region (0.4–1.0 μm) at 2-km resolution. The visible component is being specifically designed to provide data for the accurate estimation of the surface solar radiation flux and PAR.

Using eight individual pushbroom cameras, each operating at the same four visible wavelengths, the MISR instrument will provide continuous multiangle imagery of the earth. With this instrument it will be possible to obtain the angular distribution functions needed to transform radiances into fluxes for all types of surface, cloud, and aerosol conditions. These data can be used to study the influence of aerosols on the surface solar irradiance and the effects of the spatial and seasonal cloud variability on the planetary solar radiation budget.

Acknowledgments

This work has been supported by grants from the Department of Energy through participation in the ARM program (Grant DOE DE-FG03-90 ER61062) and from the National Space and Aeronautics Administration through grant NASA NAGW-3180.

References

1. Budyko, M. I., *Atlas of the Heat Balance of the Earth* (Russian), Kartfabrika Gosgeoltehiydata, 1963, 41 pp.
2. WCP-92, Report of the TOGA Workshop on Sea Surface Temperature and Net Surface Radiation, 1984.
3. Suttles, J. T. and Ohring, G., Surface radiation budget for climate applications, NASA Reference Publication, 1169, 1986, 132 pp.
4. Ellingson, R. G. and Fouquart, Y., The intercomparison of radiation codes in climate models: an overview, *J. Geophys. Res.*, 96, 8925, 1991.
5. Tarpley, J. D., Estimating incident solar radiation at the surface from geostationary satellite data, *J. Appl. Meteorol.*, 41, 1172, 1979.

6. Gautier, C., Diak, G., and Masse, S., A simple physical model to estimate incident solar radiation at the surface from GOES satellite data, *J. Appl. Meteorol.*, 19, 1005, 1980.

7. Stephens, G. L. and Webster, P. J., Cloud decoupling of the surface and planetary radiative budgets, *J. Atmos. Sci.*, 41, 681, 1984.

8. Charlock, T. P. and Ohring, G., Report to the COSPAR/WCRP Workshop on surface radiation budget (SRB) for climate and global change, Wurtzburg Workshop Rep., October 30–November 3, 1990.

9. Schiffer, R. A. and Rossow, W. B., The International Satellite Cloud Climatology Project ISCCP: the first project of the world climate research programme, *Bull. Am. Meteorol. Soc.*, 64, 779, 1983.

10. Barkstrom, B. R., The earth radiation budget experiment (ERBE), *Bull. Am. Meteorol. Soc.*, 65, 1170, 1994.

11. Paltridge, G. W. and Platt, C. M. R., *Radiative Processes in Meteorology and Climatology*, Elsevier Scientific Publishing, 1976, 318 pp.

12. Tanre, D., Deroo, C., Duhaut, P., Herman, M., Morcrette, J. J., Perbos, J., and Deschamps, P. Y., Effects atmospheriques en detection-logiciel de simulation du signal satellitaire dans le spectre solaire, Proc. 3rd Int. Colloq. Spectral Signatures of Objects in Remote Sensing, Les Arcs, France, Dec 16–20, 1985.

13. Frouin, R., Lingner, D. W., Gautier, C., Baker, K. S., and Smith, R. C., A simple analytical formula to compute clear sky total and photosynthetically available solar irradiance at the ocean surface, *J. Geophys. Res.-Oceans*, 94, 9731, 1989.

14. Fouquart, Y., Bonnel, B., and Ramaswamy, V., Intercomparing shortwave radiation codes for climate studies, *J. Geophys. Res.*, 96, 8955, 1991.

15. Gautier, C. and Byers, M., Remote sensing of global surface shortwave radiation and PAR over the ocean, Proc. Conf. on the Optics of the Air-Sea Interface, SPIE Vol. 1749, San Diego, CA, July 23–24, 1992.

16. Fouquart, Y., Bonnel, B., and Chaoui Roquai, M., Santer R., and Cerf, A., Observations of Saharan aerosols: results of ECLATS Field Experiment. I. Optical thicknesses and aerosol size distributions, *J. Clim. Appl. Meteorol.*, 26, 28, 1987.

17. Stephens, G. L., Radiation profiles in extended water clouds. I. Theory, *J. Atmos. Sci.*, 35, 2111, 1978.

18. Stephens, G. L., Radiative profiles in extended water clouds. II. Parameterization schemes, *J. Atmos. Sci.*, 35, 2123, 1978.

19. Stephens, G. L., Paltridge, G. W., and Platt, C. M. R., Radiation profiles in extended water clouds. III. Observations, *J. Atmos. Sci.*, 35, 2133, 1978.

20. Davies, R. and Ridgway, W. L., Kim, K.-E., Spectral absorption of solar radiation in cloudy atmospheres: a 20 cm-1 model, *J. Atmos. Sci.*, 41, 2126, 1984.

21. Durkee, P. A. and Coakley, J. A., Jr., and Bernstein, R. L., Effect of shiptrack effluents on cloud reflectivity, *Science*, 237, 1020, 1987.

22. Coakley, J. A., Jr., Bernstein R. L., and Durkee, P. A., Effect of shiptrack effluents on cloud reflectivity, *Science*, 237, 1020, 1987.

23. Reynolds, D. W. and Vonder, Haar, T. H., and Cox, S. K., The effect of solar radiation absorption in the tropical troposphere, *J. Appl. Meteorol.*, 14, 433, 1975.

24. Drummond, A. J. and Hickey, J. R., Large-scale reflection and absorption of solar radiation by clouds as influencing earth radiation budgets, *New Aircraft Measurements, Prepr., Int. Conf. Weather Modification*, Canberra, American Meteorological Society, 1971, 267.

25. Herman, G. F., Solar radiation in summertime Arctic stratus clouds, *J. Atmos. Sci.*, 34, 1423, 1977.

26. Foot, J. S., Some observations of the optical properties of clouds. I. Stratocumulus, *Q. J. R. Meteorol. Soc.*, 114, 129, 1988.

27. King, M. D., Nakajima, T., Radke, L. F., and Hobbs, P. V., Cloud absorption properties as derived from airborne measurements of scattered radiation within clouds, Proc. Int. Radiation Symp., Lille, France, 1989, 14.

28. Rawlins, F., Aircraft measurements of the solar absorption of broken cloud fields: a case study, *Q. J. R. Meteorol. Soc.,* 115, 365, 1989.

29. Stephens, G. L. and Tsay, S. C., On the cloud absorption anomaly, *Q. J. R. Meteorol. Soc.,* 116, 671, 1990.

30. McKee, T. B. and Cox, S. K., Scattering of visible radiation by finite clouds, *J. Atmos. Sci.,* 31, 1885, 1974.

31. Aida, M., Scattering of solar radiation as a function of cloud dimensions and orientation, *J. Quant. Spectrosc. Radiat. Transfer,* 17, 303, 1976.

32. Davies, R., The effect of finite geometry on the three-dimensional cloud fields, *J. Atmos. Sci.,* 35, 1712, 1978.

33. Welch, R. M. and Wielicki, B. A., Stratocumulus cloud field reflected fluxes: the effect of cloud shape, *J. Atmos. Sci.,* 41, 3085, 1984.

34. Harshvardhan and Thomas, R. W. L., Solar reflection form interacting and shadowing cloud elements, *J. Geophys. Res.,* 89, 7179, 1984.

35. Welch, R. M. and Wielicki, B. A., A radiative parameterization of stratocumulus cloud fields, *J. Atmos. Sci.,* 42, 2888, 1985.

36. Joseph, J. H. and Kagan, V., The reflection of solar radiation from bar cloud arrays, *J. Geophys. Res.,* 93, 2405, 1988.

37. Kobayashi, T., Parameterization of reflectivity for broken cloud fields, *J. Atmos. Sci.,* 45, 3034, 1988.

38. Bréon, F.-M., Reflectance of broken cloud fields: simulation and parameterization, *J. Atmos. Sci.,* 49, 1221, 1992.

39. Coakley, J. A. and Kobayashi, T., Broken cloud biases in albedo and surface insolation derived from satellite imagery data, *J. Climate,* 2, 721, 1989.

40. Schmetz, J., On the parameterization of the radiative properties of broken clouds, *Tellus,* 36A, 417, 1984.

41. Schmetz, J., Towards a surface radiation climatology: retrieval of downward irradiances from satellites, *J. Atmos. Res.,* 23, 287, 1989.

42. Ramanathan, V., Scientific use of surface radiation budget data for climate studies., Surface Radiation Budget for Climate Applications, Suttles, and Ohring, Eds., NASA Publication No. 1169, Washington, D.C., 1986.

43. Cess, R. D. and Vulis, I. L., Inferring surface solar absorption from broadband satellite measurements, *J. Climate,* 2, 974, 1989.

44. Dedieu, G., Deschamps, P. Y., and Kerr, Y. H., Solar irradiance at the surface from Meteosat visible data, Proc. Machine Processing of Remotely Sensed Data Symp., 1983.

45. Le Borgne, P. and Marsouin, A., Determination du flux rayonnement ondes courtes incident la surface: mise au point d'une methode operationelle partir des donnes du canal visible de METEOSAT, *Meteorologie,* 20, 9, 1988.

46. Moeser, W., Raschke, E., Incident solar radiation over Europe from METEOSAT data, *J. Climate Appl. Meteor.,* 23, 166, 1984.

47. Gautier, C., O'Hirok, W., Landsfeld, M., and Shiren, Y., Bidirectional Reflectance Effects on the Computation of Surface Solar Flux from Satellite Radiance, ARM Science Team Meet. Rep., Charleston, SC, 1994.

48. Pinker, R. T. and Laszlo, I., Modeling surface solar irradiance for satellite applications on a global scale, *J. Appl. Meteorol.,* 31, 194, 1992.

49. Darnell, W. L., Staylor, W. F., and Gupta, S. K., Estimation of surface insolation using sun synchronous satellite data, *J. Climate,* 1, 820, 1988.

50. Frouin, R. and Gautier, C., Variability of photosynthetically available and total solar irradiance at the surface during FIFE, Symp. 1st ISLCP Experiment-FIFE, Anaheim, CA, February 7–9, 1990.

51. Pinker, R. T. and Laszlo, I., Global distribution of photosynthetically active radiation as observed from satellites, *J. Climate,* 5, 56, 1992.

52. Wang, T., Satellite-derived long term net solar radiation over the global ocean surface: its relationship to low frequency SST variation and El Nino, Masters thesis, University of California, Santa Barbara, 19.

53. Chertock, B. and Frouin, R., A technique for global monitoring of net solar irradiance at the ocean surface. I. Model. *J. Appl. Meteorol.*, 31, 1956, 1990.

54. Li, Z. and Leighton, H. G., Global climatologies of solar radiation budgets at the surface and in the atmosphere from five years of ERBE data, *J. Geophys. Res.-Atmos.*, 98, 4919, 1992.

55. Bréon, F.-M., Frouin, R., and Gautier, C., Global shortwave energy budget at the Earth's surface from ERBE observations, *J. Climate*, in press.

56. Payne, R. E., Albedo of the sea surface, *J. Atmos. Sci.*, 29, 959, 1972.

57. Gautier, C. and Frouin, R., Sensitivity of satellite-derived net shortwave irradiance at the earth's surface to radiometric calibration, Proc. of 4th Int. Colloq. Spectral Signatures of Objects in Remote Sensing, Aussois, France, January 18–22, 1988 (ESA SP-287, April 1988).

58. Frouin, R. and Gautier, C., Calibration of GOES-5 and GOES-6 VISSR/VAS short wavelength channels, Proc. 3rd Int. Colloq. on Spectral Signatures of Objects in Remote Sensing, Les Arcs, France, December 16–20, 1985, 327.

59. Slater, P. N., Biggar, S. F., Holm, R. G., Jackson, R. D., Mao, Y., Moran, M. S., Palmer, J. M., and Yuan, B., Reflectance- and radiance-based methods for the in-flight absolute calibration of multispectral sensors, *Remote Sensing Environ.*, 22, 11, 1987.

60. Staylor, W. F., Reflection and emission models for clouds derived from Nimbus 7 earth radiation budget scanner measurements, *J. Climate*, 2, 419, 1990.

61. Whitlock, C. H., Staylor, W. F., Smith, G., Levin, R., Frouin, R., Gautier, C., Teillet, P. M., Slater, P. N., Kaufman, J. Y., Holben, B. N., and LeCroy, S. R., AVHRR and VISSR satellite instrument calibration results from both cirrus and marine stratocumulus IFO periods, Fire Science Rep., NASA CP 3083, 1989.

62. Frouin, R. and Gautier, C., Calibration of NOAA-7 AVHRR, GOES-5, and GOES-6 VISSR/VAS solar channels, *Remote Sensing Environ.*, 22, 73, 1987.

63. Gautier, C., Daily shortwave energy budget over the ocean from geostationary satellite measurements, *Oceanography from Space*, Plenum Press, Marine Series, New York, 1981, 201.

64. Gautier, C. and Katsaros, K., Insolation during STREX. I. Comparisons between surface measurements and satellite estimates, *J. Geophys, Res.*, 89, 11779, 1984.

65. Frouin, R., Gautier, C., Katsaros, K., and Lind, R., A comparison of satellite and empirical formula techniques for estimating insolation over the oceans, *J. Appl. Meteorol.*, 27, 1016, 1988.

66. Bates, J. and Gautier, C., Interaction between net surface shortwave flux and sea surface temperature, *J. Appl. Meteorol.*, 28, 43, 1989.

67. Oberhuber, J. M., An atlas based on the "COADS" data set: the budgets of heat, buoyancy, and turbulent kinetic energy at the surface of the global ocean, Max Plank Institute for Meteorology, 1988, 28 pp.

18

Simultaneous Assimilation of Satellite Radar Data in Coupled Wind and Wave Forecast Models

Frederic W. Dobson

18.1 Introduction

A number of new sensors on satellites, aircraft, and ships have or will become available that will be capable of remootely monitoring the marine environment in all weather (Table 18.1). These include both active microwave sensors, such as radars, and passive devices, such as microwave radiometers. In addition, upwelling radiation data will be available from a variety of sensors in both the visual and infrared bands. Geophysical parameters derived from the satellite data are now extensively used as input data to a wide variety of operational forecast, hindcast, and research-oriented numerical models of the lower atmosphere and upper ocean. Two processes are involved in the utilization: the translation of the remotely sensed

0-8493-4525-1/95/$0.00+$.50

Table 18.1   Radar Satellites Available in the Mid-1990s and Beyond

		Launch	Instrument Packages			
Satellite	Agency	Date	SAR	SCATT	RAlt	Other
ERS-1	ESA	July 1991	AMI[a]	AMI	Yes	ATSR[b]
JERS-1	JSA	February 1992	L-band			Oc col.
Topex-Poseidon	NASA	June 1992			Yes	Yes
SIR-C	NASA	1994–1995	C, L, X-band			
ERS-2	ESA	Mid-1995	AMI	AMI	Yes	ATSR-2, GOME[c]
Radarsat	CSA	Mid-1995	C-band			
Geosat	NASA	1995			Yes	
ADEOS	JSA	1996		NSCAT[d]		Oc col.
POEM	NASA	1998	C-band		Yes	AATSR[e], Oc col.

[a]   Advanced Microwave Imager.
[b]   Along-Track Scanning Radiometer.
[c]   Global Ozone Monitoring Experiment.
[d]   NASA Scatterometer.
[e]   Advanced Along-Track Scanning Radiometer.

signals into geophysical parameters, and the assimilation of the information thus derived into physical (numerical) models that provide continuous, global coverage.

This chapter is concerned with techniques for analyzing the raw returns from satellite radar sensors and assimilating them in the context of numerical models in which air-sea coupling processes are acknowledged and used as part of the assimilation process to optimize the information content of the satellite systems. There are two important payoffs: (1) the satellite information is global and the sensors extremely accurate, so that only by utilizing the algorithms by which the raw returns are converted to physical paramaters can their full accuracy and information content be achieved; and (2) since there are typically a wide variety of satellite sensors providing information useful as model input data, and since the raw data–physical parameter algorithms typically include dependencies of raw data on more than one physical parameter (e.g., both wind speed and wave height for the radar altimeter), the information content of each satellite sensor is enhanced if it is utilized in combination with others in the numerical model.

One of the most important customers for satellite-derived ocean surface information is research focused on understanding earth climate. The World Climate Research Program has called[1-4] for a program to utilize all the satellite information that is available to the World Meteorological Organization (WMO) Global Telecommunications System (GTS) in the coming years. The program calls specifically for design and testing of coupled numerical models forecasting waves and surface winds over the ocean that can assimilate all the satellite information available to them. These models can be used, not only to provide a better climatology of marine surface conditions, but also to provide estimates of the air-sea fluxes of momentum, heat, and water vapor.[5]

Without the satellite information, (which is global in extent, all weather in the case of the radar sensors, and uniform in accuracy) the models over the vast extent of oceans of the world — 70% of the earth surface — must rely on poor quality *in situ* marine data[6] and from land stations and climatology to predict changes in global weather and climate. To make use of the satellite data two conditions must be met: they must be calibrated and validated in open ocean conditions (nowhere else does the sea respond to the winds as it does there); and all types of information available must be assimilated into the same numerical model, so that the cross-coupling information among the different satellite data algorithms can be included.

18.2 Progress to Date

Some progress already has been made in understanding the relations between satellite microwave signals and the condition of the sea and ice surfaces. The algorithms used to relate satellite data to geophysical variables are given in the individual chapters (e.g., for synthetic aperture radars, see Chapter 15 and Appendix E; for radar altimeters, see Chapters 12 and 14 and Appendix D; for scatterometers, see Chapters 13 and 22 and Appendix F). However, the algorithms remain little more than empirical fits to *in situ* (or model) data and are only partially based on the actual physics of air-sea interaction processes: their validity is suspect when conditions lie outside the range of the defining measurements.[7]

Each active microwave sensor views the surface of the ocean as a reflector/backscatterer of microwave energy. The most common mechanisms (see Appendix E) are specular reflections from facets at normal incidence to the electromagnetic radiation, Bragg backscatter from groups of short and steep wind-generated waves, and surface discontinuities caused by sharp edges and breaking. All (except specular reflection) are caused in some way by the action of the wind. However, these same features, or wave systems closely coupled to them but at different wave numbers, determine the drag of the wind-roughened sea surface, that is, the downward rate of transfer/unit area, or flux, of horizontally directed momentum: the air sea momentum flux or wind stress.[8] The sea surface aerodynamic roughness and mean wave slope are dominated by the high-wave number (short-wavelength: typically 10–100 cm) components of the wave field; whereas in the open sea, the waves we observe, the so-called dominant waves (those that determine the significant wave height, for example), are at relatively long wavelengths (100–200 m). To understand the radar signals in terms of observable quantities, we must understand the connections between the long and short waves. One of the most effective concepts is the idea of a modulation transfer function, which allows the electromagnetic signals scattered by the small waves to be related to the dominant waves that modulate them (for the Synthetic Aperture Radar [SAR], see Chapter 15 and to the winds for the scatterometer, see Chapter 13).

For validation purposes,[9] a new generation of *in situ* sea surface sensors are being developed: winds can be measured from meteorological buoys much more accurately and reliably than from ships, and without the complication of allowing for the large measurement heights normally associated with drilling platforms. New, easily deployed directional wave buoys with satellite reporting capabilities have become available. Marine radars can provide wave direction and wavelength with precisions matching the satellite radars, and land-based, long-range high-frequency (HF) radars have been developed that provide surface current velocity, wave height, and wind speed. The dissipation technique[10] can now be used from ships at sea to provide estimates of the wind stress and heat flux in all sea states.

Numerical models attempt to describe the complexities of the ocean-atmosphere system with equations couched in terms suitable for computer algorithms, which attempt to address the basic variabilities of the system while providing approximations for processes too complex for the numerics (that occur, for example, at scales smaller than the grid scale or at timescales shorter than the time step). In an operational setting, for example, a meteorological forecast model will use as its initial state the set of atmospheric observations available at map time; if none are available at a given grid point, the last forecast will be substituted. The equations of the model are then stepped forward in time to produce the forecast.

Since the 1960s models have become increasingly dependent on satellite-derived information. The geostationary satellites provide views of the clouds that are used to derive cloud top heights (the air temperature varies with height, thus cloud top radiation temperatures give height) and, from cloud motion, the vertical profile of wind velocities. In the 1970s and 1980s a variety of satellite atmospheric sounders have come into use. In order to make optimum use of the satellite data, it has been necessary to restructure the numerics of the forecast models. The satellite data are assimilated into the models at the time step and grid point they refer to. This process has become known as 4-D data assimilation (assimilation in the 4 dimensions: 1 time, 3 space).[11] Along with the assimilations have come statistical procedures known as optimal interpolation,[11] which allow all data types to be assigned areas of influence based on empirically determined correlation scales[11] and to be given relative weights based on both forecast experience and inherent time and space variability. These developments, when included in models that cover the entire globe (the globally available satellite data has made this possible), have led to an increase in weather forecasting skill that enables useful forecasts out to 5 d.

The next important increase in the skill of numerical forecasts over the ocean will be due to two factors. First, the radar sensors — such as the SAR, scatterometer, and radar altimeter on European Remote Sensing Satellite-1 (ERS-1) and ERS-2 — provide global, all-weather coverage of the condition of the sea surface. Satellite-borne SAR, for instance, provides the user with an estimate of the height, wavelength, and direction of the longer waves (wavelength $>\sim100$ m with a short-wavelength cutoff associated with orbit geometry). The radar altimeter provides an estimate of the large-scale (10–10,000 km) sea surface slope, an overall average wave height, and a wind speed, but does not provide wind direction and cannot distinguish between sea and swell. The scatterometer signal provides an estimate of the density of Bragg-scattered waves on the sea surface at any moment in time, averaged over the swath of its averaging system. The density of Bragg waves is then related[7] to the wind velocity, or, if validations can be made, to the wind stress. All three sensors, taken together, provide cross-checks on each other; and extra information (such as air-sea temperature difference, derived from the requirement that surface winds predicted by the forecast model agree with both scatterometer and radar altimeter, thus demanding knowledge of both the wind stress and air-sea heat flux, which together determine the air stability and hence the air-sea temperature difference) can be derived by combining them in a model.[3]

Second, there is good evidence[12] that the ability of surface winds to transfer momentum to the water surface (which is determined by the aerodynamic roughness of the surface) is related to the state of development of the wave field, the wave age Cp/Uc, where Cp is the phase velocity of the dominant waves in the locally wind-driven spectrum and Uc is the component of the wind speed in the direction of the dominant wind-driven waves. That is, the winds and waves are coupled. (The swell is not involved, even if the swell waves dominate, since by definition the swell has nothing to do with the local wind.)

In what follows, descriptions will be provided of the three active radar systems used on satellites: SAR, scatterometer, and radar altimeter, and the way the information provided by each may be assimilated into numerical models that forecast sea surface parameters. SAR will be considered first; next the scatterometer with SAR, and then the radar altimeter with both scatterometer and SAR will be discussed. Because the three provide complementary sea surface information, assimilating the three together provides more benefit to the model than that provided by each separately. These sections will be followed by a final one that describes in some detail how the assimilation of sea surface information from the three satellite sensors together is achieved in a combined wind-wave numerical forecast model in which the marine surface wind field is coupled to the wave field. Only through their assimilation in such coupled wind-wave models can the benefits from satellite radar data be optimally achieved.

18.3 SAR Assimilation

The Synthetic Aperture Radar (see Reference 13, Chapter 15, and Appendix E) is best suited to the estimation of the wavelength and wave direction spectrum from the image it forms of the sea surface. Estimation of wave height is less direct, and hence less precise: the technique[14] for wave height determination is independent of the calibration of the radar. It involves forming the ratio of the wave-modulated fraction of the image spectrum to the background: the so-called clutter spectrum.[15] Even the estimation of wavelength and direction is complicated by a number of factors. The SAR image itself contains no information that allows the user to choose the correct wave direction from the 180° ambiguity presented (each SAR image produces a single snapshot of the sea surface modulation). Furthermore, because of the radar geometry and the asymmetry of the returns from the range and azimuth directions, no information on the sea surface wave field is available at wave numbers beyond an azimuth cutoff caused by the nonlinearity of the velocity bunching imaging mechanism, which is most severe for satellite SARs with large range-to-velocity ratios (see Chapter 15, Figure 4).

In order to resolve the 180° wave direction ambiguity and to supply the missing information to the wave spectrum at wave numbers beyond the azimuth cutoff, the SAR image spectrum is converted to a wave spectrum using an inversion algorithm in a wave model.[16,17] The technique is to use the known SAR imaging geometry and transfer function to forward map a first-guess wave spectrum, normally the output of a forecast model, into that which the SAR would image given the first-guess spectrum. The forward map is compared with the observed SAR spectrum and modified to agree with it in areas in wave number space where the SAR is believed to provide valid information, and the resulting composite SAR spectrum is inverted. The inversion[13] is a nonlinear process involving iterations to minimize a cost function that describes differences between the observed spectrum and the forward map and that determines the penalties to be charged up to the various processes incorporated in the fit. The final outcome is a best-guess wave spectrum fully consistent with all parameters that have been included in the cost function minimization procedure.

In order to make effective use of the SAR data, it is necessary[5] to consider the effect of the assimilation on the wave prediction system. Waves are generated by the wind: if the assimilation only changes the predicted wave field of the model in the vicinity of each SAR observation, then the wave field and the input wind field (which, since it is not directly estimated in most SAR algorithms, is not explicitly included in the cost minimization) will be inconsistent and the effect of the assimilation will disappear rather quickly, that is, within a few time steps of the initial insertion of the SAR information.[5] It therefore becomes necessary to adjust the input wind field used by the model to be consistent with the SAR observation. Note that such adjustments are not helpful unless both the model forecast and the SAR data are sufficiently close to the real ocean surface conditions that the adjustments are small — on the order of 10–20% in a given physical parameter is a good rule of thumb. There are a number of strategies for accomplishing this. The most straightforward one is to use the fetch-limited wave growth laws[18] to determine the correct wind based on an estimate of the fetch and duration of the existing sea. This technique will only work well if the fetch/duration can be defined — not an easy task in the open ocean for propagating storm systems. Another technique is to obtain the wind from the SAR image by computing the area-integrated backscattered energy and provide an empirical formula to relate that to the mean wind speed (that is, by using the SAR as a scatterometer), obtaining the wind direction from the model. It will be necessary in this case to include the wind algorithm in the cost function minimization, giving the consistency of wave field with wind field some weight in determining the optimal fit of the model forecast to the SAR data, to ensure that the wind and wave fields remain consistent with the model physics. When this SAR estimate of wind speed and the wave consistency constraints are applied, the model provides an independent estimate of the wind speed (or of the wind stress, depending on the algorithm used). The same applies when wind estimates are assimilated from other radar sensors, such as the scatterometer and the radar altimeter (see below). This idea is in the research stage at this time.[19]

18.4 Scatterometer Assimilation

The scatterometer (see Reference 13 and Chapter 13) is designed to provide an estimate of the wind velocity (or stress, depending on validation) to within an instrument-determined direction ambiguity.[7] It is an excellent tool to use in conjunction with SAR, since taken together the two instruments provide independent information on both the wind field used by models to predict the waves and on the waves themselves. Hence the great significance for wave forecasting of the ERS-1 Advanced Microwave Imager; this has been designed[20] to provide in its wind/wave mode of operation continuous scatterometer coverage in a 500-km swath beneath the satellite, with every 200 km a SAR imagette, or a 10×6 km square where the SAR is used to produce a wave directional spectrum. Using the assimilation techniques outlined in the previous section, it becomes possible to provide the forecaster with a wind-compatible wave field (and conversely a wave-compatible wind field) and an independent estimate of the wind velocity (or stress) over the ocean,[21] an important enhancement considering the general paucity and lack of reliability of marine wind measurements.[6]

The development of algorithms relating radar backscatter (sigma naught [σ^0]: the ratio of returned power to transmitted power) observations to the sea surface wind stress, velocity, and temperature is still the subject of debate.[7] Although it is safe to assume that a relationship does exist between the radar backscatter from a wind-roughened sea surface and the wind stress that roughens it, it is by the same token clear that the relationship is a highly complex one.[22] The present approach[23] of relating measurements of marine wind velocities directly to observations of radar backscatter at a range of incidence angles, although suited to operational use in forecast models, reveals little of the causal physics.

Present research points to three general approaches to determine a suitable scatterometer wind algorithm. The first approach — to develop a physics-based algorithm with what is known and attempt to provide reasonable estimates of what is not known — is typified by References 22 and 24. These two approaches contain the basic physics for all geophysical scatterometer algorithms.

The second approach is to form algorithms based on fits of scatterometer σ^0 observations with wind observations of various kinds (see Chapter 15). Many such algorithms exist.[23,25]

The third approach is to seek, through experiment and theory, to better understand the physics of the variety of processes in play as the wind generates the waves. A clear exposition of the many residual problems with using this approach is given in Reference 7. We have grave difficulty in sampling the wind and the wave fields to permit accuracies high enough for scatterometer validation The scattering processes are not, as is universally assumed in the existing algorithms, entirely due to Bragg scattering from the high-wave number components of the wave field; and the high-wave number behavior of the wave field (i.e., the area in wave number space where the wind stress roughens the water surface and from which the majority of the backscattered radar energy emanates), is the most poorly understood of all. We must allow[26] for the kinematic effect of the dominant waves (which causes frequency shifts in the observed spectrum) before we can characterize the high wave number region of the wave spectrum and determine its mean slope. In addition,[12,27] we must establish a solid relationship, validated in the open sea, free from the many influences of land (aerodynamic roughnesses larger by two orders of magnitude than those at sea, and not related to the wind; limited fetches for wave development as opposed to the limited durations typical of moving meteorological disturbances, etc.) between mean wave slope, the state of development of the sea (the wave age), and the aerodynamic roughness of the sea surface (which defines its ability to extract horizontal momentum from the wind, and is itself changed in the process).[21]

The best approach is to begin with the empirically based scatterometer wind algorithms and add to them as much as is known up to the present of the sensitivities of the radar backscatter to the physical parameters of the wind-roughened sea surface, (e.g., Reference 23), updated in areas of stated ignorance with the most recent findings from the field experiments.[12]

The scatterometer wind algorithm must, as for SAR, then be assimilated into the numerical forecast model. Since the algorithm estimates sea surface winds at intervals along a swath of ocean (interval 50 km, swath width 500 km for ERS-1), some care is required in their assimilation. The meteorological models typically determine surface winds by downward extension of geostrophic winds from surface pressure measurements using a boundary layer model, at grid points much more widely separated than the scatterometer measurements, but over the entire ocean surface rather than in swaths. The assimilation problem must therefore include some averaging of the scatterometer winds and distribution of the information they supply over an empirically determined area of influence. The satellite winds must influence the model wind profile, not only at the surface, but also over the entire atmospheric surface boundary layer. As for SAR, some consideration must be given to making the winds from the assimilation consistent with the wave field from the assimilation of SAR wave data in the wave forecast model.[28] Since the forecast models of the 1980s allow for little or no coupling either of surface winds to geostropic winds or of the wind field to the wave field, some major adjustments are now underway in research institutes (e. g., the European Center for Medium-Range Weather Forecasts[28a] and the Canadian Atmospheric Environment Service).[28b]

18.5 Radar Altimeter Assimilation with SAR and Scatterometers

The radar altimeter (see Reference 13 and Chapters 12 and 14) provides estimates along the satellite track of mean sea surface elevation, significant wave height and wind speed (but not wind direction), and total wave activity only, with sea and swell lumped together (it is normally calibrated in significant wave height Hs: the mean height of the highest one third of the waves in the observed field). When assimilated in a wave model in combination with SAR and scatterometer, the wave field and the concomitant wind field along the satellite track are specified completely: that is, the information passed to the forecast model includes wind speed and direction, a wave directional spectrum so that the spectrum may be partitioned into sea and swell, and Hs.

The scatterometer and the altimeter provide independent estimates of the wind speed (the altimeter algorithm is for radar incidence angles near normal with backscatter primarily specular reflections, while that for the scatterometer is for 45° incidence angles with backscatter primarily Bragg). Since both instruments provide a wind speed estimate and since assimilation of these estimates by the model demands an atmospheric boundary layer model needing to know both the lower atmosphere momentum balance and its stability, the extra degree of freedom provided by the two independent wind speed estimates can be used to provide the air-sea temperature difference, which through its influence on the air stability and hence the turbulence in the boundary layer influences the rate of growth of the wave field for a given wind.[29] The accuracy with which the air-sea temperature difference can be determined is not yet known. It will depend on the sensitivity of the data-to-model fit to the air stability, and this in turn depends on the ability of both the satellite data and the model prediction to estimate the true wind (and wave) conditions at a given time and grid point location of the model.

A radar altimeter looks at the oceans at incidence angles near normal (see Figure E.1 in Appendix E), which produces specular reflections of the radar pulses from the wind-roughened sea surface.[30] As the intensity of the surface roughness increases, the radar pulse is further scattered, and the power backscattered to the antenna decreases. In Reference 31, the intensity of the sea surface roughness is compared with the mean square surface slope:

$$\langle s^2 \rangle = \int F(\mathbf{k})\mathbf{k}^2 d\mathbf{k}$$

where $\mathbf{k}$ is the vector wave number and $F(\mathbf{k})$ is the directional wave number spectrum of the waves. The reflected radar power has been found[31] to be inversely related (or at least approximately so) to the mean square slope; in addition[32,30] for mean wind speeds up to 20 m/sec, the mean square slope is (once again, approximately) linearly related to the wind speed. These relations form the basis for existing algorithms that derive wind speed from the altimeter backscattered power σ^0. An algorithm has been developed[33] that extends the wind speed range to 40 m/sec; it is a fit of the Geosat σ^0 observations to data from typhoons and hurricanes viewed by the altimeter combined with modeling of the storm winds in the vicinity of their cores. Because both the normal-incidence angle radar reflectivity of the sea surface and the physical configuration of the sea surface itself (i.e., the mean square slope) at high sea states (i.e., wind speed >20 m/sec) is essentially unknown,[25,34,35] there is no physics-based algorithm available for obtaining wind velocity from σ^0 that is validated at high wind speeds in the open sea.

The processes active in determining wave growth and decay change rapidly with wind speed at high sea state, as wave breaking becomes more dominant and bubbles and spray become more prevalent. Neither the incidence angle dependence nor the sea state dependence of σ^0 are known in conditions above sea state 8 (full gale: wind speeds 18-20 m/sec). At high sea states or in young seas, wave breaking triggers flow separation at the crests of steep, short-wavelength components,[36] leading to enhanced rates of energy input to the short waves and supporting a large fraction of the total wind stress. The rate at which the short waves transfer their energy and momentum to other parts of the spectrum — upward to dissipative scales, downward toward the peak in the dominant waves — is related to the age of the wave field, that is, to the ratio of the speed of the dominant waves to wind speed in the wave direction.[27] Since the overall steepness of the waves, that is, their mean slope, is also related to the wave age, it is a safe assumption that σ^0 can be related to wave age.[37]

Of the variables estimated by the altimeter, the wave height is thought to be the most accurate.[37a] Due to a combination of sampling errors and a paucity of marine wind measurements from buoys, the wind speed estimates are not thought to reach the accuracy levels claimed for them in validations against the winds from forecast models.[38]

18.6 Coupled Wind-Wave Assimilation

The modeling of wind waves has progressed from empirical wave height prediction during World War II to physically realistic descriptions of the full directional spectrum of the wave field based on wind fields from operational global forecast models. The next important step in improving the estimation of wave fields in the open ocean, away from the influence of land, is the idea of coupled wind-wave models[5] combined with assimilation of the global wind and wave information available from satellites such as ERS-1.[13] In a coupled model a relation is specified between the wind stress and the state of development of the wave field.[5,27]

Note that the state of development of the wave field includes a time history, so that earlier model runs must be kept in order to predict and propagate the swell: waves that are in the model domain at map time but were generated earlier and elsewhere by the wind.[19] The idea is that the wind stress is greater over young seas (dominant waves travel at speeds much less than the wind speed) than it is over more fully developed seas (dominant waves travel at or faster than the wind speed). Therefore, it is necessary to iterate the model prediction. The process begins with a prediction of the wave field from a first-guess wind field — normally the standard output of a forecast model — and a determination of its state of development by estimation of the wave age Cp/Uc (see definition in Section 18.2). From the wave age ratio, as determined in the wave model, a better estimate of the wind stress is formed and passed back to the meteorological boundary layer model; the waves are predicted again, and so on until the change between successive iterations is

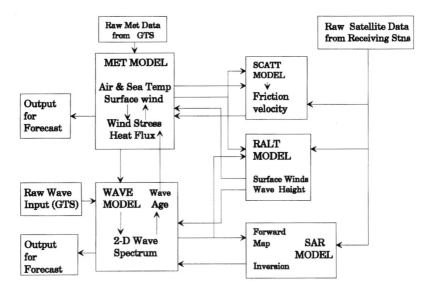

Figure 18.1   An idealized schema for a full-blown model-satellite coupled data assimilation system.

negligible. The result is a more accurate wave field and an estimate of the wind stress that is fully consistent with the state of development of the wave field instead of one determined by processes at the top of the atmospheric boundary layer, as is the case in most existing models.

The advent of multisensor radar satellites such as ERS-1 and -2 leads to a further enhancement. The ERS-1 Advanced Microwave Imager has been designed[1] to allow simultaneous estimates of wind velocity from the scatterometer, of wind speed and total wave height from the radar altimeter, and of wave direction and length from the SAR to be assimilated in coupled models. This has led to the possibility of providing from the models global fields of wind stress values that are based on a combination of satellite measurements and model physics. There will normally be enough degrees of freedom in the system to allow estimates of the air-sea temperature difference[29] and hence of the air-sea heat flux.[39] Both the wind stress and the air-sea heat flux are important and as yet poorly known[40] parameters in predictive numerical models of the global climate.

Figure 18.1 is an idealized schema for a full-blown model-satellite coupled data assimilation system. It begins with a meteorological forecast model that produces, from an analysis of data available at map time and an embedded planetary boundary layer model, a first-guess forecast of the sea surface wind field. This wind field is converted to a wind stress field using bulk formulas[39] and used as input to a wave forecast model. The wave model uses the first-guess wind stress field and any available wave height/period/direction data available on the GTS — normally the available ship data are extremely sparse — to produce a first-guess wave field; this includes propagating ahead in time and space any swell components left from the previous model run.[28]

The first-guess wind, wind stress, air and sea surface temperature, and wave fields are sent to the algorithms that convert raw backscattered radar power to physical data values for the scatterometer (SCAT model: wind stress vector), altimeter (RALT model: mean wind speed, significant wave height), and SAR (SAR model: full directional wave spectrum, and possibly mean wind speed). At grid points in the forecast model where satellite data are available at map time, each first-guess parameter is combined with the estimates from the satellite data algorithms in a (normally iterative and often nonlinear) fitting process whereby a cost function is defined that specifies the weight to be given to each of the parameters of the fit, and the weighted differences are minimized in a least squares sense.[13] The products of the fits will be optimal estimates of each parameter: wind velocity and stress, the full two-dimensional wave spectrum, and air and sea surface temperature. These are fed back into the meteorological and wave forecast models, and further iterated to account for the model coupling (that is, the state of development of the wave field is defined by the wind stress plus swell from previous runs; and the wind stress is determined by the state of development of the wave field). The final output of the procedure is a set of forecast wind and wave fields that are fitted with optimal procedures to be as consistent as possible with each other within the constraints of the assimilation and of the physics of the model. At grid points containing satellite-derived information, confidence in the accuracy of the forecast will be high, but will drop in areas where the model physics have simply carried forward an earlier analysis in space and time. It is not known by how much the skill and quantitative accuracy of the forecasts available from coupled wind-wave forecast models capable of assimilating satellite radar data will be enhanced.

From the point of view of an oceanographer interested in the climate-scale dynamics and thermodynamics of the oceans, such exercises conducted on a global scale by large numerical models may be the only way of specifying the global fluxes of momentum and heat across the sea surface.[4]

References

1. Hasselmann, K., Assimilation of microwave data in atmospheric and wave models, Proc. Conf. on Use of Satellite Data in Climate Models, Alpbach, Austria, June 10-12, 1985, *ESA Spec. Pub.*, 244, 47, 1985.
2. WCRP, Assimilation of Satellite Wind and Wave Data in Numerical Weather and Wave Models, Workshop Rep., March 25–26, 1986, Shinfield Park, U.K., WMO Tech. Doc. 148, WCP-122, Geneva, Switzerland, 1986.
3. WCRP, Global Data Assimilation for Air-Sea Fluxes, Workshop Rep., JSC/CCCO Working Group on Air-Sea Fluxes, October 1988, WMO Tech. Doc. 257, WCRP-16, Geneva, Switzerland, 1988.
4. WCRP, World Ocean Circulation Experiment, WOCE Surface Flux Determinations — a Strategy for *in situ* Measurements, WMO Tech. Doc. 304, WCRP-23, Geneva, Switzerland, 1989.
5. De Las Heras, M. M. and Janssen, P. A. E. M., Data assimilation with a coupled wind-wave model, *J. Geophys. Res.*, 97, 20261, 1992.

6. Pierson, W. J., Jr., Examples of, reasons for, and consequences of the poor quality of wind data from ships for the marine boundary layer: implications for remote sensing, *J. Geophys. Res.*, 95(C8), 13,313, 1990.

7. Pierson, W. J., Jr., Dependence of radar backscatter on environmental parameters, in *Surface Waves and Fluxes*, Geernaert, G. L. and Plant, W. J., Eds., Kluwer, Dordrecht, 1990, 173.

8. Phillips, O. M., Radar returns from the sea surface — Bragg scattering and breaking, *J. Phys. Oceanogr.*, 18, 1065, 1988.

9. Dobson, F. W., Smith, S. D., Anderson, R. J., Vachon, P. W., Vandemark, D., Buckley, J. R., Allingham, M., Khandekar, M., Lalbeharry, R., and Gill, E., The Grand Banks ERS-1 SAR Wave Experiment, *Eos*, 74, 41 and 44, 1993.

10. Dobson, F. W., Hasse, L., and Davis, R. E., *Air-Sea Interaction: Instruments and Methods*, Plenum Press, New York, 1980, 801 pp.

11. Daley, R., *Atmospheric Data Analysis*, Cambridge University Press, 1991, 457 pp.

12. Donelan, M. A., Dobson, F. W., Smith, S. D., and Anderson, R. J., On the dependence of sea surface roughness on wave development, *J. Phys. Oceanogr.*, 23, 2143, 1993.

13. Komen, G. J., Cavaleri, L., Donelan, M., Hasselmann, K., Hasselmann, S., and Janssen, D. A. E. M., *Dynamics and Modelling of Ocean Waves*, Cambridge University Press, Cambridge, England, 1995, 532.

14. Bruening, C., Hasselmann, S., Hasselmann, K., Lerner, S., and Gerling, T., On the extraction of ocean wave height spectra from ERS-1 SAR wave mode image spectra, *Proc. 1st ERS-1 Workshop, Cannes, November 1992*, ESA Publications Division, ESA SP-359, (c/o ESTEC), Noordwijk, The Netherlands.

15. Alpers, W. and Hasselmann, K., Spectral signal to clutter and thermal noise properties of ocean wave imaging synthetic aperture radars, *Int. J. Remote Sensing*, 3, 423, 1982.

16. Hasselmann, K. and Hasselmann, S., On the nonlinear mapping of an ocean wave spectrum into a SAR image spectrum and its inversion, *J. Geophys. Res.*, 96(C6), 10,713, 1992.

17. Krogstad, H. E., Samset, O., and Vachon, P. W., Generalizations of the non-linear ocean-SAR transform and a simplified SAR inversion algorithm, *Atmos.-Ocean*, 33, 61–82, 1994.

18. Perrie, W. and Toulany, B., Fetch relations for wind-generated waves as a function of wind stress scaling, *J. Phys. Oceanogr.*, 20, 1666, 1990.

19. Hasselmann, S., Bruening, C. and Lionello, P., Towards a generalized optimal interpolation method for the assimilation of ERS-1 SAR retrieved wave spectra in a wave model, Proc., 2nd ESA ERS-1 Symp., Hamburg, Germany, October 11–14, 1993, 21–26.

20. Attema, E. P. W., The active microwave instrument on board the ERS-1 satellite, *Proc. IEEE*, 79, 791, 1991.

21. Caudal, G., Self-consistency between wind stress, wave spectrum, and wind-induced wave growth for fully-rough air-sea interface, *J. Geophys. Res.*, 98, 22,743, 1993.

22. Donelan, M. A. and Pierson, W. J., Jr., Radar scattering and equilibrium ranges in wind-generated waves with application to scatterometry, *J. Geophys. Res.*, 92(C5), 4971, 1987.

23. Stoffelen, A. and Anderson, D. L. T., Wind retrieval and ERS-1 scatterometer radar backscatter measurements, *Adv. Space Res.*, 13(5), 53, 1993.

24. Plant, W. J., A two-scale model of short wind-generated waves and scatterometry, *J. Geophys. Res.*, 9, 10,735, 1986.

25. Quilfven, P., ERS-1 off-line wind scatterometer products, *Proc. IGARSS 1993, Tokyo, Japan*, 1750–1752, 1993.

26. Banner, M. L., Equilibrium spectra of wind waves, *J. Phys. Oceanogr.*, 20, 966, 1990.

27. Donelan, M. A., The Mechanical coupling between air and sea: an evolution of ideas and observations, *Strategies for Future Climate Research*, Latif, M., Ed., Max-Planck Institut für Meteorologie, Hamburg, Germany, 1993, 77.

28. Bauer, E., Assimilation of ocean wave spectra retrieved from ERS-1 SAR into the wave model WAM, Proc. 2nd ESA ERS-1 Symp., Hamburg, Germany, October 11–14, 1993, 27.

28a. Hasselmann, S., personal communication, 1994.

28b. Wilson, L., personal communication, 1994.

29. Janssen, P. A. E. M. and Komen, G. J., Effect of atmospheric stability on the growth of surface gravity waves, *Boundary-Layer Meteorol.*, 2, 85, 1985.

30. Jackson, F. C., Walton, W. T., and Hines, D. E., Sea surface mean slope from Ku-band backscatter data, *J. Geophys. Res.*, 97, 11,411, 1992.

31. Moore, R. K. and Fung, A. K., Radar determination of winds at sea, *Proc. IEEE*, 67, 1505, 1979.

32. Cox, C. and Munk, W., Statistics of the sea surface derived from sun glitter, *J. Mar. Res.*, 13, 198, 1954.

33. Young, I. R., An estimate of the Geosat altimeter wind speed algorithm at high wind speeds, *J Geophys. Res.*, 98(C11), 20,275, 1993.

34. Banner, M. L. and Fooks, E. H., On the microwave reflectivity of small-scale breaking water waves, *Proc. R. Soc. London A*, 399, 93, 1985.

35. Jaehne, B. and Reimer, K. S., Two dimensional wavenumber spectra of small-scale water surface waves, *J. Geophys. Res.*, 95, 11,531, 1990.

36. Banner, M. L. and Melville, W. K., On the separation of air flow over water waves, *J. Fluid Mech.*, 77, 825, 1976.

37. Glazman, R. E. and Pilorz, S. H., Effects of sea maturity on satellite altimeter measurements, *J. Geophys. Res.*, 95, 2,857, 1990.

37a. Pierson, W. J., personal communication, CUNY, 1994.

38. Hansen, B., Bruening, C., and Staabs, C., Global Comparison of Significant Wave Heights Derived from ERS-1 and SAR Wave Mode Altimeter and TOPEX Altimeter Data, *Proc. 2nd ERS-1 Symp. October 11–14, 1993, Hamburg, Germany*, ESA, Paris, 1994, 33–36.

39. Smith, S. D., Coefficients for sea surface wind stress, *J. Geophys. Res.*, 93, 15,467, 1988.

40. Bretherton, F., Burridge, D., Crease, J., Dobson, F., Kraus, W., and Vonder Haar, T., The 'CAGE' Experiment: a Feasibility Study, WCP-22, WMO, Geneva, 1981, 95 pp.

Part IV

Application to Sea Ice

19

AVHRR Applications for Ice Surface Studies

Konrad Steffen

19.1 Introduction

The Advanced Very High-Resolution Radiometer (AVHRR) is a satellite sensor with strong potential for studying sea ice and snow-covered ice surfaces. Its visible and infrared channels offer data with direct applicability to problems of heat balance and climate studies. Its frequent polar coverage and broad swath are sufficient to resolve synoptic and even diurnal changes of surface temperature, ice concentration, and lead patterns. The decade long availability and planned continuation of this sensor make it a primary source of data that can be used to monitor global climate.

Despite this promise, AVHRR has not been broadly applied to polar scientific problems. A number of studies have been undertaken to show how geophysical important surface parameters can be inferred using AVHRR data. Most studies, however, have worked in a demonstration and/or case study mode, and have not been developed into a monitoring capability. A reason for less extended applications to the polar regions has been and continues to be cloudiness that obscures the surface for most of the summer and much of the winter. Although there are holes in the cloud cover through which the surface can be viewed, differentiating cloud from surface over snow and ice is far more difficult than at lower latitudes where the albedo and temperature of land and water are quite dissimilar from those of clouds. Another problem is the diverse number of institutions and agencies where AVHRR data are archived, in multiple formats and media, for various periods, and all without interrelated catalogs. For this reason, the utilization of polar AVHRR data has not been used to its full potential. Present snow and ice applications from AVHRR data for polar regions are summarized in Steffen et al.[1]

19.2 Data and Processing

Resolution

AVHRR data are available in two different resolutions: high-resolution picture transmission (HRPT) and local area coverage (LAC) with 1.1-km ground resolution at nadir, 2.5-km resolution at the swath edge, and global area coverage (GAC) at 4-km resolution. The swath width is 2580 km with 2048 data samples per scan line and 10-b radiometric resolution (1024 digital levels). For additional technical information the reader is referred to Appendix A.

Navigation and Geolocation

The AVHRR data in level 1b format include information for navigation and calibration.[2] The navigation of AVHRR data can be used to depict the actual motion of the satellite.[3] Parameters in an orbit model are modified to give correct coordinates of known ground points in an image. The geolocation accuracy of AVHRR LAC data is on the order of 1.5 km in the vicinity of the land area.[4] The procedure and method are discussed in Chapter 10.

Calibration

Calibration of the AVHRR data is channel dependent. Prelaunch radiometric sensitivity of AVHRR visible and near infrared channels (1 and 2) is determined with an integrating sphere by the manufacturer of the sensors to indicate their precision.[5] The prelaunch calibration values convert the instrument counts into surface. The integrating sphere used for the prelaunch calibration, however, also has a degrading sensitivity with time. Therefore, prelaunch calibrations, which are based on the intercomparison with the integrating sphere, will always have a relative error of up to a few percent (i.e., the sphere calibration showed a drop in radiance of 10–16% between 1983 to 1987).[5] For more information on the prelaunch calibration the reader is referred to Appendix A.

A second and probably more important error source is the rate of degradation of the sensors over time. For the visible channel of National Oceanic and Atmospheric Administration (NOAA)-9 satellite, degradation rates of up to 6%/year were found.[6] Haefliger et al.[7] derived degradation rates of 0.6%/year for channel 1 and 4%/year for channel 2 for NOAA-11 based on a single comparison of *in situ* measured and modeled (AVHRR) reflectance values from the Greenland ice sheet.

The emitted infrared channels (4 and 5) require several steps for calibration. First, continuous onboard sensing of space and blackbody targets every few scan lines provides information for computing slope and intercept values for the infrared channels.[2] Energy values are calculated; and then using Planck's equation with prelaunch wave numbers, brightness temperatures can be found. Nonlinear corrections that are dependent on the scene temperature are then applied.

19.3 Ice Concentration

For some time AVHRR data have played an integral role in general ice forecasting operations that produce maps of the ice edge and chart the location of large polynyas (open water and thin ice area) and wide leads.[8] Although effective automated means have not been developed to extract ice concentration from the AVHRR data stream, analysts who prepare operational ice charts through conventional manual interpretation methods use AVHRR images to map the ice edge on a routine basis.

More extensive use of AVHRR data to map ice concentration within the pack and to extract ice-type information (whether through manual analysis, subjective classification techniques, or objective automated routines) has been limited in large part by two factors. First, discrimination of thin ice from open water is next to impossible, particularly in the vicinity of open leads and polynyas where significant atmospheric water vapor and ice crystals bias radiances are sensed from space. Second, interpretation of pixels that include mixtures of ice types that occur at subresolution scales, such as in leads or in areas where brash or small floes are common, is ambiguous.

Algorithms

AVHRR channel 2 (0.725–1.1 μm) is most useful for the determination of ice concentration, since it displays the greatest difference in reflectance between sea ice and open water. Two different methods can be used for the ice-type classification based on channel 2 depending on the season. The threshold method relates average brightness in a pixel to one ice type, and the tie point method relates the brightness value to a ratio of two types such as open water and ice.

Threshold Techniques

Using training areas, the brightness value ranges for different ice types and open water are determined. By selecting appropriate brightness thresholds, an image can be classified into three different ice types such as open water/nilas, gray ice/gray-white ice, and white ice, corresponding to a categorization of ice thickness and stages of development commonly used in sea ice research.[9] Following classification, ice concentration is then calculated as the sum of all surface types with the exception of open water.

Tie Point Algorithm

The algorithm procedure for sea ice concentration calculation was developed for Landsat imagery, and applied for different validation studies for both the Arctic and the Antarctic.[10,11] It is based on the idea that during periods where no new ice formation occurs (late spring and summer), the spectrum of classes is reduced to open water and white ice. If open water and white ice are the only two classes that are present, the assumption can be made that all brightness values in between those classes must represent ice concentrations at subresolution. Locations where a known state is assumed (i.e., 100% ice, 100% open water) are known as tie points. This algorithm thus more realistically accounts for the presence of ice floes smaller than the AVHRR pixel resolution:

$$Ic = \left(Dx - Dl/Dh - Dl\right) \cdot 100 \qquad (1)$$

where Ic = ice concentration, Dx = brightness value representing ice concentration, Dl = brightness value for open water, and Dh = brightness value for white ice.

Tie points can be found using training areas for open water and large white ice floes, where Dh represents the mean brightness for that floe. The attempt to account for subresolution size ice floes is physically more sound in areas where white ice and open water are the only surface types present. This was also found by Comiso and Zwally[10] in comparison sea ice concentrations derived from the Electrical Scanning Microwave Radiometer (ESMR). An example of the tie point ice concentration is shown in Figure 19.1* for the northern Baffin Bay.

Issues related to the derivation of ice concentration and ice-type information from AVHRR imagery were addressed by Key et al.[12] The analysis is based on the radiometric contrast of water and ice under different atmospheric conditions. Different atmospheric conditions were derived by numerical simulations with the LOWTRAN-7 model. Empirical approaches that degrade fine resolution Landsat images of leads to the coarser AVHRR resolution form the basis for analysis of scale relationships. The results of this study have broad implications with respect to feature detection in general and with respect to derivation of ice-type and ice concentration information.

Massom and Comiso[13] demonstrated that frequency distributions formed by values of pixels in an AVHRR image are polymodal and carry information about the aerial extent of different ice types that are present in the scene. Ice types were

---

*Color plates follow page 404.

assigned to individual modes by interpreting relative brightness and temperatures that define pixels within the modes. The areal extent of each ice type in the scene then can be estimated as a function of the number of pixels within each mode. The method depends on reliable cloud-masking procedures to identify cloud pixels and remove them from consideration.

Derivation of ice concentration from AVHRR imagery requires that image pixels be classified as water, ice, or cloud. Continuing progress in development of cloud-masking algorithms will effectively reduce the problem of discrimination between ice and water in the near future.[14] For clear cloud scenes, ice concentration can be achieved during winter months when good thermal contrast exists between ice and water surfaces. So far there has been no accuracy assessment for the winter case.

Errors

The question of errors introduced by the limitations of the AVHRR sensor to spectrally and spatially resolved surface features is addressed in the following. Unresolved ice features such as leads and ice floes smaller than the AVHRR field of view will be misinterpreted, causing an error in the derived ice concentration with the threshold method. These errors have to be considered in light of the fact that in some cases, for example, during breakup, the relative proportion of ice features that are smaller than 1.1 km is relatively large. The threshold method classifies different ice types based on their respective gray levels. Ice concentration is then calculated as the sum of all surface types with the exception of open water. Subresolution ice features (e.g., open water lead in white ice) would appear with a gray level between open water and white ice and therefore be classified as a young ice type. Other subresolution ice features might be too small even to be detected and will therefore be missed entirely by the threshold classification scheme.

To investigate this problem further, a theoretical error is calculated for different floe sizes and different lead frequencies. In the model, ice concentrations are represented by leads of variable width and floes of variable size. For simplicity, it is assumed that the leads are always parallel to the AVHRR scan line; and the ice floes are rectangular, also parallel to the AVHRR scan line. If we assume that the border pixel of ice floes or leads is misclassified as young ice instead of open water, the total error becomes a function of the lead spacing and the average lead width. For the following discussion, an AVHRR pixel size of 1.1 km is assumed. The ice concentration error (Δice) for leads can be parameterized as a function of lead widths (Lw), lead spacing (Ls), and AVHRR pixel resolution (Re) as following:

$$\Delta\text{ice}_{\text{lead}}\left(\%\right) = \left(100 - \left(\left(Ls - Lw\right)/Ls\right) \cdot 100\right)\left(Q/Lw\right) \qquad (2)$$

with $Q = Lw - n\,Re$; $n = 1,2,3,\ldots$; and $0 < Q < Re$.

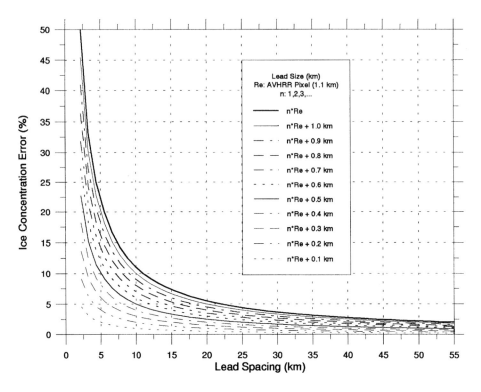

Figure 19.2 Theoretical ice concentration error due to AVHRR pixel resolution for different lead widths (greater than or equal to one AVHRR pixel) and different lead spacing (distance between leads in kilometer). The calculation is based on Equation 2.

Figure 19.2 shows the theoretical error in ice concentration for different lead widths and lead spacing. It is interesting to note that the maximum error occurs with pixel size (AVHRR resolution at nadir) and lead width equal or a multiple of the resolution. For example, at 10 km, lead spacing for lead widths of 1.1, 2.2, and 3.3 km ($n \cdot Re$) results in an ice concentration error of 11%. The error decreases for lead width smaller or bigger than the multiple AVHRR resolution.

Figure 19.3 shows the percent ice concentration error for different floe sizes and floe spacing. To account for the additional boundaries of water and ice for an ice floe compared to a lead, Equation 2 has to be modified as following:

$$\Delta \text{ice}_{\text{ice floe}}(\%) = \left(100 - \left((Ls - Lw)/Ls\right) \cdot 100\right) \cdot$$
$$(Q/Lw)\left(1 + (Ls - Lw)/Ls\right) \tag{3}$$

For example, if the floes are 5 km apart, the expected maximum error of the threshold algorithm can be as large as 32% due to the AVHRR pixel size of 1.1 km.

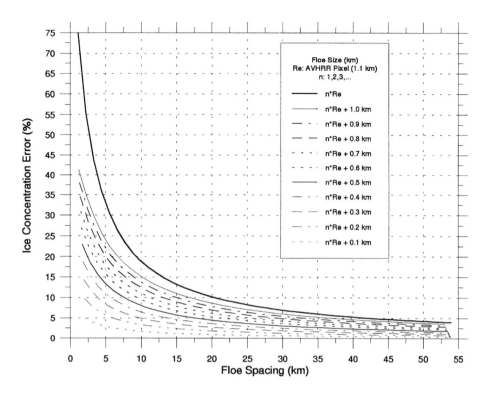

Figure 19.3 Theoretical ice concentration error due to AVHRR pixel resolution for different ice floe sizes (greater than or equal to one AVHRR pixel) and different ice floe spacing (distance between ice floes in kilometers). The calculation is based on Equation 3.

For the tie point method, the error is much smaller because this method includes subresolution information. Comparisons between various AVHRR-derived ice concentrations based on the tie point method and Landsat Multispectral Scanner (MSS) concentrations for the Bering Sea showed mean differences of −2.2 to 8.4% in ice concentration.[15]

AVHRR ice edge locations for the Bearing Sea in March 1988 were validated with Landsat imagery by Emery et al.[15] The AVHRR ice edge location was found using a subjective threshold method to classify between water and ice. Results show the AVHRR edge to be consistently seaward of the Landsat edge. Mean deviation is 1.8 km in channel 2 data, and 2.9 km in channel 4 data with root mean square (rms) errors of 3.2 and 4.4 km.

19.4 Ice Surface Temperature

The surface temperature is an important parameter for the estimation of radiative and turbulent heat fluxes. However, we have limited knowledge of atmospheric temperature, humidity, and aerosol distribution, temporal as well as spatial. This information is needed for the derivation of the surface temperature of ice and snow with a meaningful accuracy for climate studies. The sea surface temperature as derived from AVHRR is also discussed in Chapter 10).

Algorithms

Surface temperatures can be derived by using the multispectral thermal infrared data of the AVHRR sensor, channel 4 (10.5–11.5 μm) and 5 (11.5–12.5 μm). Since the atmosphere will affect different spectral channels in different ways, the difference between the brightness temperature in the separate channels can be used to provide an empirical correction. This method is called the split window method and can be applied for the uneven numbered NOAA satellite AVHRR sensors. Another approach is to calculate the various contributions (absorption and emission) to the satellite-observed thermal infrared radiance. For this method all the constituents of the atmosphere have to be known.

The general approach to estimating surface temperatures from thermal channels is to relate satellite observations to surface temperature observations with a regression model. Over Arctic sea ice, however, *in situ* surface temperature observations are rare; and therefore satellite radiance or brightness temperatures have to be modeled by forward calculation of the radiative transfer model. This approach is commonly used for the retrieval of sea surface temperatures (SST) of the ocean (see Chapter 10). Over sea ice, Key and Haefliger[16] have parameterized humidity and temperature sounding data from Soviet ice islands for different seasons; and they derived characteristic atmospheric profiles as input data for the radiative transfer model. They used an energy balance model, and modeled directional emissivities of snow to simulated AVHRR radiances. A regression model was then applied that includes the two split-window channels and the satellite scan angle vs. the simulated AVHRR radiances in order to retrieve the coefficients of Equation 4:

$$T_{\text{ice}} = a + bT_4 + cT_5 + d\left(\left(T_4 - T_5\right)\sec(\theta)\right) \tag{4}$$

where T_4 and T_5 are the satellite-measured brightness temperatures for channel 4 and 5 in Kelvin and θ is the sensor scan angle (0°–55°). The coefficients were determined through a least squares regression procedure. The rms error in the estimated ice surface temperature is stated to be less than 0.1 K in all seasons.

Errors

The rms error mentioned above is based on the regression analysis only of the modeled data; however, it does neglect the measurement errors, emissivity difference due to mixed pixels, or atmospheric effects such as ice crystal precipitation or change in ozone concentration from the mean standard profile. Lindsay and Rothrock[17] have reported an AVHRR ice surface temperature bias of +4 K for the Arctic Ocean in January 1989 based on an intercomparison with drifting buoys.

A similar method was used by Haefliger et al.[7] for the snow surface temperature retrieval of the Greenland ice sheet. Radiosonde humidity and temperature profile measurements were used as input data for the radiative transfer modeling for the AVHRR radiance at the surface, and regressed with the split window method. These AVHRR-derived snow surface temperatures were compared with *in situ* measurements that were estimated from upwelling longwave radiation measurements on the ice sheet. The comparison showed that surface temperatures were estimated to within 0.5 K in the worst case, with a mean difference of –0.2 K.

19.5 Surface Albedo

Reflectance or reflectivity is the ratio of reflected energy to the incident energy. It is a measure of the ability of a surface to reflect energy, and is specially limited to a narrow wave band or single wavelength. Reflectance is affected not only by the nature of the surface itself, but also by the angle of incidence and the viewing angle. Albedo should imply integration over the complete sphere of all directions and wavelength (e.g., 0.28–5 μm). The albedo should be distinguished from the reflectivity, which refers to one specific wavelength. The planetary albedo is the ratio of reflected and incoming solar radiation at the top of the atmosphere. The planetary albedo is a mixed signal of cloud amount, cloud type, atmospheric path length, aerosol loading in the atmosphere, and surface types, all of which make a contribution to the reflected shortwave radiation at the top of the atmosphere. The planetary albedo is an important boundary condition for the energy and heat balance of the Arctic and has to be known within a certain accuracy for the modeling of the large-scale energy exchange between the midlatitudes and the polar regions. Any long-term changes in the Arctic climate during the daylight period should be detectable as a change in the planetary albedo. As a long-term goal, the monthly mean planetary albedo as derived from AVHRR GAC data should have a relative accuracy of 0.5% and an absolute accuracy of approximately 1%.

The shortwave radiation and the surface as well as the planetary albedo can be derived from the Advanced Very High-Resolution Radiometer (AVHRR) satellite sensor. The AVHRR channels 1 and 2 (0.58–0.68 μm and 0.725–1.1 μm) measure daily the radiance at the top of the atmosphere with global coverage. The global area coverage (GAC) data set with 4-km pixel resolution would meet the objective to monitor the shortwave radiation parameters for the polar regions. This data set is readily available from the Goddard Space Flight Center (GSFC) Distributed Active Archive Center (DAAC) as a pathfinder product.

Satellite-derived albedo values are useful for the estimation of the shortwave radiation balance, an important term in surface energy balance studies. For the study of spatial and temporal changes of surface albedo, satellites provide the only data source due to the inherent dynamic change in the sea ice cover. Snow and ice surfaces have relatively high albedo values (50–90%), whereas the water surface has one of the lowest albedo values (5–10% for high solar elevations). Snow and ice albedo depend on the solar incidence angle, snow grain size and geometry,

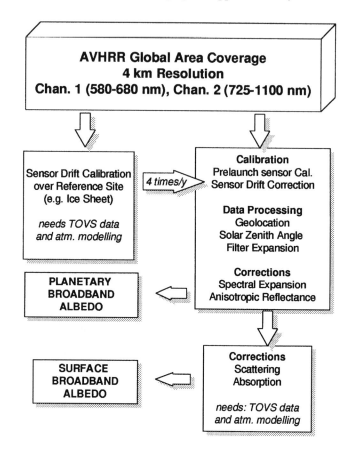

Figure 19.4   Surface and planetary broadband albedo retrieval from AVHRR GAC data.

impurities in the snow, water content in the snow cover, surface topography, and reflection angle.

Algorithms

A schematic diagram of AVHRR GAC-derived surface and planetary albedo values is given in Figure 19.4. The AVHRR sensors are only calibrated prior to the launch and therefore have no in-flight calibration capability to monitor the degradation of the sensors with time. Degradation rates of 0.3–0.6%/year for channel 1 and 1–4%/year for channel 2 were reported in the literature (Table 19.1). To monitor the AVHRR radiometric degradation over time, large-scale reference areas of well-known spectral reflectance properties such as deserts or dry snow areas of ice sheets should be used as calibration targets several times a year. Further, the temporal and spatial aerosol loading of the polar atmosphere has to be known for accurate radiative transfer modeling of scattering, absorption, and reflection of solar

Table 19.1 AVHRR Albedo Accuracy Assessment as of Today

Error Sources	Planetary Albedo (%)	Surface Albedo (%)
Prelaunch calibration	1–3	1–3
Sensor degradation per year[a]		
Channel 1	0.3–0.6	0.3–0.6
Channel 2	1–4	1–4
Atmospheric interaction		
Scattering absorption		2
Aerosol concentration		1–5
Other gases		<1
Spectral expansion	1	1
Anisotropic reflectance	Unknown	Unknown
Cloud clearing algorithm	Unknown	Unknown

[a] Sensor degradation can be monitored.

radiation. Currently, there is no aerosol climatology available, and it is believed that *in situ* data are needed for an accurate retrieval algorithm due to the large variability of aerosols with time.

The AVHRR multispectral narrowband radiometric scanners with small fields of view need proper adjustments so that their data can be used to estimate top-of-the-atmosphere albedo values (planetary albedo). The conversion of shortwave narrowband data into average planetary albedo requires adjustment of the spectral expansion from filtered narrowband to unfiltered broadband radiance of the short-wave spectrum from 0.24 to 4.2 μm.[18] The planetary albedo for snow surfaces using AVHRR narrowband satellite data can be derived through a multivariant regression analysis of Nimbus-7 earth radiation broadband data and AVHRR narrowband data.[19] The error due to the spectral expansion is in the order of 1% (Table 19.1).

To derive surface albedo values from AVHRR narrowband radiance measurements, the scattering and absorption of the atmosphere has to be included. This was successfully demonstrated by Koepke[20] through numerical simulation that used a linear relationship between clear sky planetary albedo and surface albedo. The coefficients he derived are given as a function of solar zenith angle, optical depth of the atmosphere due to aerosols, integrated ozone content of the atmosphere for AVHRR channel 1, and water vapor content for AVHRR channel 2. An alternate way to derive surface albedo values from AVHRR radiance measurements is to use a radiative transfer model such as LOWTRAN to correct for scattering and absorption as demonstrated by Haefliger et al.[7] Table 19.1 summarizes the possible errors due the atmospheric interaction, and Figure 19.1 shows the different processing steps involved to derive planetary and surface albedo from AVHRR GAC data.

Surface shortwave radiative fluxes over polar oceans can be obtained on a climatological basis from joint use of AVHRR and Special Sensor Microwave/Imager (SSM/I) data.[21] The directional AVHRR radiances at the top of the atmosphere were simulated with a radiative transfer model with realistic atmospheric conditions, and compared in an iterative mode with average AVHRR visible radiances for cloudy pixels to derive optical thicknesses of the cloud cover. The SSM/I data provided the sea ice concentrations and consequently the surface reflectance. From the retrieved optical thickness, the incoming, absorbed, and reflected solar fluxes (0.2–2.8 μm) were computed by using a two-stream narrowband model.[18] The results agreed well with *in situ* measurements, but they are strongly dependent on cloud base temperature.

Errors

Accurate surface albedo measurements from satellites are difficult to achieve because of the combined effect of clouds, variable optical path length, and non-Lambertian reflectances of most natural surfaces. Satellite sensors with narrow field-of-view scanners measure the reflectance at one or a few angles. If the reflecting surface (in our case snow, ice, and water) is not Lambertian, then the measurements have to be corrected by the anisotropic reflectance factor (ARF), which is defined as the bidirectional reflectance normalized relative to the hemispheric reflectance. The first airborne bidirectional radiances of snow-covered surfaces have been reported by Hall et al.[22] For wet snow surfaces, the ARF at large solar zenith angles can be as large as 2.[23] For new snow the ARF is close to one, and most likely can be neglected for AVHRR applications.

19.6 Conclusion

Today, a major emphasis in polar research is directed toward monitoring surface parameters that describe present climate conditions and help us understanding how the climate system functions. AVHRR data have the potential to provide a rich data stream bearing on polar conditions and processes; however, these data have been underutilized, due mainly to the inaccessibility of calibrated, geolocated LAC and HRPT data sets. Many applications such as surface temperature, lead statistics, and ice concentration require multitemporal data sets for process studies and for reducing the uncertainty in derived surface parameters.

Retrieval of ice surface temperature shows promise and would represent a significant contribution to the field, considering the current lack of spatially detailed measurements in the polar regions. This might be the most important application of AVHRR data in the polar regions, since some of the other applications can be more effectively met by other satellite systems (e.g., ice concentration with passive microwave satellite data from SSM/I). However, retrieved data have not been yet validated, although the estimated accuracy is believed to be between 1 and 4 K. Atmospheric models (radiative transfer) to account for absorption and emission of thermal radiation are in place, but additional information on ice crystal precipitation and aerosol distribution are needed to reduce the uncertainty in ice surface temperature retrieval.

The same parameters are needed for an improved albedo retrieval. Currently, narrowband albedo can be retrieved for cloud-free pixels within an accuracy of

1–4% for summer conditions (no ice crystal precipitation) if coincident atmospheric profile measurements are available.

Ice concentration algorithms make use of surface reflectance and surface temperature information; and consequently also need improved atmospheric input data to characterize temporal and spatial variability of water vapor, diamond dust, and ice crystals in conjunction with improved cloud-masking algorithms. Emphasis should be placed on use of a combined sensor algorithm (i.e., passive microwave, thermal and visible, synthetic aperture radar) to circumvent present problems with cloud masking in the visible and thermal spectra.

Long-term monitoring of the shortwave radiation parameters are required for Arctic climate and process studies for both polar regions from 55° to the pole. The AVHRR GAC data will become available through the GSFC DAAC as a pathfinder product — the Earth Observation System (EOS) DAAC — in the near future.

References

1. Steffen, K., Bindschadler, R., Comiso, J., Eppler, D., Fetterer, F., Hawkins, J., Key, J., Rothrock, D., Thomas, R., and Weaver, R., Snow and ice applications of AVHRR in polar regions, *Ann. Glaciol.*, 17, 1, 1993.
2. Lauritson, L., Nelson, G.J., and Porto, F.W., Data extraction and calibration of TIROS-N/NOAA radiometers, NOAA Technical Memorandum NESS107, 1979.
3. Baldwin, D.G. and Emery, W.J., Precise AVHRR image navigation and thermal calibration, *Ann. Glaciol.*, 17, 414, 1992.
4. Emery, W.J., Brown, J., and Nowak, Z.P., AVHRR image navigation: summary and review, *Photogr. Eng. Remote Sensing*, 55(8), 1175, 1989.
5. Abel, P., Prelaunch calibration of NOAA-11 AVHRR visible and near IR channels, *Remote Sensing Environ.*, 31, 227, 1990.
6. Staylor, W.F., Degradation rates of the AVHRR visible channel for the NOAA 6, 7, and 9 spacecraft, *J. Atmos. Ocean. Technol.*, 7, 411, 1990.
7. Haefliger, M., Steffen, K., and Fowler, C.W., AVHRR surface temperature narrow-band albedo comparison with ground measurements for the Greenland ice sheet, *Ann. Glaciol.*, 17, 49, 1993.
8. Hufford, G.L., Sea ice detection using enhanced infrared satellite data, *Mariners Weather Log*, 25(1), 1, 1981.
9. Steffen, K., Atlas of sea ice types, deformation processes and openings in the ice, *Zurcher Geographische Schriften*, 20, Swiss Federal Institute of Technology, Zurich, Switzerland, 1986.
10. Comiso, J.C. and Zwally, H.J., Antarctic sea ice concentrations inferred from Nimbus 5 ESMR and Landsat imagery, *J. Geophys. Res.*, 87(C8), 5836, 1982.
11. Steffen, K. and Schweiger, A.J., A multisensor approach to sea ice classification for the validation of DMSP-SSM/I passive microwave derived sea ice products, *Photogr. Eng. Remote Sensing*, 56(1), 75, 1990.
12. Key, J., Stone, R., Maslanik, J., and Ellefsen, E., The detectability of sea ice leads in thermal satellite data as a function of atmospheric conditions and measurement scale, *Ann. Glaciol.*, 17, 227, 1993.
13. Massom, R.A. and Comiso, J.C., The determination of surface temperature and new sea ice classification using AVHRR, *J. Geophys. Res.*, 99(C3), 5201–5218, 1994.
14. Welch, R.M., Sengupta, S.K., Goroch, A.K., Rabindra, P., Rangaraj, N., and Navar, M.S., Polar cloud and surface classification using AVHRR imagery: an intercomparison of methods, *J. Appl. Meteorol.*, 31, 405, 1992.

15. Emery, W.J., Radebaugh, M., Fowler, C.W., Cavalieri, D., and Steffen, K., A comparison of sea ice parameters computed from Advanced Very High Resolution Radiometer and Landsat satellite imagery and from airborne passive microwave radiometry, *J. Geophys. Res.,* 96(C12), 22,075, 1991.

16. Key, J. and Haefliger, M., Arctic ice surface temperature retrieval from AVHRR thermal channels, *J. Geophys. Res.,* 97(D5), 5885, 1992.

17. Lindsay, R. and Rothrock, D., Surface temperature and albedo distributions of Arctic sea ice from AVHRR, *Ann. Glaciol.,* 17, 391, 1993.

18. Saunders, R.W., The determination of broad-band surface albedo from AVHRR visible and near-infrared radiances, *Int. J. Remote Sensing,* 11(1), 49, 1990.

19. Wydick, J.E., Davis, P.A., and Gruber, A., Estimation of broadband planetary albedo from operational narrowband satellite measurements, NOAA Tech. Rep. NESDIS 27, U.S. Department of Commerce, 1987.

20. Koepke, P., Removal of atmospheric effects from AVHRR albedos, *J. Appl. Meteorol.,* 28, 1341, 1989.

21. Kergomard, C., Bonnel, B., and Fouquart, Y., Retrieval of surface radiative fluxes on the marginal zone of sea ice from operational satellite data, *Ann. Glaciol.,* 17, 201, 1993.

22. Hall, D., Foster, J., Irons, J., and Dabney, D., Airborne bidirectional radiance of snow surfaces in Montana, U.S.A., *Ann. Glaciol.,* 17, 35, 1993.

23. Steffen, K., Bidirectional reflectance of snow at 500–600 nm, large scale effects of seasonal snow cover, in *Proc. IAHS Pub.,* 166, 415, 1987.

20

Sea-Ice Geophysical Parameters From SMMR and SSM/I Data

Josefino C. Comiso

20.1 Introduction

Large-scale spatial and seasonal characterization of the global sea-ice cover can best be done presently with satellite passive microwave data.[1-3] Such data provide comprehensive spatial coverage at good temporal resolution and are insensitive to darkness while contaminated only slightly in some channels by effects due to clouds and weather. Microwave data have been available almost continuously since launch of the Nimbus-5 Electrically Scanning Microwave Radiometer (ESMR) that afforded good data from 1972 to 1976. ESMR was followed by the Nimbus-7 Scanning Multichannel Microwave Radiometer (SMMR) that was in operation from 1978 through 1987 and was subsequently replaced by the DMSP-Special Scanning Microwave Imager (SSM/I), three versions of which have been launched since 1987. The series of observations from these sensors provides a consistent long-term data set that can serve as an important resource for polar and climate related studies.

The marked contrast in emissivity between ice-free and ice-covered oceans at some frequencies enables quick identification of location and evaluation of spatial

0-8493-4525-1/95/$0.00+$.50

extent of the sea-ice cover from brightness temperature data. The multichannel data can be used to obtain geophysical ice parameters that are useful for a variety of applications. However, accurate determination of these parameters requires a good understanding of the emission properties of the material at various wavelengths and polarizations and during different seasons.[4,5] The coarse resolution of the sensor (about 25×25 km) makes it imperative that mixing algorithms are used to infer the physical characteristics of the emitting surfaces. Such algorithms make the resolution immaterial in many applications, but only if certain criteria are satisfied. For example, the multichannel, dual-polarized sensors are capable of accurately discriminating at the most only three radiometrically distinct surfaces since the data are basically two dimensional.[6,7] Thus, in areas where there are three very dominant types of surfaces, including open water, accurate quantification of the fractional coverage of each surface is possible. However, there are areas where more than three radiometrically different surface types[5,8-10] exist, causing ambiguous retrieval of the fraction of each type of surface. Further complications are caused by snow cover, roughness, flooding, and surface wetness, especially during spring and summer.[4] In this section, the emissivity of sea ice, techniques for deriving ice parameters from passive microwave data, associated errors, and unresolved problems will be discussed.

20.2 Microwave Emissivities of Sea Ice

The basic property of the surface of interest that the satellite sensor detects is its emissivity, which is the ratio of the radiative flux emitted by the material per unit solid angle per frequency to that of a blackbody at the same physical temperature. The emissivity of a material depends on its dielectric property that controls both absorption and scattering coefficients. For sea ice, a major factor affecting the emissivity is salinity.[11,12] Cold saline ice floes, like first-year and young ice, are opaque and lossy because the imaginary part of their dielectric constant is relatively high. On the other hand, cold desalinated ice floes such as Arctic multiyear ice are transparent because the imaginary part of their dielectric constant is low. Because of voids (or air pockets) that scatter some of the emitted radiation, the radiative flux at the surface of multiyear ice floes is suppressed causing considerably lower effective emissivities for multiyear ice than those of first-year ice.[13]

The frequency and polarization dependence of the emissivity of three dominant types of surfaces (first-year ice, multiyear ice, and open water) in sea ice-covered oceans in winter are shown in Figure 20.1 (ranges of SMMR and SSM/I frequencies are indicated). The actual frequencies, polarizations, and incidence angles of the satellite sensors are given in Appendix C. The emissivity of first-year ice is shown to be almost frequency independent. The reasons are because the surface is opaque regardless of frequency and because effects due to internal scattering are not significant. The emissivity of multiyear ice, on the other hand, shows large inverse relationship with frequency. This is consistent with increases in internal scattering by voids as the wavelength of the radiation gets shorter and more comparable to the

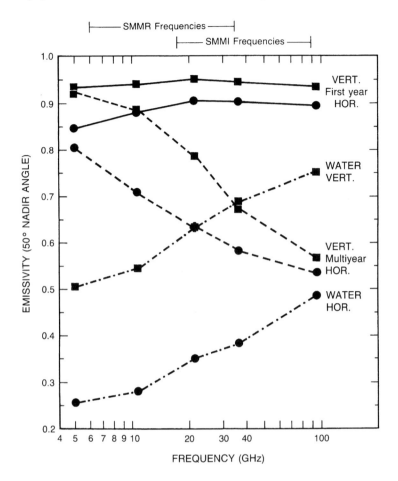

Figure 20.1 Frequency dependence of electromagnetic radiation in the microwave region over sea ice (first-year and multiyear ice) and ocean for both horizontal and vertical polarizations. (From Svendson, E. et al., *J. Geophys. Res.-Oceans*, 88, 2785, 1983. With permission of the American Geophysical Union.)

sizes of the scatterers. The emissivity of water is shown to increase with frequency on account of frequency dependence of its dielectric property and hence its emissivity.[14] For water, the emissivities at vertical polarization are also shown to be considerably higher than those at horizontal polarization.

The emissivity of first-year ice, as shown in Figure 20.1, changes as the snow cover characteristics vary due to the different thicknesses, granularities, and wetness. The emissivity decreases as the thickness (and grain size) increases especially at high frequencies when the wavelength of the radiation is comparable in size to snow ice crystals. The presence of ice lenses and snow ice at the snow-ice interface further increases scattering effects, especially at horizontal (H) polarization.[12] Also, very heavy snow load on relatively thin first-year ice cover causes flooding because of negative freeboard. Flooding causes slush to form at the snow-ice interface and thereby alters the effective emissivity of the surface.[5]

The emissivities of multiyear ice, as given in Figure 20.1, also fluctuate because the ice signature is influenced primarily by its salinity and the size of its voids (or scatterers). The latter parameters in turn depend on the thermal history of the floe that may not be spatially uniform throughout the Arctic basin. Manifestations of this are the observed variations in emissivity from one multiyear ice floe to another and also from one region to another.[10,15,16] Furthermore, second year floes have been observed to have signatures different from those of older floes.[8] Flooding and subsequent refreezing, which make the salinity profile similar to that of first-year ice, have also been observed.[10] Furthermore, regional variations in the time history of the Arctic ice cover as revealed statistically by buoy data[17] suggest a possible link to the observed regional variations in emissivity.

Three-dimensional representations of the distributions of emissivity data in the Arctic region are shown in Figure 20.2. Emissivities were derived using SMMR brightness temperature and satellite thermal infrared data.[4] Two-dimensional projections of the data points are also provided. The scatter plots show that data points in the Arctic region, representing near 100% ice cover, tend to cluster in the region between the labels A and D in the plots, while data from open water areas cluster near label H. The nonlinearity in the clustering of the data points between A and D in both 3- and 2-dimensional versions (with one exception) indicates that the emissivity of consolidated ice is not a simple combination of only two radiometrically distinct ice types. Lack of linearity suggests that there are more than two radiometrically different types of surfaces in thick ice regions. Such possibility has been reinforced by recent studies.[10,18]

The emissivity of sea ice changes with season, especially during spring and summer.[4,19-21] Enhancements in ice cover emissivity are especially obvious over predominantly multiyear ice regions during onset of spring. Such increases are caused by the presence of moisture in the ice crystals as indicated earlier. The emissivity subsequently goes down when the surface gets saturated by water and meltponding occurs.

20.3 Sea-Ice Geophysical Parameters

Ice Concentration

One of the most important parameters provided by passive microwave data is ice concentration. Ice concentration is used to estimate ice extent, actual ice area, and amount of open water within the ice pack. The latter is in turn needed in the calculation of heat fluxes between the ocean and the atmosphere, and salinity fluxes between the ice cover and the ocean. Several techniques have been developed to obtain ice concentration from passive microwave data.[1,7,20,22,23,27] A review of these

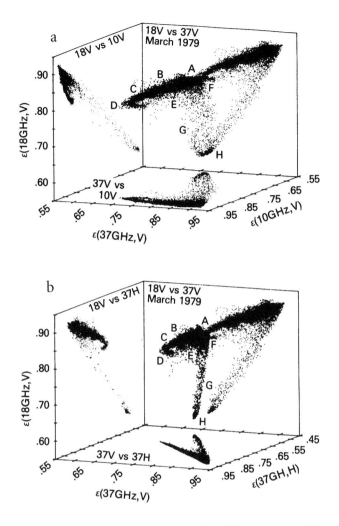

Figure 20.2 Three-dimensional emissivity scatter plots of (a) 10 GHz vs. 18 GHz vs. 37 GHz at vertical polarization, and (b) 18 GHz vs. 37 GHz at vertical polarization vs. 37 GHz at horizontal polarization. (From Comiso, J. C., *J. Geophys. Res.-Oceans*, 91, 985, 1986. With permission of the American Geophysical Union.)

techniques and a comparison of some ice concentration results are presented by Steffen et al.[24] Although the techniques are consistent in finding the location of the ice edge, the derived fractions of open water within the ice pack are not always consistent, partly on account of the use of different sets of SSM/I channels. Comparative studies with Landsat images have not yielded conclusive evaluation of the merit of each technique, since coverage of such images is usually limited and is not available during winter when ice is most expansive because of persistent clouds and long durations of darkness each day. However, current work with Synthetic

Aperture Radar (SAR) data may provide new insights about the differences in the retrieved values, especially in flooded and snow-covered rough areas.

The bootstrap technique as described by Comiso[6] and Comiso and Sullivan[25] will be used to illustrate how ice concentration is derived. The technique uses basic radiative transfer equations and at the same time takes advantage of unique multichannel distributions of the emissivity of sea ice as illustrated in Figure 20.2. In the central Arctic region, which is covered predominantly by multiyear ice, the 2-dimensional projections of the 3-dimensional plots show nonlinear clustering of consolidated ice data except for the 37 H vs. 37 V version (Figure 20.2b). A similar scatter cluster plot in Figure 20.3a indeed shows very high correlation of the horizontal (H) with the vertical (V) polarization ice data at 37 GHz. Data from this set of channels are used primarily to obtain ice concentration in the central Arctic region. The virtues for the utilization of these two channels are (1) it provides the best resolution from the set of channels (i.e., 19 and 37 GHz at both polarizations) used for ice algorithms; (2) the standard deviation of the brightness temperatures of consolidated ice is less than ± 2.5 K[6] and may provide the accuracy suitable for studies in the Arctic region where the percentage of open water has been noted to be as small as 2% in winter;[26] (3) surface temperature variations are taken into account (see Figures 20.3b and c); and (4) the slope of the line AD, which is used to find the tie point for consolidated ice, is consistently close to 1.0 in every winter data from either SMMR or SSM/I. In the seasonal sea ice region, however, the sole use of the 37 H and 37 V channels do not provide consistent identification of consolidated ice data points. Sometimes consolidated ice is difficult to discriminate from mixtures of ice and water likely on account of flooding, and snow and roughness effects. The use of 19 and 37 GHz data at vertical polarization was found to be more consistent in this regard.[25] This set of data was thus used primarily in the seasonal ice region, including the Antarctic region.[25]

The brightness temperature of open water within the ice pack is usually uniform because of smooth surface and relatively constant temperature. In the open ocean, atmospheric effects, waves, higher temperatures, and foam cause the brightness temperatures to increase substantially in some areas. Application of the algorithm in these areas, produces unrealistic ice concentration values in the open ocean. Fortunately, data from the open ocean regions can be separated from those in ice-covered areas on account of predictable changes in brightness temperatures in the region when atmospheric forcing and ocean conditions change. This is especially evident in scatter plots of 18 vs. 37 GHz brightness temperature SMMR data at vertical polarization. The brightness temperatures of open ocean in these channels are approximately linearly related and are distinctly clustered in these plots for easy separation of these data from those of ice-covered oceans. The utilization of this set of channels to mask out open ocean areas was applied in Comiso et al.,[27] but was first documented in Comiso and Sullivan.[25] A similar technique that uses gradient ratios as the threshold has been used by Gloersen and Cavalieri.[28] The use of the same methods was not as effective for SSM/I data because the 19 GHz-SSMI channels are closer to the water vapor line (22 GHz) than the 18 GHz-SMMR

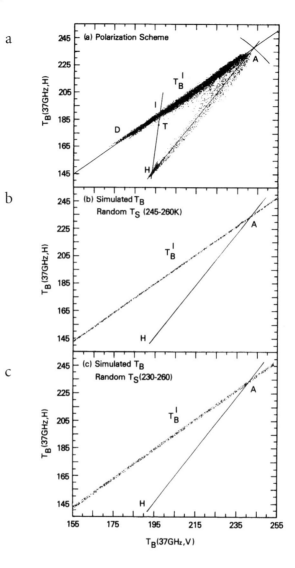

Figure 20.3 (a) Schematics for ice concentration determination, (b) sensitivity of data to variations in physical temperature of the ice from 250 to 265 K generated at random, and (c) sensitivity of data to variations in physical temperature of the ice from 240 to 265 K generated at random. (From Comiso, J. C., *J. Geophys. Res.-Oceans*, 91, 989, 1986. With permission of the American Geophysical Union.)

channels. As a consequence, in the 19 vs. 37 GHz plots, the SSM/I data points for open ocean are less linear, and some data points are in the same general location as those from ice-covered areas. However, scatter plots of 19 GHz vs. 22 GHz (vertical polarization) data show a better separation between data points in ice covered regions from those in the open ocean. This set was therefore used instead of 19 GHz vs. 37 GHz data set. Also, most of the open ocean data that cannot be masked by the 19 GHz vs. 22 GHz data, can be masked using the difference of 22 GHz and 19 GHz data at vertical polarization. The data is masked if the difference is greater than 16 K in the Antarctic regions (for all seasons), or if it is greater than 14 K during the winder (or greater than 23 K during the summer) in the Arctic region.

To obtain ice concentration, it is generally not necessary to invoke the ice type. The line AD in the scatter plot in Figure 3 can be parametrized to represent near 100% ice. Also, a data point H represents open water within the ice pack. Ice concentration for any data point at T can be derived from the radiative transfer equation and a mixing formulation[6] and is given by:

$$C_I = (T_T - T_H)/(T_I - T_H) \qquad (1)$$

where T_T, T_H, and T_I are brightness temperatures of the data point, of open water, and of consolidated ice, respectively. Identical results are derived using the brightness temperatures of one or the other channel (x or y axis). However, there is a singularity when $T_I = T_H$, which can be avoided by using corresponding values from the other channel. Since the ratio of the distances HT and HI (in Figure 3) provides the same answer and no singularity, this ratio is used in the algorithm (instead of Equation 1) for convenience.[25]

Errors in the determination of ice concentration are difficult to establish because of the difficulty in obtaining comprehensive validation data that can be used to test retrievals in different regions and seasons. However, some crude estimates can be made partly based on known physical and radiative characteristics of the surface. In the central Arctic during winter where ice concentration is expected to be very high, the accuracy in the determination of ice concentration is relatively good because the standard deviation in the signature for consolidated ice, when the 37 GHz channels are used, is only about 2.5 K. If the signature of open water is well known, such standard deviation would lead to an uncertainty in ice concentration calculation of about 3%. If the major sources of error (including the uncertainty in the identification of the reference points for open water and sea ice that are inferred from the data, radiometer noise, variations in temperature, and atmospheric effects) are taken into consideration, an accuracy of 5–10% in the central Arctic is possible.[4,6] Such accuracy would already be useful in identifying areas of large divergences in the Arctic region.

Errors are higher in the seasonal ice region than in the central Arctic region because of more unpredictable signatures. This is partly because of spatial changes in surface temperature that are not as effectively accounted for when the 19 and 37 GHz (vertical polarization) set of data is used.[29] However, in this region, spatial variations in the temperature of the snow/ice interface, which is the primary source of the signal, are minimal because the snow cover is such an effective insulation material. A key source of error is the constantly changing emissivity of some surfaces. For example, when leads open up during winter, the open water gets exposed to the cold atmosphere and grease ice forms at the surface in a few minutes. The surface then goes through a metamorphosis from grease ice to nilas, to young ice, and then to first-year ice with snow cover. During these transitions, the emissivity of the surface can change considerably from one stage to another.[5,29,31] Since such changes in emissivity are not taken into account in ice concentration algorithms, the derived fractions of open water are therefore not strictly those of open water and may include some mixtures of grease ice, and new ice. For example, the algorithm would provide an ice concentration of 62% in an area covered by 100% gray nilas.[30] This problem has been studied in the Bering Sea area[31,32] for the development of possible algorithms for new ice, but additional work is needed and it is not clear whether ambiguities can be resolved with current sensors.

In the summer, meltponding also causes underestimates in the ice cover. Preliminary results of comparative study with SAR images suggest that the ice concentration values derived from SSM/I data can be as much as 30% less than those from SAR.[32] Another factor is flooded floes, which have been observed in the Arctic[33] and can represent as much of 30% of the ice cover in areas such as the western Weddell Sea. Not much is known about the signature of flooded ice, but the effect is believed to be significant especially at channels (i.e., 19 GHz or lower) in which the signal reflects mainly emission from either the snow/ice interface or slush.[5]

Coded ice concentration maps, derived using the bootstrap technique, in both hemispheres and for all seasons are shown in Figure 20.4. The maps illustrate the large extent of the ice cover and its strong seasonality in both hemispheres. The marginal ice zones are also well identified. However, the ice concentration gradients in this region may partly reflect changes in emissivity from the ice edge to the inner pack. Low ice concentration values within the pack during the summer may also partly reflect effects of meltponding and flooding.

Several field and aircraft experiments have been performed in both polar regions,[9,10,15,35-38] and in some of them the basic assumptions about ice types and interpretation of the cluster plots have been confirmed. However, validation of satellite ice concentration data using data from these experiments has not been easy. Field data are difficult to use because of limited coverage compared with the large footprint of the satellite sensors (about 30 × 30 km). While generally easier to interpret because of fine resolution and availability of ancillary measurements, aircraft data are useful but also need to be validated by ground measurements. The other strategy has been to utilize high-resolution satellite (or space shuttle) data for validation.[39-42] While the use of high-resolution data has its advantages, such strategy degenerates into comparative analysis because the other satellite data also need to be validated. For example, unambiguous discrimination of open water, grease ice, small pancakes, and gray nilas in both visible and microwave channels may be impossible even with high-resolution sensors. Generally, however, the passive microwave data provide valuable information about large-scale characteristics of the ice cover as well as locations of ice edges, polynyas, and extensive leads. It is also useful to note that some of the comparative studies yielded high correlation coefficients.[43]

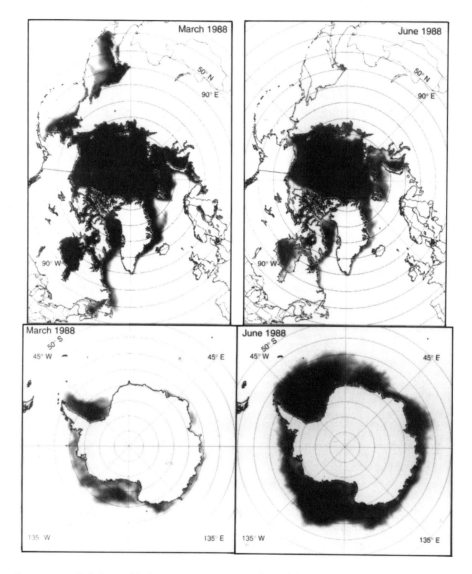

Figure 20.4   Coded monthly ice concentration maps derived from SSM/I brightness temperature data for March, June, September, and December 1988 in both Northern and Southern Hemispheres.

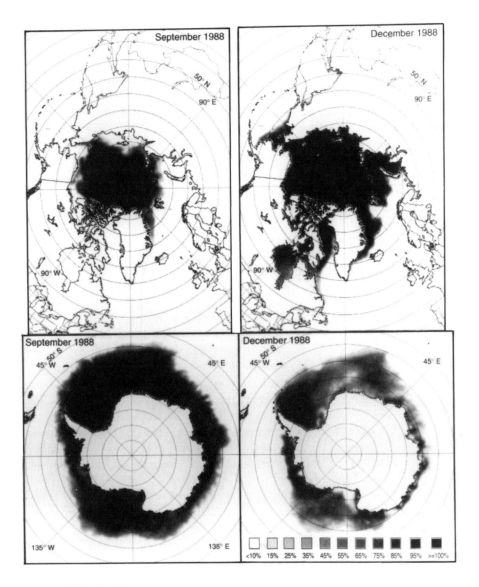

Figure 20.4 (Continued)

Ice Surface Type

With the bootstrap technique, ice concentration can be derived without an *a priori* assumption about the type of ice surface within the satellite footprint. This is possible only because the consolidated ice data points appear linear in the scatter plot of the set of channels used in the algorithm. Some algorithms[20,22,23] calculate multi-year ice and first-year ice concentrations by assuming that there are only three types of surfaces, including open water. The bootstrap technique described previously can be slightly modified to derive the fraction of each ice type if there are only two radiometrically different ice surfaces, the average emissivities of which are known. In a 2-dimensional scatter plot, the three radiometrically different surfaces, including open water surface, would be represented by three corners of a triangle. The surface type fraction can then be linearly interpolated from this triangle, assuming pure type at each corner point. However, as a rule, it is not easy to find the corner points on account of effects due to snow cover, wetness, temperature, and flooding. Furthermore, the number of radiometrically different ice types are usually more than two,[5] which is beyond what the set of channels from the radiometers can effectively discriminate.

The ambiguities in the classification of surface types may be minimized by applying the algorithm on a regional basis. Previous analysis of subsets of the Arctic Ocean and vicinity indicated the merit of regional studies.[4] However, finding a standard method of establishing the boundaries of each of these subsets for global studies is not trivial because the polar regions are so vast and because of the dynamic character of the sea ice cover. This problem can be addressed by using a combination of geographic boundaries and physical property boundaries. Different bays, seas, and oceans may be regarded as different geographic regions. The use of geographic boundaries provides an ability to take advantage of the time history of the ice pack. For example, if a bay (e.g., Hudson Bay) is ice free during the summer, it is unlikely that it would have multiyear ice during the subsequent winter. In such a region, the primary ice types would be first-year ice and new ice.

Physical property boundaries can be associated with the different ice regimes, such as the outer, inner, divergence, convergence, and perennial zones. The outer zone, which includes the marginal ice zone, is perhaps the easiest to identify in the passive microwave data because of the marked contrast in emissivity between the ocean and the ice-covered regions at some frequencies. It is also characterized as the region where loose pancakes are located and where the effect of waves dominate. The inner zone, on the other hand, is the region where the ice cover is practically a continuous sheet of ice except where there are leads and polynyas. The divergence zone is the area of extensive lead formation while the convergence zone is the area of extensive rafting and ridging. The perennial zone is the area where ice floes survive during the summer. The problem is how to separate these different regimes consistently, especially as applied to the Arctic region where the history of the ice cover can be so varied and complex.

An unsupervised technique for identifying the ice regimes/types has been developed[44] and uses multichannel cluster analysis (called isoclass). Isoclass utilizes all

SSMI channels and finds data that are clustered together within specific standard deviations. Data in each cluster are expected to have similar radiative and therefore physical characteristics. Physically distinct regions such as the marginal ice zone and the seasonal ice region, for example, form different clusters. Different clusters in the perennial zone have also been identified.[44] However, although the procedure is a viable option, good field data to establish correct interpretation of each of these clusters are required and have been difficult to obtain. Before an ice-type algorithm can be applied in a region, the typical types of ice and their emissivities must be known. Further research, including the concurrent use of SAR data and the use of neural net for surface classification, is in progress.

Snow Cover, Melt Effects, and Flooding

Although snow cover, melt effects, and flooding are generally regarded as sources of errors in the determination of ice parameters, the quantification of their areal coverage would be useful, especially for climate and process studies. Quantification of the radiative characteristics of snow cover over sea ice is especially promising over first-year ice floes because of the opacity of this ice type. If snow is the only factor that affects the spatial variability in emissivity, then the data can be parameterized to quantify snow thickness/granularity effects through concurrent use of radiative transfer modeling and field measurements. In areas not affected by extreme weather conditions, the use of the 85-GHz channels may be especially useful as demonstrated previously[9,45] because it is the frequency most sensitive to snow cover effects. However, the snow cover is not a simple homogeneous layer,[46] and some effects on observed satellite measurements may not be easy to separate from parameters of interest. Further research dedicated for this purpose is needed before quantitative estimates of snow cover parameters can be made.

The passive microwave data also show strong evidence of surface melt during spring.[47] Effects of short term refreezing after melt[48] and meltponding[33] have also been studied. Quantification of the areal extent of these effects would be very useful because they are closely linked with changes in albedo and ice structure. However, ambiguities are difficult to resolve and may require concurrent use of ancillary data (e.g., SAR, infrared, or Landsat).

Ice Extent and Actual Ice Area

Estimates of the ice extent and actual ice area may provide the key to a good understanding of the state and future of the ice cover. The time dependence of these parameters may also provide an early signal for a possible change in climate. Ice extent is the ocean area that is partly or fully covered by ice. With satellite data, however, there is a lower limit in the discrimination of partly ice-covered ocean and fully open ocean because of overlap in the radiative characteristics of these two surfaces. This limit usually ranges from 10 to 15% when SMMR and SSM/I data are used and is also usually the limit imposed by the use of open ocean mask (or weather filter), as described earlier. Thus, the ice extent as measured by passive microwave data is usually defined in terms of >n% extent where n represents the lower limit. Because of the magnitude of ice-covered area and relatively high concentrations in most areas, global ice extent can be calculated at a very good statistical accuracy.[1-3] However, there are errors associated with land mask and antenna patterns at the ice edges that should be taken into consideration. Sensitivity analysis also indicates a significant change in ice extent when n = 10 is used instead of n = 15, which implies the importance of having consistent value of n for long-term analysis.

Actual ice area is the total area actually covered by ice and can be obtained by taking the sum over the entire region of the products of the area and ice concentration in each data element. The actual ice area, in combination with ice thickness distribution, provides the information needed for mass balance, meltwater, and salinity flux studies. Accurate determination of actual ice area requires accurate retrieval of ice concentration and removal of some of the biases in ice concentration such as those associated with known fluctuations of the emissivity of 100% ice. The use of fixed brightness temperatures for 100% ice (tie point) causes overestimates of ice concentration in some areas and underestimates in others. Thus, ice areas calculated from ice concentration maps where values greater than 100% have been truncated to 100% may cause an underestimate of the real area because underestimated values are not compensated by overestimated values. Biases associated with underestimates of ice concentration in predominantly new ice areas and the use of open ocean mask that causes the exclusion of very low concentration ice should also be removed if possible or reported otherwise.

20.4 Discussion and Conclusions

Passive microwave satellite data will continue to be the most important resource for studying the large-scale distribution of the sea ice cover for many years to come. The data set is noted for comprehensive and consistent spatial coverage at good temporal resolution. They have been very useful for assessing the ice extent and locations of the ice edges, polynyas, and lead formations. Also, the historical data have already been converted into digital polar maps that can be conveniently used for time series studies. These maps are already available in a convenient polar stereographic format.[49] For a long-time history of the ice cover, a blending technique has to be developed to enable consistent characterization of the ice cover when combining data from different satellite sensors. Where there is overlap in coverage (as, for example, between SMMR and SSM/I in 1987, and between two SSM/Is in 1992), reference points, masking, and calibration can be adjusted to ensure that geophysical products are consistent.

Several decomposition techniques have been developed to calculate ice concentration. Generally, they provide results that are useful, especially for large-scale global studies. However, there are unresolved problems. For example, these techniques do not provide an accurate assessment of fractional areas covered by new and young ice. Such assessments are required in various applications such as the calculation of heat fluxes between the ocean and the atmosphere and salinity fluxes between the ice cover and the ocean. Furthermore, only two ice types (i.e., first-year and multiyear ice) are assumed in ice-type algorithms, making other surface types[5] unaccounted for. Refinements to better take into account melt and meltponding effects are also highly desirable.

The application of ice-type algorithms on a local and seasonal basis would minimize ambiguities in the results. Subsets for local analysis may be obtained using both geographic and physical properties boundaries. The latter can be determined using an unsupervised multidimensional cluster analysis and neural network. With fewer ice types in each local region, there would be less error in the determination of ice-type fraction. Estimates for ice concentration would also be improved. However, problems in active ice growth regions cannot be fully resolved. In this regard, it would be useful if ice concentration maps are generated together with ice-type maps where the latter indicates which surface type is prevalent in each data element. The latter could be generated using passive microwave in conjunction with existing temperature, pressure, and wind data.

A complementary or alternate approach would be to make use of known time variability of the ice cover. The use of a physical ice model and a Kalman filter, as has been done in some studies,[50] is a good first step in this direction. Ancillary data (e.g., wind and temperature) comparable in spatial resolution to that of SSM/I would be highly desirable for this purpose. Because of good time resolution and spatial coverage, the time-variational information that can be inferred from SSM/I data is invaluable and should be fully utilized.

References

1. Zwally, H. J., Comiso, J. C., Parkinson, C. L., Campbell, W. J., Carsey, F. D., and Gloersen, P., Antarctic sea ice, 1973–1976: satellite passive microwave observations, NASA Spec. Publ.-459, 1983.
2. Parkinson, C. L., Comiso, J. C., Zwally, H. J., Cavalieri, D. J., Gloersen, P., and Campbell, W. J., Arctic sea ice 1973–1976 from satellite passive microwave observations, NASA Spec. Publ. 489, 1987.
3. Gloersen P., Campbell, W., Cavalieri, D., Comiso, J., Parkinson, C., and Zwally, H. J., Arctic and Antarctic sea ice, 1978–1987: satellite passive microwave observations and analysis, NASA Spec. Publ. 511, 1992.
4. Comiso, J. C., Sea ice effective microwave emissivities from satellite passive microwave and infrared observations, *J. Geophys. Res.*, 88, 7686, 1984.
5. Eppler, D., Anderson, M. R., Cavalieri, D. J., Comiso, J. C., Farmer, L. D., Garrity, C., Gloersen, P., Grenfell, T., Hallikainen, M., Lohanick, A. W., Maetzler, C., Melloh, R. A., Rubinstein, I., Swift, C. T., and Garrity, C., Passive microwave signatures of sea ice, *Microwave Remote Sensing of Sea Ice*, Carsey, F., Ed., American Geophysical Union, Washington, D.C., 1992, 47.
6. Comiso, J. C., Characteristics of Arctic winter sea ice from satellite multispectral microwave observations, *J. Geophys. Res.*, 91, 975, 1986.
7. Rothrock, D. A., Thomas, D. R., and Thorndike, A. S., Principal component analysis of satellite passive microwave data over sea ice, *J. Geophys. Res.*, 93, 2321, 1988.

8. Tooma, S. G., Mannella, R. A., Hollinger, J. P., and Ketchum, R. D., Jr., Comparison of sea-ice type identification between airborne dural-frequency passive microwave radiometry and standard laser/infrared techniques, *J. Glaciol.*, 15, 225, 1975.

9. Comiso, J. C., Grenfell, T. C., Bell, D. L., Lange, M. A., and Ackley, S. F., Passive microwave *in situ* observations of winter Weddell Sea ice, *J. Geophys. Res.*, 94, 10891, 1989.

10. Grenfell, T. C., Surface-based passive microwave studies of multiyear ice, *J. Geophys. Res.*, 97, 3485, 1992.

11. Vant, M. R., Gray, R. B., Ramseier, R. O., and Makios, V., Dielectric properties of fresh sea ice at 10 and 35 GHz, *J. Appl. Phys.*, 45, 4712, 1974.

12. Matzler, C., Ramseier, R. O., and Svendsen, E., Polarization effects in sea ice signatures, *IEEE J. Ocean. Eng.*, OE-9, 333, 1984.

13. Gloersen, P., Wilheit, T. T., Chang, T. C., and Nordberg, W., Microwave maps of the polar ice of the earth, *Bull. Am. Meteor. Soc.*, 55, 1442, 1974.

14. Swift, C. T., Passive microwave remote sensing of the ocean — a review, *Boundary-Layer Meteorol.*, 18, 25, 1980.

15. Comiso, J. C., Wadhams, P., Krabill, W., Swift, R., Crawford, J., and Tucker, W., Top/bottom multisensor remote sensing of Arctic sea ice, *J. Geophys. Res.*, 96(C2), 2693, 1991.

16. Carsey, F., Summer Arctic sea ice character from satellite microwave data, *J. Geophys. Res.*, 90, 5015, 1985.

17. Colony, R. and Thorndike, A., Sea ice motion as a drunkard's walk, *J. Geophys. Res.*, 90, 965, 1985.

18. Thomas, D. R., Arctic sea ice signatures for passive microwave algorithms, *J. Geophys. Res.*, 98, 10037, 1993.

19. Carsey, F., Arctic distribution at end of summer 1973–1976 from satellite microwave data, *J. Geophys. Res.*, 87, 5809, 1982.

20. Cavalieri, D. J., Gloersen, P., and Campbell, W. J., Determination of sea ice parameters with the Nimbus 7 SMMR, *J. Geophys. Res.*, 89, 5355, 1984.

21. Grenfell, T. and Lohanick, A. W., Temporal variations of the microwave signatures of sea ice during the late spring and early summer near Mould Bay, NWT, *J. Geophys. Res.*, 90, 5063, 1985.

22. Svendsen, E., Kloster, K., Farrelly, B., Johannessen, O. M., Johannessen, J. A., Campbell, W. J., Gloersen, P., Cavalieri, D., and Matzler, C., Norwegian Remote Sensing Experiment: evaluation of the Nimbus 7 Scanning multichannel microwave radiometer for sea ice research, *J. Geophys. Res.-Oceans*, 88, 2781, 1983.

23. Swift, C. T., Fedor, L. S., and Ramseier, R. O., An algorithm to measure sea ice concentration with microwave radiometers, *J. Geophys. Res.*, 90, 1087, 1985.

24. Steffen, K., Cavalieri, D. J., Comiso, J. C., St. Germain, K., Gloersen, P., Key, J., and Rubinstein, I., The estimation of geophysical parameters using passive microwave algorithms, *Microwave Remote Sensing of Sea Ice*, Carsey, F., Ed., American Geophysical Union, Washington, D.C., 1992, 201.

25. Comiso, J. C. and Sullivan, C. W., Satellite microwave and *in situ* observations of the Weddell Sea ice cover and its marginal ice zone, *J. Geophys. Res.*, 91(C8), 9663, 1986.

26. Wittman, W. I. and Schule, J. J., Comments on the mass budget of Arctic ice pack, in Proc. Symp. Arctic Heat Budget and Atmospheric Circulation, Fletcher, J. O., Ed., RM5233-NSAF, Rand Corp., Santa Monica, CA, 1966, p. 215.

27. Comiso, J. C., Ackley, S. F., and Gordon, A. L., Antarctic Sea ice microwave signature and their correlation with *in situ* ice observations, *J. Geophys. Res.*, 89, 662, 1984.

28. Gloersen, P. and Cavalieri, D. J., Reduction of weather effects in the calculation of sea ice concentration from microwave radiances, *J. Geophys. Res.*, 91, 3913, 1986.

29. Grenfell, T. and Comiso, J. C., Multifrequency passive microwave observations of first year sea ice grown in a tank, *IEEE Trans. Geosci. Remote Sensing*, GE-24, 826, 1986.

30. Comiso, J. C., Grenfell, T. C., Lange, M., Lohanick, A. W., Moore, R. K., and Wadhams, P., Microwave remote sensing of the southern ocean ice cover, *Microwave Remote Sensing of Sea Ice*, Carsey, F., Ed., American Geophysical Union, Washington, D.C., 1992, 233.

31. Weneshan, M., Maykut, G. A., Grenfell, T. C., and Winebrenner, D. P., Passive microwave remote sensing of thin sea ice using principal component analysis, *J. Geophys. Res.*, 98, 12,453, 1993.

32. Cavalieri, D., A passive microwave technique for mapping new and young sea ice in seasonal sea ice zones, *J. Geophys. Res.*, in press.

33. Comiso, J. C. and Kwok, R., Summer Arctic ice concentration and characteristics from ERS1-SAR and SSM/I data, Proc. 1st ERS-1 Symp.: Space at the Service of our Environment, Cannes, France, November 4–6, 1992, ESA SP-359, 1993, 367.

33a. Meese, D., private communication, CRREL, Hanover, NH, 1989.

34. Tucker, W. B., III, Gow, A. J., and Weeks, W. F., Physical properties of summer sea ice from Fram Strait, *J. Geophys. Res.*, 92(7), 6787, 1987.

35. Campbell, W. J., Wayenberg, J., Ramseier, R. O., Vant, M. R., Weaver, R., Redmond, A., Arsenault, L., Gloersen, P., Zwally, H. J., Wilheit, T. T., Chang, T. C., Hall, D., Gray, L., Meeks, D. C., Bryan, M. L., Barath, F. T., Elachi, C., Leberl, F., and Fan, T., Microwave remote sensing of sea ice in the AIDJEX main experiment, *Boundary Layer Meteorol.*, 13, 309, 1978.

36. Ramseier, R. O., Gloersen, P., and Campbell, W. J., Variation of the microwave emissivity of sea ice in the Beaufort and Bering Seas, in *Proceedings of the URSI Commission II*, Schanda, E., Ed., Institute of Applied Physics, University, Berne, Switzerland, 1974, 87.

37. Johannessen, O. M., Campbell, W. J., Shuckman, R., Sandven, S., Gloersen, P., Johannessen, J. A., Josberger, E. G., and Haugan, P. M., Microwave study programs of air-ice-ocean interactive processes in the seasonal ice zone of the Greenland and Barents Seas, *Microwave Remote Sensing of Sea Ice*, Carsey, F., Ed., American Geophysical Union, Washington, D.C., 1992, 261.

38. Cavalieri, D. J., Crawford, J. P., Drinkwater, M. R., Eppler, D. T., Farmer, L. D., Jentz, R. R., and Wackerman, C. C., Aircraft active and passive microwave validations of sea ice concentration from the DMSP SSM/I, *J. Geophys. Res.*, 96, 21989, 1991.

39. Comiso, J. C. and Zwally, H. J., Antarctic sea ice concentrations inferred from Nimbus-5 ESMR and LANDSAT imagery, *J. Geophys. Res.*, 87, 5836, 1982.

40. Burns, B. A., Cavalieri, D. J., Keller, M. R., Campbell, W. J., Grenfell, T. C., Maykut, G. A., and Gloersen, P., Multisensor comparisons of ice concentration estimates in the marginal ice zone, *J. Geophys. Res.*, 92, 6843, 1988.

41. Martin, S., Holt, B., Cavalieri, D. J., and Squire, V., Shuttle imaging radar B (SIR-B) Weddell Sea ice observations: a comparison of SIR-B and scanning multichannel microwave radiometer ice concentrations, *J. Geophys. Res.*, 92, 7173, 1987.

42. Steffen, K. and Schweiger, A. J., NASA team algorithm for sea ice concentration retrieval from Defense Meteorological Satellite Program Special Sensor Microwave Imager: comparison with Landsat satellite imagery, *J. Geophys. Res.*, 96, 21971, 1991.

43. Cavalieri, D. J., The validation of geophysical products using multisensor data, *Microwave Remote Sensing of Sea Ice*, Carsey, F., Ed., American Geophysical Union, Washington, D.C., 1992, 233.

44. Comiso, J. C., Arctic multiyear ice classification and summer ice cover using passive microwave satellite data, *J. Geophys. Res.*, 95, 13411, 13593, 1990.

45. Lohanick, A., Passive microwave signatures of laboratory-grown undeformed first year ice with an evolving snow cover, *J. Geophys. Res.*, 98, 4667, 1993.

46. Garrity, C., Characterization of snow on floating ice and case studies of brightness temperature changes during the onset of melt, *Microwave Remote Sensing of Sea Ice*, Carsey, F., Ed., American Geophysical Union, Washington, D.C., 1992, 233.

47. Gogineni, S. P., Moor, R. K., Grenfell, T. C., Barber, D. G., Digby, S., and Drinkwater, M., The effects of freeze-up and melt processes of microwave signatures, *Microwave Remote Sensing of Sea Ice*, Carsey, F., Ed., American Geophysical Union, Washington, D.C., 1992, 233.

48. Anderson, M. R., The onset of spring melt in first-year ice regions of the Arctic as determined from Scanning Multichannel Microwave Radiometer data for 1979 to 1980, *J. Geophys. Res.*, 92, 13153, 1987.

49. Barry, R. G, Maslanik, J., Steffen, K., Weaver, R. L., Troisi, V., Cavalieri, D. J., and Martin, S., Advances in sea ice research based on remotely sensed passive microwave data, *Oceanography*, 6, 4, 1993.

50. Thomas, D. R. and Rothrock, D. A., The Arctic Ocean ice balance: a Kalman smoother estimate, *J. Geophys. Res.*, 98, 10053, 1993.

21

Airborne and Satellite SAR Investigations of Sea-Ice Surface Characteristics

Mark R. Drinkwater

21.1 Introduction

The logistical difficulty of making measurements in polar oceans during winter makes remote sensing techniques attractive. Microwave Synthetic Aperture Radar (SAR) offers day and night imaging, without impact from atmospheric conditions. SAR satellite receiving stations located in Fairbanks, Alaska; Tromsø, Norway; Kiruna, Sweden; West Freugh, Scotland; and Prince Albert and Gatineau, Canada, form a chain of station-receiving masks that cover all but the eastern Arctic basin. Similar Antarctic stations are operated by the Germans at the Chilean General Bernardo O'Higgins base; and by the Japanese at Syowa.[1] A further Antarctic station is currently being built at the U.S. McMurdo base[2] to be operational in 1995–1996, and will complete coverage of the Southern Ocean around the Antarctic continent.

0-8493-4525-1/95/$0.00+$.50

This bipolar network forms the basis for over a decade of continuous satellite observations of the polar ice cover.

Sea ice plays a key role in climate through its interactions with and feedbacks to the atmosphere and ocean.[3] As ice covers on average 10% of the global ocean area (rising to a maximum of 13%), this high-albedo insulating layer acts as an intermediary in the way in which the local atmosphere and ocean communicate. Sea-ice characteristics reflect and respond to the balance of fluxes of momentum, heat, water vapor, and salt at the ocean surface, by adjustments in thickness and salinity distribution. Through surface albedo and fraction of leads, ice surface conditions impact the net heat flux at the surface. Similarly, winter sea-ice growth preconditions the mixed layer, due to salinization by salt rejection.[4] It influences global ocean characteristics from the perspective of participating in formation of water masses such as Antarctic bottom water or the high-salinity shelf water found along the shelves of the Weddell and Ross Seas[5] and the Beaufort and Chukchi Seas. Thus sea ice has an important impact beyond locally regulating the exchange of heat, momentum, and water vapor between ocean and atmosphere.

In recent years SAR evolved and matured into an operational tool,[6] but the data have barely been exploited to their full scientific potential. This chapter points toward some of the insight SAR can give to sea-ice surface conditions, while identifying drawbacks and difficulties with using data or applying them in geophysical investigations. Chapter 24 develops and extends some of these themes with specific geophysical applications of the surface information obtained from SAR.

21.2 SAR and the Study of Sea Ice

From Seasat to the Present Day

Seasat laid foundations for SAR remote sensing of sea ice, returning high-quality data from the Beaufort and Chukchi Seas.[7] This short-lived satellite mission (see Appendix E) prevented planned validation experiments to understand the impact of sea-ice geophysics upon L-band backscatter. Since 1978, aircraft measurements were the main method of studying microwave interactions with snow and sea ice. These were conducted with various instruments with unique operating characteristics, their specifications given in Appendix E. Varying viewing geometry, frequency, and polarization strongly impacts sensitivity to surface phenomena, making it necessary to interpret resulting data with care. Airborne SAR campaigns conducted during field experiments enabled simultaneous measurements of sea-ice surface properties. The following studies continued development of geophysical applications between Seasat and 1991.

One of the most intensive, long-term applications of airborne SAR has been throughout a series of experiments to study air-sea interaction in the seasonal ice zone. Johannessen et al.[8] describe results of the 1979 Norwegian Remote Sensing Experiment (NORSEX); the Marginal Ice Zone Experiments (MIZEX) conducted in 1983, 1984, and 1987; and the Seasonal Ice Zone Experiment (SIZEX) in 1989. Early versions of the Jet Propulsion Laboratory (JPL) AIRSAR, the CCRS/Environmental Research Institute of Michigan (ERIM) SAR, and the CCRS SAR were used in this series of experiments. Results from these data defined the role of SAR in monitoring the morphology and structure of marginal ice zones in the Greenland Sea, Fram Strait, and Barents Sea with application to monitoring mesoscale oceanographic activity and sea-ice dynamics along ice edges. Such SAR observations led to considerable interest in modeling ocean processes such as ice edge upwelling, eddy formation,[8] and deep convection,[9] all of which directly result in surface expressions traced by the SAR-imaged sea-ice drift.

In parallel to experiments described above, similar seasonal ice zone experiments were being conducted in the Labrador Sea in preparation for the use of C-band ERS-1 and Radarsat data. The Labrador Ice Margin Experiments (LIMEX) were conducted in 1987 and 1989 with support from the CCRS aircraft.[10-13] These experiments were unique because they were the first with a C-band SAR instrument. Results led to developments in understanding wave imaging in marginal ice zones, the evaluation of C-band backscatter models for sea ice, and the influence of different ice rheologies on marginal ice zone dynamics.[14]

A number of experiments took place prior to the launch of the European Remote Sensing Satellite-1 (ERS-1), in preparation for the use of satellite data in sea-ice monitoring. The first was the Bothnian Experiment in Preparation for ERS-1 (BEPERS-88) in the Gulf of Bothnia in February 1988.[15] Following this, a series of experiments also began in the Canadian archipelago. The Seasonal Ice Monitoring Site (SIMS) experiment was first conducted in Resolute Passage in May and June 1990.[16] Continuation experiments have subsequently been conducted in 1991 and 1992 with the CCRS SAR to monitor seasonal change in Lancaster Sound. The latter was conducted under the new name Seasonal Ice Monitoring and Modeling Site (SIMMS '92). This new name reflects the evolution of this annual experiment toward utilizing time series SAR and field data to model the snow and sea-ice response to short- and long-wave radiation dynamics.[17]

ERS-1 Validation Experiments

After failure to capture simultaneous field measurements during the Seasat mission, various experiments were conceived with the object of calibration of the radar or validating approaches to extract sea-ice information from ERS-1 SAR data during the early lifetime of the satellite. These were ARCTIC '91, conducted in the late summer to early fall period in the high Arctic; the Baltic Experiment for ERS-1 (BEERS-92) from January to March 1992 in the Gulf of Bothnia; the Seasonal Ice Zone Experiment (SIZEX-92) in the Barents Sea in March 1992; and the Winter Weddell Gyre Study (WWGS '92) in the Weddell Sea, Antarctica, from May to August 1992. For reports on the preliminary findings of each of these individual studies, the reader is referred to papers presented at the 1st ERS-1 Results Symposium.[18]

Validation activities have been focused on the capability of SAR to image, differentiate, and monitor different types of sea ice. For the most part ERS-1 data are shown applicable to the problem of calculating areal fractions of different ice types, and especially to calculating the regional fraction of multiyear ice in the Arctic. Perhaps the most promising validation result is that SAR images can be used effectively to track ice floes under different conditions. Ice tracking opens doors to future scientific investigations, because kinematic information is the key to measuring ice divergence or convergence. Estimates of the thin ice fraction, the heat and salt fluxes into the upper ocean, and thus the ice growth rate are then possible (see Chapter 24). As physical modeling goes hand in hand with the development of scientific applications of these data, it is necessary then to point out drawbacks associated with utilizing SAR data.

21.3 Cm-Wavelengths and Sea-Ice Geophysics

To date a large number of studies have been conducted to understand interactions of microwaves with sea ice.[19-21] Rather than describe each result in detail, a number of important findings are summarized in this section to identify restrictions in using data with known parameters under certain snow and ice conditions or seasons. A more detailed review of the physical basis for microwave interactions with sea ice is provided by Reference 22, and a breakdown of major results from microwave radar studies is also given in Reference 23.

Impact of Frequency

Microwave image content depends on the proportion of the transmitted power reflected or scattered back to the radar. One key to using SAR for studying sea-ice geophysics is that backscattering and the penetration depth through the snow cover into the ice are frequency dependent. At shorter centimeter wavelengths such as X-band, electromagnetic radiation barely penetrates beyond the surface of higher salinity sea ice, scattering largely at the surface. One argument is whether enough information can be gleaned from the characteristics of this scattering for fundamental geophysical differences within ice body to be recognized. The converse strategy is to employ L-band or longer wavelengths to penetrate into the ice and to sense the structure and morphology of the ice from the volume scattering that originates from internal inhomogeneities.

Microwave Signatures of Ice Types

Recognition of various components of an ice cover by way of unique frequency-dependent backscatter signatures has long been considered the best route toward recovering proxy information on ice thickness. As microwave techniques have proved unsuccessful in deriving ice thickness directly, the best alternative was considered to be to map ice classes reflecting age or thickness through their salinity or roughness related backscattering signature. Here we briefly describe the success or drawbacks in recovering information using this approach.

Figure 21.1 shows a heuristic model of the annual growth of sea ice and provides an indication of the relative importance of various geophysical parameters on C-band SAR backscattering. The model represents thermodynamically influenced changes in the relative importance or efficiency of the snow and ice scattering on components of the total backscattered signal (at a typical incidence angle of 25°). Individual panels represent significant factors in ice or snow cover development, together with accompanying changes in the relative importance of components of the microwave backscatter. Important transitions are indicated in each panel together with a curve that shows the general seasonal progression in that parameter. In the lowermost panel a series of periods are indicated that describe general geophysical applications making best use of the combined information provided by these data.

First-Year Ice

As sea ice grows and ages, its backscatter signature changes. Provided ice grows thermodynamically without deformation or surface roughening, it would follow a growth sequence similar to that depicted in Figure 21.1. Obviously ice growth can begin in a given location at any time of year, but Figure 21.1 simply shows an uninterrupted and complete growth cycle of first-year ice. From its origin as new ice between 10 and 20 cm thick, it is an extremely efficient reflector due to its high salinity. If smooth, thin ice appears as the lowest power target in a SAR image, since surface roughness also determines how strong backscatter occurs. Thus the amount of deformation and surface roughening of the thin ice types is critical to the discrimination of thin ice in SAR images.[22,23] Pancake ice, which undergoes wave disturbance during growth, can in contrast appear extremely rough at all wavelengths from X- to L-band. Thus the ice growth environment is critical to the signature of thin ice types. An anomalous situation observed by various authors for young sea ice is documented in Figure 21.1 as a dotted line in the early part of some of the panels. The so-called frost-flower cycle may roughen thin ice to the extent that this high-salinity surface causes high backscatter values that can be confused with other ice.[24]

As ice grows through an intermediate stage known as gray ice into thick first-year (FY) ice, its surface grows colder and rougher, and acquires a snow cover. Moreover, the lower electromagnetic absorption gets with ice age, the higher backscatter becomes. Though this argument appears counterintuitive, various competing effects serve to override the reduced reflectivity caused by reduced salinity and the impedance matching effect of a snow cover. For instance, a snow cover may induce ice surface roughening, while also raising the temperature at the snow/ice interface by insulation. Thus, despite thicker FY ice being less saline than new or gray ice, its backscatter is often observed to be roughly 5 dB higher than younger ice forms in the range 1–10 GHz.[23,25]

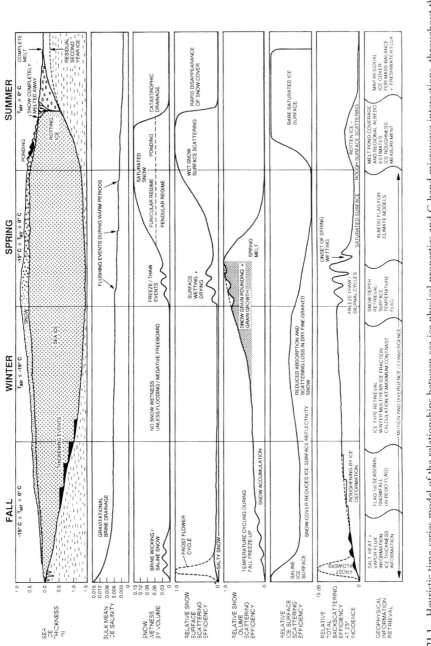

Figure 21.1   Heuristic time-series model of the relationships between sea ice physical properties and C-band microwave interactions, throughout the thermodynamic growth cycle. The lower panel shows potential geophysical information contained in the microwave backscatter record.

Multiyear Ice

First-year ice thick enough to survive the summer melt becomes multiyear (MY) or old ice (signified as a dashed line in Figure 21.1). Typically MY ice is morphologically distinctive with the upper ice consisting of freshened raised areas with a bubbly, low-density upper layer.[26] The process of melt-freeze temperature cycling and the flushing of brine produces low-salinity ice that generally supports a deeper winter snow cover. Winter SAR observations of old ice in the Arctic at frequencies of 5 GHz and above indicate that this ice has the strongest backscatter (around −10 dB at 23°) of any target other than pancake ice or thin ice with frost flowers. It appears that the lower salinity of this old ice enables greater transmission, lower absorption, and deeper penetration into the ice volume. Air inclusions and inhomogeneities in the lower density upper ice cause strong volume scattering sufficient to dominate over the corresponding levels of snow and snow/ice surface scattering.

Transmission and Absorption of Microwaves

Transmission of microwaves in sea ice is determined by scattering and absorption within the medium. These two components arise from the salinity and air inclusion content of the sea ice, as well as structural transformations that the ice undergoes. As sea ice ages it becomes desalinated,[26] and what begins as relatively high-salinity young first-year (FY) ice (>10‰ salinity) becomes less saline as it thickens. Arctic multiyear ice (MY) exceeding 1 year old normally has a lower density and salinity upper layer after experiencing summer melt processes, and is generally lower than 2–3‰ salinity. Plots of absorption and modeled penetration depths are shown in Figure 21.2 for the typical range of salinity.[20,22] First-year (FY) ice attenuates a transmitted wave rapidly within a few tens of centimeters of the surface of the ice; and of the available SAR systems, only L- and P-band can sense deeper than 50 cm under most naturally occurring sea-ice conditions. In contrast, cold MY ice experiences frequency-dependent penetration depths varying (in theory) from 1 m at X-band up to several meters at L-band. The result of penetration into MY ice at frequencies higher than C-band is that volume scattering from inhomogeneities within the ice becomes dominant.[23] This factor results in old, lower salinity MY ice having a backscatter value greater than most other ice types. This characteristic allows multiyear ice to be distinguished from lower backscatter FY ice types in C-band aircraft and ERS-1 data.

The early focus in SAR systems was on longer microwave wavelength systems such as L-band, and Seasat recovered useful mesoscale information on ice concentration, floe sizes, and shapes. The L-band wavelength is too long, however, to sense the microscopic differences between FY and MY. Notwithstanding this drawback, L-band responds most effectively to macroscopic internal deformation and structural features within the ice, such as pressure ridges and pressure zones, and leads or fractures. A shift to favor shorter wavelengths was because of the greater responsiveness to ice surface dielectric differences and roughnesses. While X-band SAR is often touted as being the best ice salinity discriminator, there are trade-offs in the

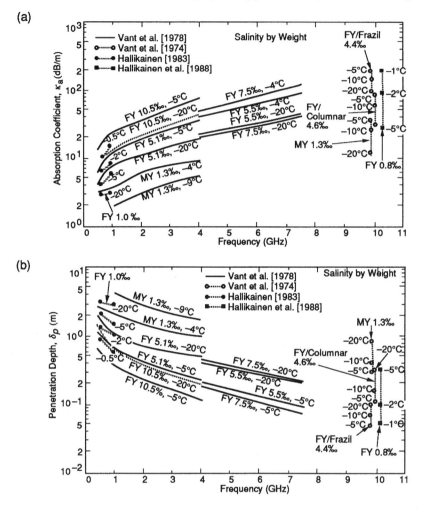

Figure 21.2   Summary of (a) experimental power absorption and (b) equivalent penetration depth in sea ice, at varying microwave frequencies. (From Hallikainen, M. and Winebrenner, D. P., AGU *Geophys. Monogr.*, 68, 36, 1992. With permission.)

information content provided by different frequencies and polarizations. The impact of the snow cover is one serious limitation to recovering information about the sea ice, due to the reduced penetration of short wavelengths in wet snow. This problem is treated later in this chapter.

Polarization Diversity

Until the planned European Space Agency Envisat polarimetric mission in the 21st century, the next decade in satellite remote sensing is restricted to single-channel instruments (Appendix E). Additional polarization information provided by polarimetric airborne systems at first sight appears irrelevant in the context of conducting current satellite SAR geophysical studies. However, results from recent studies show that the polarimetric airborne SAR is a welcome complement to single-channel SAR in terms of developing geophysical applications.

Scattering Model Development

Polarimetric SAR provides complex backscatter coefficients at different combinations of linear polarization. These enables synthesis or reconstruction of a backscatter image at any preselected polarization of the incident wave. Recent results using JPL AIRSAR data (see Appendix E) illustrate the advantage of additional polarizations in obtaining a more thorough understanding of scattering fundamentals.[27] These data are now being used in developing models and the analysis tools required for interpreting the physical basis of single-channel sea-ice signatures.[28] Development and testing of fully polarimetric backscattering models is critical because multichannel techniques are the only way to completely characterize key ice properties involved in the scattering process. Ultimately, backscatter model inversion using SAR image data will require incorporation of all essential scattering physics before realizations of solutions containing the key physical properties of sea ice can successfully be made.

Distinguishing Between Water and Ice

The main deficiency of single-channel SAR techniques is that the dielectric constant at the surface of smooth new ice can be sufficiently high that thin ice is indistinguishable from ocean water (on the basis of low backscatter magnitude). On the other hand, wind waves can generate rough surface scattering from open water in leads that can easily exceed the backscatter of the brightest MY ice target (–8 dB). Both situations cause difficulty by reversal of contrast between ice floes and their background, and this confounds automated techniques to study lead opening or ice edge location.

L-band polarimetric data and models have recently been coupled to demonstrate that it is possible to discriminate unambiguously between open water and young ice (in the range 0–30 cm) in leads. This approach requires that the incident wavelength be sufficiently long that this undeformed high salinity thin ice layer appears smooth enough that small perturbation surface scattering theory is valid.[29] The ratio of backscatter at vertical-vertical (VV)- and height-height (HH)-polarization (pol), then becomes independent of surface roughness and is instead dependent on the dielectric constant. Using an approach suggested in Reference 30, it is shown that VV/HH polarization ratios can conveniently resolve discrimination difficulties between water and new ice[31] based on order of magnitude differences in dielectric constant.

Ambiguities in Sea Ice Classification

Multichannel airborne JPL AIRSAR data can be used to remove ambiguities or difficulties in discriminating important types of ice at single C- or L-band wavelengths.[32] Polarimetric data is more adept at classifying thin ice, while also distinguishing a number of unique FY signatures, and can be used to generate a detailed ice-type chart. The value of satellite SAR data is demonstrated when these fully polarimetric data are degraded back to their single-frequency, single-polarization constituents. Comparisons of ice classification charts using C-band VV-pol (ERS-1 simulated) or L-band HH-pol data (J-ERS-1 simulated) with the fully polarimetric charts are used to quantify errors or deficiencies in classification using current spaceborne SAR. The study in Reference 32 also shows that combined L-band HH and C-band VV image data from J-ERS-1 and ERS-1 would be more powerful for studying sea ice than any single-channel data set.

Snow Cover: A Thermal Insulator and Microwave Blanket

Snow plays a critical role geophysically and in terms of microwave backscattering. Dry snow has a higher albedo than sea ice, thereby reflecting a higher proportion of incoming shortwave radiation; however, it also tends to increase the physical temperature at the sea-ice surface by virtue of its low thermal conductivity. Snow-fall on sea ice plays a significant role in determining the subsequent heat balance at the surface of the sea ice due to the insulating capacity of the snow layer. In an analogy to its thermodynamic effect, snow also regulates transmission of micro-waves. Snow depth, grain morphology, and structure, while dependent on the thermal atmospheric forcing, also play a significant role in the microwave scattering and absorption of penetrating microwaves.

Dry Snow

Winter snow is laid down with negligible melt metamorphism, and precipitated crystals become broken down and compressed by wind drift. This fine-grained dry snow is effectively transparent to microwaves, and the loss factor is on the order of 15% of the value of pure ice. Having a small dielectric constant (Figure 21.3a and b) and low absorption coefficient, it allows microwaves to propagate over long distances up to several meters before being completely absorbed or scattered (Figure 21.3c). Typical snow depths on thick FY and MY ice in the Arctic therefore present little impediment to incident microwaves. Additionally, the dielectric constant of dry snow is sufficiently low that the impedance between air and snow is almost matched. This results in negligible surface scattering or internal volume scattering and most of the wave being transmitted into the snow before being scattered at the snow/ice interface — where the largest dielectric contrast is encountered.

An assumption of a structure-free snowpack is somewhat unrealistic for most naturally occurring snow covers. After snowfall, thermal gradients through the snowpack promote changes in snow crystal shapes and sizes, and influence the backscatter. Snow metamorphism and vapor fluxes can result in internal layers causing surface scattering contributions or enlargement of snow grains and thereby Rayleigh volume scattering (at X- and C-bands). Some of these effects on

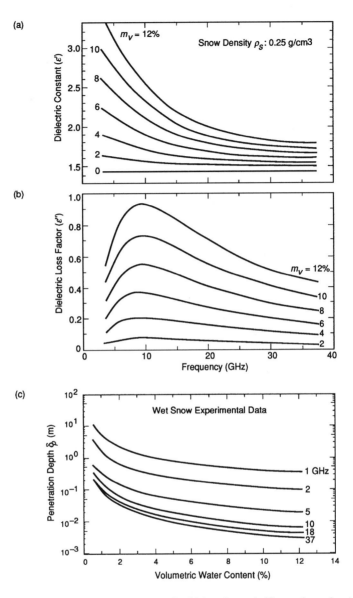

Figure 21.3 Wet snow (a) permittivity (ε'); (b) loss factor (ε'') at volume fractions of water of 0–12%, with constant snow density of 250 kg m^{-3}, and; (c) frequency-dependent penetration depth for varying snow water content. (From Hallikainen, M. and Winebrenner, D. P., AGU *Geophys. Monogr.*, 68, 40, 1992. With permission.)

microwave signatures are described in Reference 33. The most significant effect occurs when a snow cover develops layers with different density and crystal characteristics. The effects of metamorphism, seasonal melting, and refreezing that promote such layering are described later.

Wet Snow

When snow melting occurs, the liquid water that appears in the air-ice mixture dramatically changes the influence of snow on incident microwaves (Figure 21.1). In contrast to dry conditions, wet snow has a permittivity ε' that becomes frequency dependent (Figure 21.3a). Equally it has a dielectric loss ε'' between 100 and 300 times as large as dry snow and tends to 1.0 at X-band[22,34] in Figure 21.3b for saturated snow. As snow wetness increases to 2% by volume, incident microwaves at frequencies above 5 GHz are absorbed at a rate of tens of decibels per meter. Absorption coefficients of around 0.24/cm are measured in moderately wet snow, [33] thus translating to the typical penetration depths shown in Figure 21.3c.

Frost Flowers

Riming on sea ice, commonly known as frost flowering, is a process by which a snow cover may develop on the surface of the sea ice without mass input from falling snow. This was first identified in Reference 24 as a special case in terms of radar scattering because the rime crystals adsorb brine expelled onto the surface of the rapidly growing ice sheet. This creates an extremely high dielectric constant layer that is an efficient rough surface scatterer at C-band and higher frequencies.[24] Recent ERS-1 results indicate numerous situations in which frost flowers appear to be a tractable explanation for the high C-band backscatter (~−10 dB), and surface scatterometer experiments[34a] confirm that their backscatter can attain values more typically associated with the brightest MY ice floes. Their occurrence certainly causes the highest known values of C-band backscatter associated with FY ice, other than for rough pancake ice.

    Natural occurrence of such features remains dependent on a number of special atmospheric and ice growth conditions. Richter-Menge and Perovich[34b] recently studied natural forms of frost flowers and made detailed measurements of the brine content of the flowers. Laboratory measurements by Martin[34c] further identify conditions under which they may form so that their appearance in satellite data can be used as a flag for specific environmental conditions. It is clear, however, that their occurrence is closely linked to high heat flux or humid situations where a vapor source combines with an advective term over a relatively cool thin ice surface, for the growth of needle or featherlike hoar crystal growth. These features are observed to be highly ephemeral and can appear and disappear overnight depending on wind or snowfall conditions, abruptly raising or lowering backscatter values by up to 15 dB. Their appearance in ERS-1 data indicates that they may remain on young sea ice in leads for up to 10 d[34d] before additional precipitation or wind destroys their effect.

Seasonal Considerations

The main seasonal drivers are air and ocean temperatures and summer insolation, and together these modulate thermal conditions within the sea ice. Ambient air temperatures and surface humidities control the sensible and latent heat fluxes and through the energy balance control the ice growth: sunlight is the major agent of melt. The most significant indicators of the thermal balance of the sea ice are the snow- or ice-surface temperature (in the absence of snow) and the snow/ice interface temperature. These enable the thermal state of the sea ice and surface snow to be determined. In the Arctic, seasonal changes due to thermodynamic forcing from the atmosphere bring about the most significant changes in the microwave response of the sea ice. Figure 21.1 depicts some of these changes by way of the introduction of liquid water into the snowpack during spring melting and disappearance of snow during summer. The effect of microwaves encountering wet snow has been described in detail, but its consequences for information retrieval have not. Snow wetness and the seasonal impact on sea-ice properties have a considerable impact on the backscatter contributions from the snow and ice.

Figure 21.1 indicates that for the most part data on ice growth during fall freezeup signify a rapid stabilization of backscatter signatures when air temperatures fall below $-10°C$.[35] Until this point in time early snowfall can impact scattering by absorbing surface brine on the young ice sheet and increasing snow microwave absorption. Temperature cycling during diurnal cycles also has an impact on the total backscatter if liquid water appears in the snow, and this is represented in Figure 21.1 as ripples during the early fall. Generally during the fall, the ice and snow cover reflect the net heat flux environment, and the rapid stabilization of Arctic microwave signatures shown in Reference 35 indicate the transition to a negative heat balance.

Ice growth continues steadily into the winter, with an accompanying increase in mean snow depth. The winter snow cover in the Arctic and Antarctic has been observed to be an extremely complex medium, with layering occurring as a result of natural radiative processes under atmospheric forcing. This results in layers often of significantly different densities, and salinities, resulting in some internal scattering when the gradients in properties are strong enough. Stratification of naturally evolving snow cover is characterized by pronounced vertical density variations at a scale height comparable to the microwave wavelength.[33] The superposition of waves reflected at various interfaces can produce noticeable interference and polarization effects in ground-based scatterometer data. However, for the most part, winter snow-structure effects are limited to frequencies higher than C-band while the spatial variations in the snow properties on the scale of the satellite SAR resolution incoherently average out such effects.

The springtime appearance of moisture in the snow has a dramatic effect on both the snow cover and its microwave properties. In the pendular situation (Figure 21.1) where the snowcover begins to melt (i.e., below ~3% wetness) free water is retained at grain boundaries by capillary suspension. Microwaves may still penetrate through the damp surface layer, but with some attenuation of the resulting backscatter from the snow/ice interface.[36] Figure 21.1 reflects the transition to a saturated snow layer together with the corresponding reduction in the ice surface scattering. Once the snowpack becomes isothermal in the late spring, liquid water builds up until the point (>3% wetness) where pore spaces open and liquid begins to drain. This wet snow layer completely masks the sea ice from incident microwaves, preventing sensing of the sea ice beneath. In this situation SAR measurements can only provide information on the snowpack on top of the ice layer, or indeed roughness-related properties of the ice.[37,38] A dramatic reduction in multiyear sea-ice backscatter occurs at the onset of spring melt, and this rapid change in the snowcover allows melt detection in ERS-1 SAR images.[39]

In early summer the absorption in the remaining surface snow is great, and internal scattering within the snow is extinguished. In late summer, when no further snow is present on the surface of the sea ice, the backscattered signal is dominated by ice surface roughness and the density and wetness at the surface of the floe. During late summer conditions, surface roughness dominates the scattering situation and the morphological characteristics of ridges and structure of ridging zones become more clear. Under these conditions FY and MY ice become indistinguishable.

Validation Measurements and Surface Proof

In situ or field geophysical data collection is the accepted form of validatory data for remote sensing techniques and the term ground truth is applicable while field experiment data are still used to revise geophysical algorithms. In many applications remote sensing now leads acquisition of basin-wide measurements with temporal and spatial coverage and accuracy superior to surface-measured data. The term ground truth (in reference to *in situ* data) is thus outdated and requires revising now that some SAR techniques such as ice-velocity tracking have become accepted as the best available within the accuracy and precision bounds of existing measurement techniques. It is proposed that for SAR to make the transition to becoming an accepted form of quantification of certain sea-ice geophysical parameters, the term for *in situ* data be renamed from ground truth to some other term such as surface proof. This term then implies that the satellite technique is equally accurate, and that surface measurements will confirm or deny rather than supercede their accuracy.

To make more powerful scientific use of SAR products, physical models for sea ice and snow must be successfully married to backscatter models to understand the thermal or dynamic cause or effect of observed signature changes. In accordance, the style of making surface validation measurements must be more rigorously linked to the requirements of these models to directly support this association. Continuing development of microwave scattering models is necessary to understand ice signature variability, but the key to making geophysical measurements with SAR data is to realign their development with accepted geophysical models explaining dynamics or thermodynamics of sea ice. Surface experiments must make associated measurements of variables characterizing the forcing behind changes in physical properties (such as the radiation balance) in order that these relationships can be exploited. Chapter 24 builds on this theme.

Infrequent point measurements in space and time are the main limitation of surface measurements, and thus the whole approach to providing validatory data must be revisited. In light of the fact that future Radarsat data will provide entire weekly coverage of sea ice in both hemispheres, it is difficult to conceive of a scheme for comparing surface-proof measurements and geophysical products from SAR data. The answer probably lies in the judicious use and careful positioning of instrumented buoys, and the continuation of well-crafted and coordinated surface measurement programs. Such efforts must endeavor to support those scientists developing physical/scattering models by accurate quantification of the most relevant geophysical characteristics of the sea-ice cover. As such these validation experiments will continue to be a necessary part of scientific utilization of SAR images.

21.4 Microwave Scattering Models and Inversion

In the previous section, the basis for observing sea-ice surface conditions is discussed. An example is considered here that couples SAR and surface measurement data in a microwave model backscatter simulation exercise, using polarimetric SAR data from the JPL AIRSAR.

To date many theoretical models developed to simulate backscattering from snow and sea ice could not account for many geophysical situations in snow-covered sea ice because they were poorly related to the physics of snow and ice. In many cases this is due to assumptions inconsistent with naturally occurring ice, or to the fact they try to match abstract internal parameters with realistic or naturally occurring properties. Model development is proceeding at a rate soon to catch up with geophysical applications.[29] Polarimetric models such as that developed in Reference 28(a,b) are being validated using polarimetric SAR data described earlier. One advantage is that the frequency and polarization sensitivity of the model can be fully tested.

The problem with most scattering models is that they are only valid within a particular range of frequency, ice roughness, or ice salinity. Examples of testing model capability under well-defined and characterized surfaces are proving most successful.[29] An example of L-band results from model tests are illustrated for a thin ice sheet in Figure 21.4.[28] Results of matching model calculated values with measured conventional backscatter coefficients indicate a good comparison in Figure 21.4a. This simulation explains that backscattering from thin ice requires a high-salinity surface (expressed in Reference 28 as a brine skim or slush layer) in order to explain differences between VV- and HH-pol data. Behavior of the complex correlation between HH- and VV-pol backscatter is expressed as a magnitude ($|\rho|$) and phase ($\angle\rho$) in Figure 21.4b and c. The value of $|\rho|$ clearly expresses a decrease with incident angle while $\angle\rho$ remains close to zero. This trend is explained by the relative contributions of scattering from the surface and volume over this incidence angle range. While surface scatter dominates up to angles of around 30° incidence, volume scatter becomes dominant beyond this point. Waves penetrating the ice sheet that undergo internal scattering become decorrelated, hence the reduction in $|\rho|$. A slight positive shift in modeled $\angle\rho$ reflects the anisotropic scattering effect of

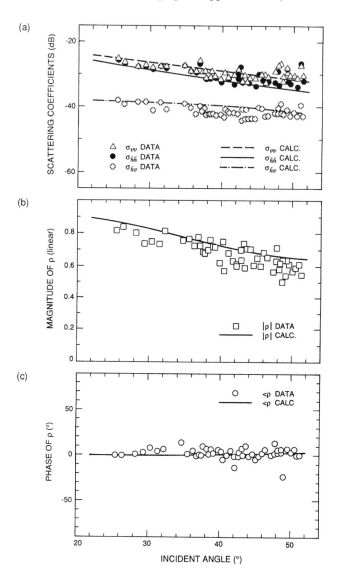

Figure 21.4 Fully polarimetric microwave scattering model results from L-band simulations of thin ice in March 1988 in the Beaufort Sea. (a) Compares measured VV-, HH- and HV-pol backscatter and simulated data; (b) indicates the magnitude $|\rho|$ of the complex correlation between VV and HH linearly polarized returns; (c) shows the corresponding phase of ρ. (From Nghiem, S. V., Kwok, R., Yueh, S. H., and Drinkwater, M. R., *J. Geophys. Res.* In press.)

tilted brine inclusions in the ice: though scattered, the data points appear to reflect a similar trend.

Modeling efforts are necessary to clarify the physics and electromagnetics governing interactions of microwaves with sea ice at various wavelengths and polarizations.[29] Recent fully polarimetric models must be tested, refined, and validated so that they can be used to provide information about ice fabric. By obtaining knowledge of the structure and dielectric properties of sea ice through microwave scattering models, we form a basis for measurement of indicators of ice salinity, thickness, and strength, together with the flux environment regulating ice conditions.

21.5 Discussion and Conclusions

SAR is ready to be exploited as a competent sea-ice measurement tool. This requires that we expend time validating approaches and confirming the geophysical utility of the extracted data. Ground proof is a necessary step in establishing SAR-derived data as a credible supplement to scattered surface point measurements.

The application of rapidly evolving algorithms to problems of monitoring sea ice geophysics is a necessary step to establishing the credibility of microwave SAR. We must all use and validate products of these applications so that the remote sensing SAR tool becomes accepted in sea ice geophysics. To make best use of this rapidly evolving tool, however, one must recognize the drawbacks involved in using aircraft or satellite SAR data. The researcher must appreciate the difficulties of using these techniques so that they can interpret the observations confidently. This chapter describes the problems and difficulties associated with use of microwave SAR instruments both from airborne and spaceborne platforms in the context of the snow and ice surface characteristics. It outlines the impact of various microwave parameters, seasonal changes, and snow cover conditions on geophysical interpretation of the data so that the limitations of these data can be recognized. Applications are described that exploit the advantages of these data in investigations of ice surface characteristics. Snow and ice surface information obtained using microwave techniques is described and applied further in Chapter 24 in a variety of geophysical studies.

Acknowledgment

This work was performed at the Jet Propulsion Laboratory, California Institute of Technology under contract to the National Aeronautics and Space Administration.

References

1. Fea, M., The ERS-1 ground segment, *ESA Bull.*, 65, 49, 1991.
2. Jezek, K. and Carsey, F. D., Eds., McMurdo SAR Facility, BPRC Technical Rep., No. 91-01, Byrd Polar Research Center, Ohio State University, Columbus, OH, 1991, 31 pp.
3. Ebert, E. E. and Curry, J. A., An intermediate one-dimensional thermodynamic sea ice model for investigating ice-atmosphere interactions, *J. Geophys. Res.*, 98(C6), 10085, 1993.
4. Gordon, A. L., Two stable modes of southern ocean winter stratification, in Chu, P. C. and Gascard, G. C., Eds., *Deep Convection and Deep Water Formation in the Oceans*, Elsevier Science Publishers B. V., Amsterdam, 1991, 17.
5. Gordon, A. L. and Huber, B. A., Southern Ocean winter mixed layer, *J. Geophys. Res.*, 95, 11655, 1990.
6. Carsey, F. D., Barry, R. G., and Weeks, W. F., Introduction, Microwave remote sensing of sea ice, in Microwave Remote Sensing of Sea Ice, Carsey, F., Ed., *Geophys. Monogr.*, American Geophysical Union, 68, 1992, 1.
7. Fu, L. and Holt, B., Seasat views oceans and sea ice with synthetic aperture radar, *JPL Publ.*, 81(120), 1982, 200 pp.
8. Johannessen, O. M., Campbell, W. J., Shuchman, R., Sandven, S., Gloersen, P., Johannessen, J. A., Josberger, E. G., and Haugan, P. M., Microwave study programs of air-ice-ocean interactive processes in the seasonal ice zone of the Greenland and Barents Seas, in Microwave Remote Sensing of Sea Ice, Carsey, F., Ed., *Geophys. Monogr.*, American Geophysical Union, 68, 1992, 261.
9. Roach, A., Aagaard, K., and Carsey, F. D., Coupled ice-ocean variability in the Greenland Sea, *Atmos.-Ocean*, 31(3), 319, 1993.
10. McNutt, L., Argus, S., Carsey, F., Holt, B., Crawford, J., Tang, C., Gray, A. C., and Livingstone, C., The Labrador ice margin experiment, March 1987 — A pilot experiment in anticipation of radarsat and ERS-1 data, *Eos Trans., AGU*, June, 634, 1987.
11. Carsey, F. D., Digby-Argus, S. A., Collins, M. J., Holt, B., Livingstone, C. E., and Tang, C. L., Overview of LIMEX '87 ice observations, *IEEE Trans. Geosci. Remote Sensing*, 27(5), 468, 1989.
12. Drinkwater, M. R. and Digby-Argus, S. A., LIMEX '87; International Experiment in the Labrador Sea Marginal Ice Zone, *Polar Rec.*, 25(155), 335, 1989.
13. Raney, R. K., Argus, S., and McNutt, L., Executive Summary: LIMEX '89, RADARSAT Program Office Publication, August 30, 1989, 13 pp.
14. Carsey, F. D. and Raney, R. K., Eds., LIMEX Special Issue, *IEEE Trans. Geosci. Remote Sensing*, 27, 5, 1989.
15. Askne, J., Ed., The Bothnian Experiment in Preparation for ERS-1, 1988 (BEPERS-88), Special Issue, *Int. J. Remote Sensing*, 13, 2373, 1992.
16. Barber, D., Johnson, D. D., and LeDrew, E. F., Measuring climatic state variables from SAR images of sea ice: the SIMS SAR validation site in Lancaster Sound, *Arctic*, 44(1), 108, 1991.
17. Barber, D., Assessment of the Interaction of Solar Radiation (0.3 to 3.0 mm) with a Seasonally Dynamic Snow Covered Sea Ice Volume, from Microwave (2.0 to 5.0 cm) Scattering, Earth Observations Laboratory Rep., ISTS-EOL-TR93-002, University of Waterloo, Waterloo, Ontario, Canada, N2L 3G1, 1993, 266 pp.
18. ESA, Proc. ERS-1 Results Symp., ESA Special Rep., SP-359, 275, 1992.
19. Ulaby, F. T., Moore, R. K., and Fung, A. K., *Microwave Remote Sensing: Active and Passive*, Vol. 2, Addison Wesley, Reading, MA, 1982, 1064 pp.
20. Ulaby, F. T., Moore, R. K., and Fung, A. K., Eds., *Microwave Remote Sensing: Active and Passive*, Vol. 3, Addison Wesley, Reading, MA, 1982, 2162 pp.
21. Carsey, F. D., Ed., Microwave Remote Sensing of Sea Ice, *Geophys. Monogr.*, 68, American Geophysical Union, 1992, 462 pp.
22. Hallikainen, M. and Winebrenner, D. P., The physical basis for sea ice remote sensing, in Microwave remote sensing of sea ice, *Geophys. Monogr.*, 68, American Geophysical Union, 1992, 29.

23. Onstott, R., SAR and Scatterometer Signatures of Sea Ice, in Microwave remote sensing of sea ice, *Geophys. Monogr.*, 68, American Geophysical Union, 1992, 73.

24. Drinkwater, M. R. and Crocker, G. B., Modelling changes in the dielectric and scattering properties of young snow-covered sea ice at GHz frequencies, *J. Glaciol.*, 34(118), 274, 1988.

25. Livingstone, C. E., Singh, K. P., and Gray, A. L., Seasonal and regional variations of active/ passive microwave signatures of sea ice, *IEEE Trans. Geosci. Remote Sensing*, GE-25(2), 159, 1987.

26. Tucker, W. B., Perovich, D. K., Gow, A. J., Weeks, W. F., and Drinkwater, M. R., Physical properties of sea ice relevant to remote sensing, in Microwave remote sensing of sea ice, *Geophys. Monogr.*, 28, American Geophysical Union, 1992, 9.

27. Drinkwater, M. R., Kwok, R., Rignot, E., Israelsson, H., Onstott, R. O., and Winebrenner, D. P., Potential applications of polarimetry to the classification of sea ice, in Microwave remote sensing of sea ice, *Geophys. Monogr.*, 28, American Geophysical Union, 1992, 419.

28a. Nghiem, S. V., Kwok, R., Yueh, S. H., and Drinkwater, M. R., Polarimetric Signatures of Sea Ice Part I: Theoretical Model, *J. Geophys. Res.*, in press.

28b. Nghiem, S. V., Kwok, R., Yueh, S. H., and Drinkwater, M. R., Polarimetric Signatures of Sea Ice Part II: Experimental Observations, *J. Geophys. Res.*, in press.

29. Winebrenner, D. P., Bredow, J., Drinkwater, M. R., Fung, A. K., Gogineni, S. P., Gow, A. J., Grenfell, T. C., Han, H. C., Lee, J. K., Kong, J. A., Mudaliar, S., Nghiem, S., Onstott, R. G., Perovich, D., Tsang, L., and West, R. D., Microwave sea ice signature modelling, in Microwave remote sensing of sea ice, *Geophys. Monogr.*, 28, American Geophysical Union, 1992, 137.

30. Winebrenner, D. P., Tsang, L., Wen, B., and West, R., Sea ice characterization measurements needed for testing of microwave remote sensing models, *J. Oceanic Eng.*, 14(2), 1989, 149.

31. Drinkwater, M. R., Kwok, R., Winebrenner, D. P., and Rignot, E., Multi-frequency polari- metric SAR observations of sea ice, *J. Geophys. Res.*, 96(C11), 1991, 20679.

32. Rignot, E. and Drinkwater, M. R., Winter sea ice mapping from multi-parameter synthetic aperture radar, *J. Glaciol.*, in press.

33. Mätzler, C., Applications of the interaction of microwaves with the natural snow cover, in *Remote Sensing Reviews*, Vol. 2, Harwood Academic Publishers, 1987, 259.

34. Dierking, W., Sensitivity Studies of Selected Theoretical Scattering Models with Applications to Radar Remote Sensing of Sea Ice, Rep. 33, Alfred Wegener Institute for Polar and Marine Science, Bremerhaven, Germany, November 1992, 113 pp.

34a. Onstott, R. O., personal communication, 1994.

34b. Perovich, D. K. and Richter-Menge, J. A., personal communication, 1994.

34c. Martin, S., personal communication, 1994.

34d. Ulander, L. M. H., Carlström, A., and Askne, J., Effect of frost flowers: rough saline snow and slush on the ERS-1 SAR backscatter of thin Arctic ice, *Int. J. Remote Sensing*, in press.

35. Drinkwater, M. R. and Carsey, F. D., Observations of the late-summer to fall transition with the 14.6 GHz SEASAT scatterometer, Proc. IGARSS '91 Symp., Vol. 3, IEEE Catalog No. CH2971-0, June 3–6, 1991, Espoo, Finland, 1991, 1597.

36. Drinkwater, M. R., LIMEX '87 ice surface characteristics; implications for C-band SAR backscatter signatures, *IEEE Trans. Geosci. Remote Sensing*, 27(5), 501, 1989.

37. Livingstone, C. E. and Drinkwater, M. R., Springtime C-band SAR backscatter signatures of Labrador Sea marginal ice: measurements vs. modelling predictions, in *IEEE Trans. Geosci. Remote Sensing*, 29(1), 1991, 29.

38. Livingstone, C. E. and Drinkwater, M. R., Correction to springtime C-band SAR backscatter signatures of Labrador Sea marginal ice: measurements vs. modelling predictions, *IEEE Trans. Geosci. Remote Sensing*, 29(3), 1991, 472.

39. Winebrenner, D. P., Nelson, E. D., and Colony, R., Observation of melt onset on multiyear Arctic sea ice using the ERS-1 SAR, *J. Geophys. Res.*, submitted.

22

Sea-Ice Studies with Scatterometers

Alain Cavanié and Francis Gohin

22.1 Introduction

Satellite-borne scatterometers, such as those of Seasat, ERS-1, or NSCATT, are flown with the principal objective of determining ocean surface winds (in force and direction) with sufficient accuracy to contribute useful data to meteorological models; furnish stand-alone climatologies with which to compare such models; and offer detailed descriptions of meteorological phenomena such as cyclones, which are difficult to model with accuracy. Such instruments, which vary in their choice of frequency, polarization, and footprint size, must in any case share certain common characteristics. They must have at least two or much preferably three antennas having different azimuth angles in order to separate out wind directions; the error in their backscattering measurements is necessarily of only a few percent, and they must cover large areas of the ocean surface in limited time.

These qualities will make them particularly suited to monitor the evolution of sea ice at polar ocean scales. Offering footprint sizes on the order of 25–50 km, they are certainly no match for Synthetic Aperture Radars (SARs) as instruments for detailed scrutiny of the regional behavior of sea ice; however, they may help to integrate such regional studies in a global view.

Much work has been conducted in past years with airborne scatterometers and some with that of Seasat over sea ice to investigate the behavior of the backscattering coefficient, σ^0 (ratio of the returned power per unit square angle in the direction of the transmitting antenna to the power incident on the surface per unit area).[1,2] Since this parameter is roughness and snow cover as well as incidence angle dependent, an attempt to comprehensively predict σ^0 without *in situ* information appears at present far from reach.[3] Rather, based on the extensive data set so far obtained by the Advanced Microwave Instrument (AMI)-WIND which is the C-band (5.3 GHz) VV-polarized scatterometer of the European Space Agency (ESA) ERS-1 satellite, we will discuss different remote sensing applications of satellite-borne scatterometers over sea ice, based on the empirical knowledge acquired by the analysis of such data.

22.2 Detecting Sea Ice Before Wind Processing

Backscattering from open water areas of the ocean, at a given incidence angle, varies considerably (>10 dB) as a function of wind speed and wind direction; it also varies over different types of sea ice at different seasons in about the same proportions.[1] Thus it is not surprising that backscattering signatures from sea ice and open water, even when data from two or more antennas are used in combination, may at times look very much alike. Seasat wind data suffered from such erroneous wind estimations and, as of May 1993, ERS-1 fast delivery product (FDP) data was not screened for such errors although the algorithm described hereafter, based on the anisotropy coefficient, is being evaluated to exclude sea ice zones from the wind processing and could be used in the near future.[4] Off-line processing of the ERS-1 scatterometer data at the French Processing and Archiving Facility Centre ERS-1 d'Archivage et de Traitement [CERSAT] presently uses sea surface temperature files that determine sea ice zones mainly from climatological data; the feasibility of using an ice-detection algorithm based on the important difference between derivatives of σ^0 with respect to incidence angle over sea ice and open water is under study. It might be thought that the difference in backscattering between water and ice surfaces could be used directly to detect the water-ice transition. Although this method can be applied locally in time and space, it requires human interpretation in view of the important variability of backscattering over both ice and water, particularly in summer months, when large puddles of water are formed on the ice. This has led to the development of the two methods hereafter described, which draw on other properties of backscattering from ice and water.

For fast delivery products that are to be produced in near-real time, throughput is essential; and the ice-detection algorithm employed must be simple to implement and maintain, and quick to execute. First, this leads to determining from analysis of the Electrically Scanning Microwave Radiometer (ESMR) data, monthly geographic zones in both hemispheres where sea ice occurs.[4] Only inside these zones will a test for possible sea-ice cover be performed.

This test is based on the fact that backscattering from sea ice is quite isotropic with respect to azimuth angle; since ERS-1 is yaw steered, the fore and aft beams of the scatterometer, which point at azimuth angles of 45° and 135° to the right of the ground track, reach the sea surface at incidence angles, which are identical for all practical purposes.[5] The anisotropy coefficient, A, is defined as:

$$A = (\sigma_1 - \sigma_2)/(\sigma_1 + \sigma_2)$$

where σ_1 and σ_2 are the σ^0 of the fore and aft beams. It would be extremely small if measurement noise, due mainly to speckle, could be eliminated; since measurement noise varies between 4 and 6% of the measured σ^0, it can be shown that the standard deviation of A varies correspondingly from about 2.8 to 4.3%. Having stored the absolute values of A measured during several previous days in a polar grid mosaic having roughly 25- × 25-km elements, the decision is taken to accept as open water a point of measurement if at least one of the values in the corresponding mosaic element is superior to a given threshold. Test results indicate that about 80% of the points over open water are detected as such, and less than 4% over zones of concentrated ice.[6] That results are not better over open water is due to the fact that fore and aft beams give nearly identical σ^0 when winds blow either in along-track or cross-track directions. Elimination of nearly all erroneous wind measurements over sea ice would have been possible if data from July, the month during which puddles of water form at the ice surface, had been eliminated.

The off-line algorithm presently under evaluation is quite different from that previously described. It is based on the fact that for moderate incidence angles, around 253, the derivative of σ^0 with respect to incidence angle is much smaller (i.e., σ^0 decreases much more rapidly with increasing incidence angle) over open water than over sea ice. Since off-line processing is conducted over several days of data at a time, it is possible to first use the data set to map this parameter determined by measurements at a given distance offtrack. Since ERS-1 has three antennas, pointing at 45°, 90° and 135° to the right of the ground track, incidence angles of the central beam are some 10° smaller than those of the side beams, which allows a direct estimation of the derivative. Kriging, an objective interpolation method, is applied to the available data to transform information, initially on line segments, into a quasi-continuous mapping.[7] A region is declared open water if the resulting estimate of the σ^0 derivative with respect to incidence angle is less than a given threshold value. Comparison with passive microwave data of the ATSR-M, also flown on ERS-1, gives quite comparable results concerning sea ice boundaries.

22.3 Detection of Different Ice Types

It is well known that σ^0 are generally higher over multiyear than over first-year sea ice, exceptions to this rule occurring during the summer thaw.[8] This is due to the reworking of ice as it ages by several different processes: decrease of its salinity

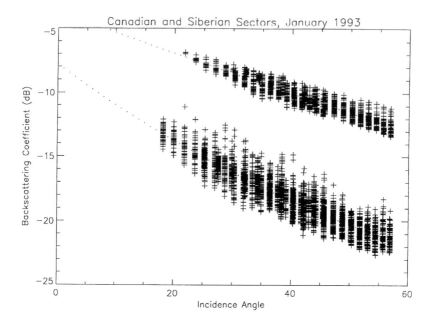

Figure 22.1   Shows σ^0 as a function of incidence angle in January 1993, over the Canadian (multiyear ice) and Siberian (first-year ice) sectors. Measurements from the three-beam data of ERS-1 AMI-WIND, as given by ESA fast delivery products.

through the downward drift of brine pockets that leave air holes in their track and increase the volume backscattering, roughening of the surface at small scales by thawing and refreezing of the snow cover water, and at large scales buckling under excessive pressure forces. As an illustration of this, Figure 22.1 shows measurements made with all three antennas of the ERS-1 scatterometer in a region of multiyear ice (84°–88°N, 30°–60°W, so-called Canadian sector), as well as in a region of first-year ice (72°–75°N, 152°–170°E, so-called Siberian sector). These geographic zones were chosen after consulting the information extracted from the ESMR passive micro-wave data that show the Siberian sector to be free of ice in summer and the Canadian sector to be covered with multiyear ice at all times of the year.[4] The σ^0 from multiyear ice are some 7–8 dB above those from first-year ice; moreover, they display in decibels, a nearly linear relationship with incidence angle, while some curvature is apparent in the cloud of first-year ice measurements.

Multiyear ice measurements also show a closer grouping around their mean curve than the first-year measurements, skewed toward higher values. This is an indication that some of the mechanisms involved in the transformation from first-year to multiyear ice are already at work in the first months following the formation of ice.

Behavior of first-year and multiyear ice as described in Figure 1 are observed over the other regions of the Arctic as well as the Antarctic oceans, from the time of late summer/early autumn freezeup to that of the late spring/early summer surface

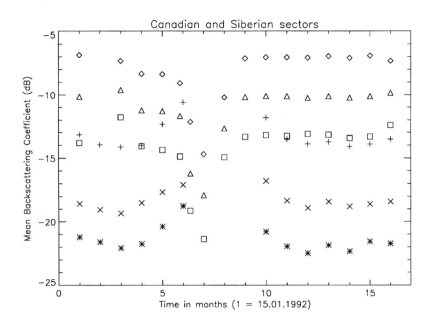

Figure 22.2   Mean values of σ^0 over the Canadian and Siberian sectors, at 20°, 40° and 60° incidence angles; time in months (1 = 15.01.1992). For the Canadian sector, diamonds, triangles, and squares represent σ^0 at 20°, 40°, and 60° incidence angles, respectively; these are replaced correspondingly by +, ×, and * signs for the Siberian sector.

thaw. During this period σ^0 at given incidence angles may be used to distinguish regional variations of ice types.

22.4 Seasonal Variations of Backscattering

Because good wind measurements require that satellite-borne scatterometers be very stable instruments, devoid of long term drift, they are ideal for observing the seasonal evolution of backscattering over sea ice at different incidence angles, on regional to basin-size scales.

Although many different techniques could be applied to arrive at similar results, we have chosen to estimate mean values of σ^0, adjusting by the least squares method, a second-order curve through the data and adopting the value on the curve at the desired incidence angle.

Figure 22.2 shows the evolution of the σ^0 thus determined from both the Canadian (multiyear ice) and Siberian (first-year ice) sectors from January 1992 to July 1993. In February 1992, too little data were available over the Canadian sector to be represented; and the presence of significant open water areas, detected by use of the anisotropy coefficient as previously described, in July through September over the Siberian sector, led to the exclusion of data during that period. Although the points generally represent monthly means, an exception was made for the month of June over the Canadian sector; in this case data from June 1 to 21 was treated separately from that from June 22 to 30, because a rapid change was observed on the 22nd. This explains the two sets of points straddling June 15. The evolutions of σ^0 over first-year and multiyear ice zones follow opposite directions: increasing in the early spring and decreasing in autumn over the Siberian sector and decreasing in late spring and increasing in late summer over the Canadian sector; time shifts observed are certainly due to the fact that the Canadian sector is 12° farther north than the Siberian, which reduces the period of relatively warm weather. The dramatic decrease of σ^0 during summer is linked to the large areas of ice covered by puddles of water during this period.[7]

The nearly linear relationship between σ^0 (in decibels) and incidence angle over multiyear ice, as well as the positive curvature detectable for first-year ice on Figure 22.1, can be followed from month to month on Figure 22.2. For the Canadian sector, the σ^0 at 40° are nearly equidistant from those at 20° and 60°, while they are much closer to the values at 60° for the Siberian sector. This difference in behavior can be attributed to the importance of volume scattering in multiyear ice; we suggest that it could perhaps be used in the calibration of sea ice backscattering models. Although only a little over a year of data has been accumulated since calibrated ERS-1 scatterometer data became available from ESA, it is clear that intercalibration of instruments of this type (of course, at the same frequency and polarization) could be readily conducted over multiyear ice regions that show very stable backscattering behavior (stable to 0.1 dB) over most of the year; this suggestion would find an ideal application if ERS-2 was launched while ERS-1 is still operational. Difficulties may otherwise arise, since it is known that multiyear ice floes drift in space and time.

22.5 Ice Drift Measurements at Basin Scales

Ice drift at different time and space scales can be, and has been, measured using successive, precisely located, images of different instruments (SAR, SLAR, AVHRR, etc.). The important dynamic range in backscattering between smooth first-year and rough multiyear ice, and the high accuracy of the ERS-1 scatterometer measures (4–6% standard error) suggest that characteristic ice features could be followed over several months if suitable images were produced.

Generating such images implies that the strong dependence of backscattering on incidence angle be removed, in order to use the data over an important part of the swath. The procedure we have used is to bin data of several days in the National Snow and Ice Data Center (NSIDC) roughly 25×25 km, polar grid.[9] In each grid element the derivative of σ^0 (in decibels) with respect to incidence angle is evaluated from three-beam measurements and a linear correction, using this derivative, applied to the individual backscattering values to correct them to a nominal incidence angle of 40°; to reduce the effect of curvature, only lines 5 to 14 of the 19-line swath are used.

Figure 22.3* shows σ^0 over the Arctic Ocean in November 1992, thus evaluated at 40° incidence angle. Open water areas, detected using the anisotropy coefficient, are marked in brown. Because the σ^0 are given in linear units and not in decibels, color scale ranges from 0.0 to 0.17. The general features of the image correspond to what can be extracted from passive microwave images made by the Scanning Multichannel Microwave Radiometer (SMMR) or Special Sensor Microwave/Imager (SSM/I) in previous years: north of the Siberian coast new ice has formed that is driven northward by prevailing winds while multiyear ice accumulates north of Canada and Greenland. Presently, awaiting SSM/I data for the same period, comparisons are somewhat subjective, but will be investigated further to see whether the passive and microwave data are redundant or complementary.

Several significant features, most of them in first-year ice on the scale of 100–300 km, have been identified and can be followed from late summer into late spring when the general thaw eliminates differences in backscattering between ice types. These features could occur, either due to buckling of the ice under compression, or from the drift of large floes of multiyear ice into the first-year ice region. From scrutiny of similar images in previous months, we believe that the roughly C-shaped feature north of Novosibirskiye Island was formed by the second mechanism. Initially spotted in mid-August touching the northern shore of the island, its drift was still followed as of May 1993. The spot of strongest backscatter, at 76.2°N, 141.7°E in early October 1992 reached 79.7°N, 158.0°E in early April 1993; this corresponds to about 440 km as the crow flies.

22.6 Directions for Future Work

Microwave brightness temperatures (such as furnished by the SSM/I) are certainly far better adapted than backscatter from satellite-borne scatterometers to measure ice concentrations; this is due to the significant brightness temperature difference between water and ice surfaces. However, backscatter data give information on surface and volume scattering, which is linked to quite different physical properties of the ice. Work must be pursued to use these two sources of data jointly. For this, long-term measurements in late spring and summer of ice surface temperatures and water puddle areal extent will be necessary.

By the same token, SARs give an excellent small-scale description of the ice and its motions; but such studies should be nested in the longer term, larger scale information coming from scatterometers and radiometers. Lagrangian drifters have, in past years, been an excellent source of data in the Arctic; no doubt their results should be compared to and integrated with the ice drift seen with successive scatterometer images. Finally, we must await impatiently the flight of NSCATT whose data, combined with that of the ERS satellites, will offer an opportunity to study different frequency and polarization combinations over sea ice.

*Color plates follow page 404.

References

1. Ulaby, F.T., Moore, R.K., and Fung, A.K., Radar measurements of sea ice, in *Microwave Remote Sensing*, Vol. 3, Artech House, Norwood, 1986, chap. 20–4.
2. Carsey, F.D. and Pihos, G., Beaufort-Chuchki seas summer and fall ice margin data from seasat: conditions with similarities to the Labrador sea, *IEEE Trans. Geosci. Remote Sensing*, 27, 541, 1989.
3. Livingstone, C.E. and Drinkwater, M.R., Springtime C-band SAR backscatter signatures of Labrador sea marginal ice: measurements versus modeling predictions, *IEEE Trans. Geosci. Remote Sensing*, 29, 29, 1991.
4. Parkinson, C.L., Comiso, J.C., Zwally, H.J., Cavalieri, D.J., Gloersen, P., and Campbell, W.J., Arctic sea ICE 1973–1976: Satellite Passive-Microwave Observations, NASA SP-489, 1987.
5. Cavanié, A. and Gohin, F., Interpretation of ERS-1 scatterometer data over sea ice, in *Proc. ERS-1 Geophys. Validation Workshop*, European Space Agency, Paris, 1992.
6. Lecomte, P., Cavanié, A., and Gohin, F., Recognition of sea ice zones using ERS-1 scatterometer data, presented at IGARSS'93, Tokyo, August 18–21, 1993.
7. Matheron, G., The theory of regionalized variables and its applications, Ecole des Mines de Paris, 1971.
8. Onstott, R.G. and Gogineri, S.P., Active microwave measurements of Arctic sea ice under summer conditions, *J. Geophys. Res.*, 90, 5035, 1985.
9. DMSP SSM/I Brightness Temperature and Sea Ice Concentration Grids for the Polar regions on CD-ROM, User's Guide, NSIDC, Special Rep. 1, Boulder, CO, 1992.

23

Satellite Remote Sensing of Ice Motion

William J. Emery, Charles W. Fowler, and J. A. Maslanik

23.1 Introduction

Displacement of sea ice over time is perhaps the most distinctive feature in satellite images of the polar ice cover. Whereas in the past the only means of determining ice motion was by direct observation from camps on the ice or from ice-bound ships and more recently using satellite-tracked buoys,[1] satellite remote sensing now permits operational tracking of ice movement over the entire Arctic and Antarctic oceans, with a spatial coverage and detail not possible using field observations.

Accurate tracking of sea ice motion on these spatial and temporal scales can play important roles in applications ranging from climate studies to ship operations. Since ice movement is an integral part of ice-ocean-atmosphere interactions in the polar regions, ice displacement is a key parameter in numerical models of polar processes and climate. Remotely sensed ice motions are a useful tool for verifying model performance. In turn, combinations of satellite data with model output improve prediction of ice movement that can assist in planning for shipping and offshore oil drilling operations in the polar regions.

Since sea ice moves relatively quickly (typically tens of kilometers per day in some areas), the key requirement of a remote sensing instrument for ice-motion mapping is frequent repeat coverage. The extensive cloud cover and low sun angles in polar regions place further limitations on remote sensing. Within these general requirements, the types of satellite instruments available offer a unique combination of advantages and limitations. Visible-band and infrared sensors onboard meteorological satellites, such as the Advanced Very High-Resolution Radiometer (AVHRR) on the Television Infrared Observing Satellite (TIROS) platforms, cover the polar regions several times per day with fields of view of approximately 1.1 km at nadir and 3.5 km at the swath edge. Spatial coverage of the polar regions with AVHRR is excellent, with the overlap of the passes at polar latitudes.

The main limitation of such visible and thermal-band sensors for ice-motion mapping is cloud cover. In polar regions where cloud cover is extensive and persistent, a cloud-free pair of images may not be available for a particular area for weeks at a time. Visible wavelength channels are of course only usable during sunlit times from spring to fall. While the thermal channels record brightness temperatures year round, they may not be useful during parts of the summer when the open-ocean and sea-ice surface temperatures are similar. In general then, only thermal images are suitable for winter mapping, while visible-band images can be used during the summer, and possibly both during parts of spring and fall months.

Active microwave instruments such as Synthetic Aperture Radar (SAR) carried onboard ERS-1 and JERS-1 and planned for Radarsat provide much higher spatial resolution (tens of meters) and can sense the surface in virtually any type of weather. Given the relatively high spatial resolution of the instrument, SAR data are particularly applicable to the study of small-scale ice movement in areas such as the marginal ice zones and deforming ice. However, due to the large data volume and the associated processing requirements, SAR imagers typically collect data along a fairly narrow swath (hundreds of kilometers vs. thousands of kilometers for the AVHRR). The frequency of sampling individual locations is therefore less, although the increase in overlap among orbits approaching the pole, as well as variations in orbit configurations, can provide relatively frequent coverage, such as the 3-d repeat cycle of ERS-1.[2]

Like radar, passive microwave sensors such as the Special Sensor Microwave/Imager (SSM/I) onboard the Defense Meteorological Satellite Program (DMSP) satellites offer the advantage of year-round operation under all atmospheric conditions. However, passive microwave images from satellites typically have spatial resolutions of tens of kilometers — insufficient to resolve the individual leads and floes needed to track the motion of specific ice features. Large masses of ice with unique emissivity properties can potentially be tracked over time, aided by the frequent temporal sampling and multispectral capabilities of such sensors.

For some applications, such as climate studies comparing the response of the ice pack over different space and timescales, it may be advantageous to combine ice motions from several of these different image types, together with ice motions determined from drifting buoys. This is particularly true for studies involving sea ice models, where model scales may vary from tens of kilometers for mesoscale models to hundreds of kilometers for general circulation models that incorporate sea ice dynamics. In these cases, SAR data may be used to define local processes, while regional and synoptic-scale processes are determined from AVHRR imagery.

With this introduction in mind, the following sections describe some of the basic considerations involved in determining ice motion from satellite data, and introduce some of the different techniques used to calculate ice velocities.

23.2 Techniques

Image Registration

Assuming that a pair of images has been assembled that cover approximately the same area over a span of a few days, the first processing issue that must be addressed to extract sea-ice motion is image registration or geolocation. All satellite imagery has some positioning error associated with the earth location of each pixel in the imagery. This error arises from uncertainty in orbital modeling, the precise time the data was collected by the satellite, modeling of the shape of the earth, and knowledge of the pointing accuracy of the instruments on the satellite. In the first stage of the geolocation process, satellite time-and-position (ephemeris) data typically obtained from government tracking sources are supplied to an orbit model to calculate the earth location of pixels in the imagery. Error is introduced by approximations within the orbital model, inaccuracies within the original ephemeris, and unmodeled changes in the satellite positioning between image acquisition time and the time the ephemeris was obtained. As this time gap between ephemeris and image acquisition increases, the error introduced by the model also increases and can become substantial. Some satellites such as the TIROS platforms use an onboard clock that is subject to drift and is corrected at intermittent intervals. Since this clock-supplied time from the satellite may be wrong, the prediction of the satellite position then will be incorrect. In addition, the pointing angles of the instrument (e.g., roll, pitch, and yaw) may not be completely known.

The shape of the earth is modeled typically as an oblate spheroid, when satellite data are projected to an approximation of the earth surface. Since the shape of the earth does not exactly fit the oblate spheriod model and map projections are local approximations of a spherical earth, additional errors are introduced in this process.

The residual error from this initial orbit-modeling geolocation can be reduced further by shifting or warping images based on specific unchanging surface features such as coastlines or islands identifiable on each image. Using these control points, registration error can typically be reduced to less than one pixel width. Even for sea ice images over open ocean, the lack of immediate control points can be compensated for, by using control points elsewhere in the orbit to compute the spacecraft attitude. These attitude parameters can then be propagated in time to produce one pixel accuracy in images without nearby control points.

A final consideration in the geolocation process is that the images should be projected to a map grid that is close to equidistant. Then, by knowing the pixel resolution, any displacement of the ice in pixel values can be simply converted into velocities over the entire image. The most common projection is the modified polar stereographic projection that, while not a true equidistant projection, is suitable for most polar applications.

Considerations of Error and Application on Choices of Imagery

The desired application of ice-displacement products determines the types of imagery used for ice-motion mapping. The choice of imagery is driven by the required precision of the ice motions, as well as the prevailing lighting and cloud cover conditions during the period of interest.

Since ice motion computed from satellite imagery represents the displacement between the acquisition times of the two images, these motions do not represent instantaneous velocities. Changes in the speed and direction of the ice can occur between the times the two images are acquired, especially when the length of time between images becomes greater. In turn, the spatial resolution of the imagery and the geolocation accuracy control the precision with which ice displacements can be mapped. Figure 23.1 describes this relationship between spatial resolution and temporal sampling. For example, for AVHRR data at 1.1 km nadir resolution and 1 d between image pairs, a velocity resolution of about 1 cm/sec is possible. The 4-km resolution global area coverage (GAC) AVHRR data allow velocity resolutions of slightly better than 5 cm/sec for images separated by 1 d. SAR data at 100-m resolution with 3 d between image pairs would give very fine velocity resolution better than 0.25 cm/sec. As can be seen, a trade-off exists between resolution constraints and time difference between images. Figure 23.1 also describes the effects of registration error. For example, using the case of AVHRR data with a 4-km resolution and assuming a coregistration error of one half pixel and 50 h between images, an error of +/− 1 cm/sec is possible for calculated vectors.

In addition to these considerations relating to required precision and temporal sampling, the initial conversion of satellite radiances or backscatter to representations of surface albedo, temperature, roughness, or emissivity properties affect the suitability of specific sensors for the desired application. Fortunately, these image calibration and postprocessing issues affect ice-motion mapping only in terms of the ability to detect sea ice features. The only processing that may be needed is to correct for sun angle, which in turn is only necessary for visible-band data acquired at different times of day. Since the radiometric relationships between spectral channels and between radiances and geophysical features are not important, image enhancement routines can be applied to assist in identifying ice leads and floes.

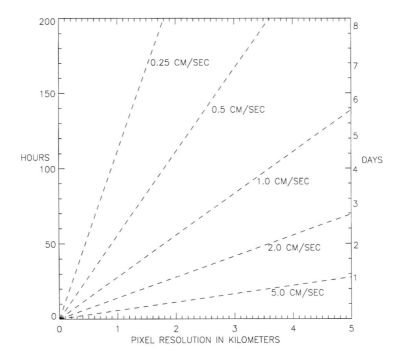

Figure 23.1 Graph showing the relationship of image resolution, time difference between images, and calculated velocity resolution.

Motion Detection: Manual Tracking Methods

Although time-consuming and limited by the patience and care of the interpreter, manual tracking of ice movement can provide a valuable overall impression of ice motion, including general direction of ice movement useful for shipping operations and initial estimates of the sea ice response to atmospheric forcings. In this approach, a feature in the ice pack such as an ice floe or lead is identified in both images. By finding the pixel displacement and knowing the pixel resolution and time span between images, velocity vectors can be computed. An early example of this approach tracked individual ice floes for about 1 month in 1973, and revealed some of the dominant circulation patterns near the east Greenland coast.[3] Using modern workstations and display software, image enhancements and image-looping or flickering procedures are particularly useful tools for manual detection and tracking of ice displacements.

Spatial Correlation Methods

While manual methods are suitable for some applications and for a first look at ice-atmosphere interactions, automated methods are clearly desirable. An effective method for tracking displacements is the use of correlations between sets of pixels

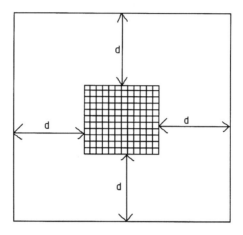

Figure 23.2   Figure showing the search and target areas for correlation.

between image pairs. A small target area in one image is correlated with many areas of the same size in a search region of the second image. The displacement of the ice is then defined by the location in the second image where the correlation coefficient is the highest. Although this process can be automated quite easily, it makes the basic assumption that the sea-ice cover can be treated as a solid, with the ice movement only a translation of this solid plane (e.g., ignoring deformation and rotation of the ice). This assumption is generally valid over short distances away from ice edges such as the marginal ice zone where the ice is less constrained.

The selection of the sizes of the target and search windows in the images depends on several factors. The size of the target window cannot be too large to negate this solid-plane assumption, but the window size must be large enough so that the correlation still has some statistical significance. The size of the search window is dependent on the expected displacement of the ice (Figure 23.2). The inside box is the target window and d is the maximum distance in pixels that the ice is expected to move. Cross-correlations are then computed at each possible location within the search window. An example of the cross-correlation values for a typical case (Figure 23.3) show a peak where the maximum correlation coefficient denotes the location of the ice movement. This method, typically referred to as a maximum cross-correlation (MCC) procedure, is widely used,[4,5] including applications to both AVHRR and SAR data.[6-8]

Since the number of correlations done in the MCC process is $(2d + 1),^2$ computation time increases exponentially as the value for d (the maximum ice movement distance) increases. If ice displacement is on the order of tens of pixels, computation time can run into several hours or days using typical workstation computers. One approach to speed the calculations is through Fast Fourier transforms (FFT). The FFT of the target window is multiplied by the complex conjugate of the FFT of the search window. The inverse FFT of the product will then show the lag or the amount of displacement of the ice. For maximum efficiency, the windows should

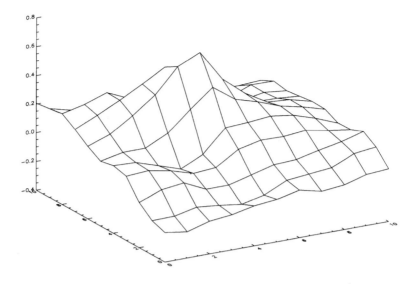

Figure 23.3 Plot of correlation values from typical cross-correlation calculations of search and target windows.

be powers of two in size and the target window must be padded to the same size as the search window. The padding and aliasing that occur with the use of FFTs can introduce some error, but are not usually significant. However, if any masking is done, such as of land masses or clouds, FFTs may not be appropriate because all values in the image are used; and the masked portions may force the ice velocity vector to a wrong value. The direct method of calculating the correlation coefficient is preferable in these cases.

A second approach to reducing computation time is by applying some prior knowledge about likely ice conditions and motion. With this knowledge in hand, the search window can be reduced in size. If such knowledge is not available initially, reduced-resolution images can be used to supply an low-resolution estimate of ice velocities that can then be used as starting values, or seed points, in the full-resolution images. In this nested correlation approach,[9,10] very small search windows can be used.

Another method of speeding up the process is to calculate cross-correlations at a reduced number of locations in the search window through an iterative process in which the maximum correlation coefficient is found and the size of the search window is reduced at each step. Looking again at Figure 23.3, the peak could be converged upon in just a few steps. However, since multiple peaks may be present, care must be taken to ensure that the method converges at the correct peak. Initial knowledge about likely ice displacement is also available from wind fields, which can explain as much as 70% of the variability in the ice movement.[11] Gridded wind fields supplied by forecast models such as the National Meteorological Center (NMC) and European Center for Medium-Range Weather Forecasts (ECMWF)

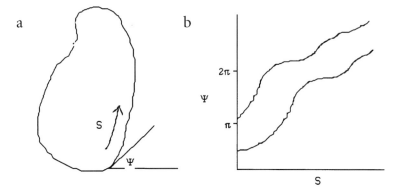

Figure 23.4   Given an arbitrary shape (a), a Ψ-S curve (b) can be generated. From these curves, rotation of shapes can then be calculated.

can be used to estimate ice displacement as a seed value to reduce the search window size as described above.

Nontranslational Ice Displacement Methods

The assumption made earlier that ice acts as a solid plane clearly does not hold for all conditions. In areas of fast moving ice, in areas where ice is deforming, or in the ice margins near open ocean, the ice may deform and/or rotate. If this more complete knowledge about the dynamics of the ice is needed, simple spatial correlation methods may not be able to find matches between images, and other more involved methods must be applied.

Most of these more complicated methods are essentially extensions of shape-recognition techniques. In general, images are first segmented into regions, and the characteristics of individual ice features are determined. A Ψ-S correlation is one method used to define a curve for the shape of the individual ice floes.[12] At point S along the boundary of the ice floe, the tangent angle Ψ is determined (Figure 23.4a). The curve of each segment in the first image is then compared with curves of all likely matches in the second image using correlations. If the shapes are the same, but rotated, the curves will only have a bias and a shift between the lines (Figure 23.4b). The rotation can then be calculated from this shift, with the distance between the centers representing the displacement. Another similar method segments the image into ice and water, computes seven invariant moments for each ice floe segment, and then proceeds with comparisons.[13]

An Example: AVHRR-Derived Ice Motions in the Beaufort Sea

To illustrate some of the basic processing steps needed in operational ice-motion detection, we consider the problem of estimating ice velocities from meteorological

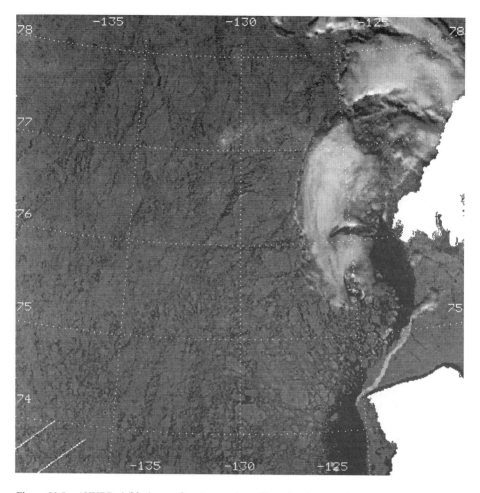

Figure 23.5 AVHRR visible-image showing portion of Beaufort Sea on July 10, 1992. Image covers area of about 500 × 500 km centered at 76°N and −130°W.

satellite imagery. Figures 23.5 and 23.6 are AVHRR thermal-channel images in the Beaufort Sea acquired 3 d apart in late fall (images are about 500 km square and centered at 76°N latitude and −130°W longitude, with a pixel spacing of 1 km). Since cloud motion must somehow be distinguished from sea-ice motion, a maximum likelihood classification is first used to detect and mask out cloud-covered areas along with land areas. Unlike applications such as sea surface temperature calculations where pure clear sky pixels are desired, cloud-contaminated pixels can be used in the ice-motion calculations as long as some surface features are visible.

The importance of cloud screening is demonstrated by applying the direct method of calculating the ice displacements (target window of 10 × 10 and a maximum displacement of 5 pixels) from the images prior to cloud detection. A fairly smooth field of ice velocities can be seen in the middle of the resulting

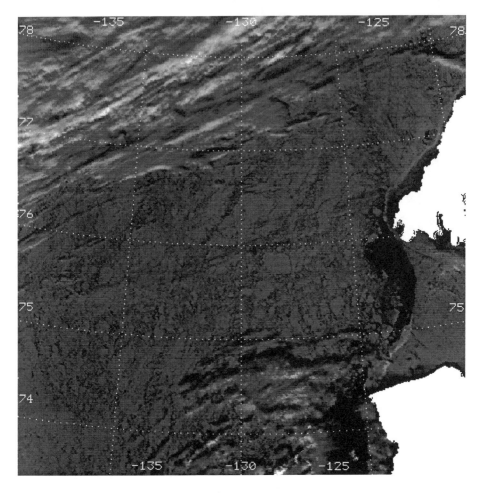

Figure 23.6   Same as Figure 5, except on July 11, 1992.

motion-vector plot (Figure 23.7), among other more chaotic vectors representing changes in cloud cover. Using the land and cloud masked images, the signal-to-noise ratio is improved, although many spurious vectors still remain along cloud edges. A spatial coherence filter[5] and thresholding of the correlation coefficients can also be applied to distinguish realistic ice-motion vectors from cloud changes (Figure 23.8). This filter assumes that the ice velocity field is fairly coherent locally. The filter compares neighboring vectors, and if displacement differences between the center vector and the eight surrounding vectors are less than two pixels, then the center vector is accepted as a good vector. The filter parameters are changed depending on the ice conditions and location. While the spatial coherence filter alone could potentially be used to remove cloud-related vectors (or alternatively, be used as a cloud-detection procedure), a prior screening of the images to mask out cloudy areas generally improves the resulting product.

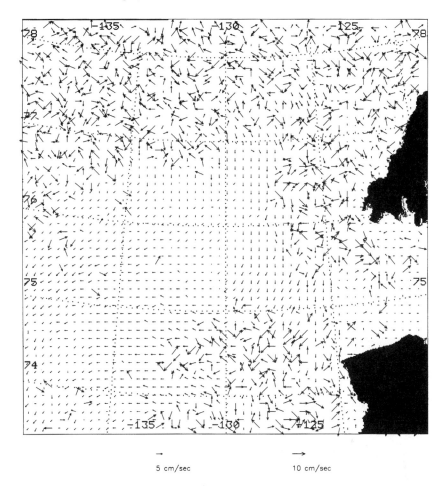

Figure 23.7   Plot of all calculated vectors from the July 10 and 11 AVHRR images.

Operational Ice-Motion Products from SAR Imagery

Given the massive data storage and processing requirements imposed by spaceborne SAR systems, centralized facilities have been established in different countries to generate and distribute SAR products to users. One such center, the Alaska SAR Facility (ASF) in Fairbanks, Alaska, routinely processes ERS-1 and JERS-1 SAR data into ice-motion vectors (as well as other geophysical products)[14-16] for distribution to data users. The ASF ice-vector product is generated using the basic spatial correlation and the Ψ-S correlation methods described above. Many different filtering techniques are used, and each vector has associated with it a confidence value. The vector product is distributed along with mean wind information for the area.

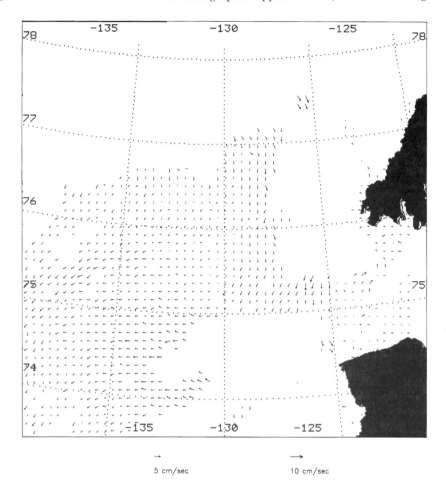

5 cm/sec   10 cm/sec

Figure 23.8   Plot of vectors after using two filtering techniques: first, a cutoff value was assigned to the correlation coefficients; and second, a nearest-neighbor filter was used.

23.3 Summary

Satellite imagery from a variety of sensors can be used to detect sea-ice motion; the choice of data depends on the precision and detail with which ice displacements must be mapped, and the prevailing environmental conditions such as cloud cover and solar illumination.

Synthetic aperture radar data are particularly useful for accurate, localized mapping of ice motion under all conditions, while imagery from meteorological sensors such as AVHRR provide broader and more frequent estimates of ice motion, but having lower precision and accuracy and being subject to cloud cover. Automated ice-displacement algorithms operate by matching the spatial relationships of leads and floes in the ice pack over time through simple cross-correlations of pixel values in moving windows, or by matching individual features. Since ice-motion mapping is based on relative contrast between open-water or thin-ice features and thicker sea ice rather than on absolute reflectances or backscatter, ice-tracking routines are fairly forgiving of data quality. Precise instrument calibration and spectral classifications are not required, and useful ice motions can be calculated from partially cloud-obscured areas and from images with cloud shadows and varying illumination angle. With the advent of centralized data processing and distribution centers such as the Alaska SAR Facility, satellite-derived ice motions are now being supplied routinely to users for tasks ranging from climate-change studies to ship operations.

References

1. Thorndike, A. S. and Colony, R., An estimate of the mean field of Arctic Sea ice motion, *J. Geophys. Res.*, 89, 10,623, 1984.
2. Dade, E. F., Rothrock, D. A., Colony, R., and Olmsted, C., Estimating Repeat Coverage of Arctic Sea Ice with ERS-1 SAR, APL Technical Rep., APL-UW TR9114, July 1991, 49 pp.
3. LaViolette, P. E. and Hubertz, J. M., Surface circulation patterns off the East Coast of Greenland as deduced from satellite photographs of ice floes, *Geophys. Res. Lett.*, 2(9), 400, 1975.
4. Ninnis, R. M., Emery, W. J., and Collins, M. J., Automated extraction of pack ice motion from advanced very high resolution radiometer imagery, *J. Geophys. Res.*, 91(C9), 10725, 1986.
5. Emery, W. J. and Fowler, C. W., Fram Strait satellite image-derived ice motions, *J. Geophys. Res.*, 96(C3), 4751, 1991.
6. Fily, M. and Rothrock, D. A., Extracting sea ice data from satellite SAR imagery, *IEEE Trans. Geosci. Remote Sensing*, GE-24(6), 849, 1986.
7. Collins, M. J. and Emery, W. J., A computational method for estimating sea ice motion in sequential seasat synthetic aperture radar imagery by matched filtering, *J. Geophys. Res.*, 93(C8), 9241, 1988.
8. Vesecky, J. F., Samadani, R., Smith, M. P., Daida, J. M., and Bracewell, R. N., Observation of sea-ice dynamics using synthetic aperture radar images: automated analysis, *IEEE Trans. Geosci. Remote Sensing*, 26(1), 38, 1988.
9. Fily, M. and Rothrock, D. A., Sea ice tracking by nested correlations, *IEEE Trans. Geosci. Remote Sensing*, GE-25(5), 570, 1987.
10. Fily, M. and Rothrock, D. A., A Fortran Program for Digital Ice Tracking by Cross-Correlation, Applied Physics Laboratory, Informal Document Series APL-UW 1-88, January 1988, 1.
11. Thorndike, A. S. and Colony, R., Sea ice motion in response to geostrophic winds, *J. Geophys. Res.*, 87(C8), 5845, 1982.
12. McConnell, R., Kwok, R., Curlander, J. C., Kober, W., and Pang, S. S., Ψ-S correlation and dynamic time warping: two methods for tracking ice floes in SAR images, *IEEE Trans. Geosci. Remote Sensing*, 29(6), 1004, 1991.
13. Zhang, H. J., Automatic Tracking Ice Floe From Satellite Imagery Via Invariant Moment Matching, IEEE Proc., IGARRS92, 1992, 582.
14. Kwok, R. and Pang, A., Performance of the Ice Motion Tracker at the Alaska SAR Facility, IEEE Proc., IGARRS92, 1992, 588.
15. Kwok, R., Curlander, J. C., McConnell, R., and Pang, S. S., An ice-motion tracking system at the Alaska SAR facility, *IEEE J. Oceanic Eng.*, 15(1), 44, 1990.
16. Kwok, R. and Cunningham, G., Alaska SAR Facility-Geophysical Processor System-Data User's Handbook, Version 1.1, February 1993, JPL, JPL D-9526, 1.

24

Applications of SAR Measurements in Ocean-Ice-Atmosphere Interaction Studies

Mark R. Drinkwater

24.1 Introduction

The difficulty in observing the polar sea-ice cover and its influence on the atmospheric and oceanic boundary layer is the ephemerality of active geophysical processes. Identifying timescales of sea-ice formation, change, and decay is the key to successful measurement of these processes. Sea ice continuously adjusts to the ocean and atmosphere, and the magnitude or scale of corresponding changes is determined by the atmospheric and oceanographic forcing terms ranging spatially from meters to thousands of kilometers, and temporally from diurnal to decadal. Table 24.1 indicates the spatial and temporal scales of key sea-ice variables of interest to sea-ice geophysicists and oceanographers.

Table 24.1   Sea-Ice Variables of Primary Interest for Atmosphere-Ocean-Ice Interaction Studies

Parameter	Horizontal Spatial Scale	Temporal Scale	Accuracy
Ice extent	100 m–10 km	1 d–10 y	<0.1 % of area
Ice thickness	0–100 km	1 d–7d	10–50 cm
Thin ice coverage	1 m–10 km	1 h–10 d	<1% of area
Ice divergence	1–100,000 km^2	1–7 d	<1 % /d
Ice type	10–1000 km	7 d–1 m	<5% of type fraction
Snow depth	0–1 km	1–7 d	5 cm
Summer melt	100 m–1 x 10^6 km^2	7 d–6 m	<5% of area
Meltponding (albedo)	0–10 km^2	7 d–6 m	<3% of area
Ice motion	0.1–10 km	1–7 d	2 cm/s
Ice growth rate	0.1–10 km	1 h–7 d	10 cm/d

A conventional concern of high-latitude oceanographers is that information gathered from space is not as accurate or as relevant as that collected first hand from research vessels or *in situ* by moorings or drifters. However, microwave radar imaging of sea ice is a young and evolving field, just as acoustic tomography and acoustic Doppler current profiling (ADCP) once were. Synthetic aperture radar (SAR) is an ideal candidate for monitoring sea-ice surface characteristics on a variety of time- and space scales. Currently operating satellites with SAR will be succeeded for at least another decade by a new generation of approved instruments (see Appendix E). This will provide more than 10 years of continuous and almost unrestricted high-resolution imaging.

Traditionally ice extent is regarded as the parameter that microwave techniques can most successfully measure (Table 24.1). This is done routinely by passive microwave instruments such as the Special Sensor Microwave/Imager (SSM/I) which frequently map large areas (see Chapter 20). Additional capability offered by higher resolution SAR (~30 m) is that individual features such as ice floes and leads can be tracked and monitored in space in time-sequential images. This capability is reviewed in Reference 1 and the ability of the ERS-1 SAR to monitor patterns of ice motion and circulation in the Beaufort Sea is demonstrated. Algorithms measuring ice motion in satellite images similar to those described in Chapter 23 enable routine motion measurements in areas such as the Fram, Denmark, and the Bering Straits. Whereas ice motion in these areas is too vigorous for buoys to be efficient or cost-effective, SAR images can provide new data on the Arctic freshwater balance by measuring ice mass flux through these regions. Equally important is the contribution of SAR to observing dynamics of leads and thin ice areas. SAR images acquired at intervals of several days enable changes in thin ice area to be quantified in relation to the motion field and spatial variations in velocity of ice floes. As subsequent sections show, when this information is coupled with physical models of ice growth and with air temperature and wind data, observations of divergence and thin ice formation in leads and polynyas can be employed in estimates of the brine flux or salt precipitation rate in the upper ocean. Models can be exploited to estimate the sensible and latent heat fluxes, and with sufficient coverage and frequency enable regional heat budgets to be monitored. With combinations of these measurements and data from judiciously placed buoys, SAR will make significant contributions toward assimilating information on variables in Table 24.1.

Experiments described in Chapter 21 provided a wealth of airborne microwave signature data. Images acquired in conjunction with surface measurements provide the raw material with which to realize the potential of contemporary satellite SAR data. The main problem now facing polar oceanographers and sea-ice geophysicists is how to translate the knowledge gained from these SAR experiments into methods of extracting useful information from satellite systems. This section outlines some new applications of SAR ice-motion and surface characteristics data in geophysical process studies.

24.2 Resolution and Coverage Issues

Of the historical spaceborne SARs described in Appendix E, the Shuttle Imaging Radar (SIR) Space-Shuttle payload is the only spaceborne sensor other than Seasat SAR to recover sea-ice data between 1978 and 1991. It recorded only limited amounts of imagery of marginal ice in the Southern Ocean during the SIR-B mission in 1984.[2] Its follow-on SIR-C mission, slated for launches in 1994, will also have low orbit inclination and therefore recover limited sea-ice data unless coinciding with maximum Southern Ocean ice extent. Of the ongoing satellite missions, ERS-1 (and J-ERS-1) SAR data currently are of most practical use in geophysical studies. These have provided a largely uninterrupted data stream, and are recording considerable amounts of SAR image data of sea ice. In this section the impact of the resolution and coverage of these systems is investigated with respect to scientific investigations. Further information about present and future microwave instruments collecting sea-ice data is given by,[3,4] and the operating characteristics of each of these instruments are further described in detail in Appendix E.

A major difference between satellite and airborne systems is their inherent revisit capability and imaging swath width. Within range, an aircraft can revisit a surface site many times offering high spatial and temporal resolution, while the polar-orbiting ERS-1 satellite SAR achieves nominally a 3-d revisit with a small fraction of the earth visited at 1-d intervals.[5] Conversely, the satellite has an indefinite revisit (over long timescales of months to years) and greater spatial coverage, but with slightly reduced spatial resolution. Each of these approaches has its drawbacks, and more often than not there is a trade-off. However, the solution to many of these problems is to have a steerable SAR antenna on a satellite platform The SCANSAR mode that Radarsat offers will solve many of these difficulties (see Appendix E).

Temporal and Spatial Resolution

The capability to revisit or repeat an image of an area on the ground has an impact on the regional monitoring of lead areas or motion-induced divergence in the ice cover. Aliasing of observations of the areal extent of open water and thin ice is a topic of considerable interest to recovering information about the salt flux and heat exchange between ocean and atmosphere. Is a repeat cycle of 3 d sufficiently high temporal resolution to monitor the process of new ice formation in leads, for instance? It is clear that the exact date of lead opening is critical for a number of reasons. Primarily ice growth is most vigorous during the early growth phase, throughout the period of the largest ocean-atmosphere heat exchange. Furthermore, concomitant salt fluxes during this phase are significantly higher, and thus knowledge of the age of a lead is essential to understanding the impact on the upper ocean. Equally importantly, the SAR repeat period is critical to observing seasonal ice zone features, as the timescales of features in the marginal ice zone are susceptible to rapid changes in thermodynamic and dynamic forcing. Consequently, repeat periods of 3 d, typical for ERS-1 SAR, are unsuitable for some MIZ monitoring and most applications must rely on snapshot images.

Orbits with longer repeat cycles often have subrepeat cycles allowing imaging of a similar area on the ground at shorter intervals. Precessing orbits such as that of the 35-d repeat orbit phase of ERS-1, allowed 3-d repeat imaging and tracking of a moving patch of sea ice over a longer period, provided that ice motion was westward and consistent with the precession of the orbit and that ice floes and features within the swath are larger than the resolution limit. Problems only occur in (1) regions where the circulation patterns dictate that the main velocity vector component is in a direction opposite to the orbit precession; and (2) in marginal ice zones where ice drift velocities exceed around 0.6 m/sec. Under these circumstances it is impossible to track the ice over long time periods and distances.

Spatial resolution in SAR data is superior to most alternative data sources and is sufficient for most applications except special cases found in marginal ice zones. Ice-floe tracking becomes almost impossible when floes decrease below the resolving power of the sensor and ice concentrations are high.[6] In spite of the problematic tracking, some information may be drawn from the patterns formed by the modulation of backscatter due to floe size or floe edge distributions. Increasing spatial resolution in order to solve these problems is a thorny issue, because while limiting the study of marginal ice zones and narrow features such as leads, data volumes are already at a level that makes image products unwieldy. Studies have shown that for the majority of cases ice pack monitoring can successfully utilize low 100-m resolution images. Such is the case that all Alaska SAR Facility image postprocessing to ice-motion velocity fields and ice type grids is currently performed on 100-m degraded resolution image data. The only short-term solution to improving on 30-m satellite SAR resolution in seasonal ice zones is to use higher resolution aircraft SAR.

Spatial Coverage

Satellite SAR has a large advantage over aircraft data when it comes to applications of the data requiring monitoring of processes responding to synoptic scale ocean and atmosphere dynamics. A typical 100-km swath from satellite data is sufficient to conduct mesoscale or regional studies, and contiguous image strips can be used to trace events over wide areas in the along-track direction. In some cases it is impossible to monitor ice motion orthogonal to the swath and in the wrong direction (counter to orbit precession). The latter problem may be overcome by a SAR system with the flexibility to image much wider areas. Radarsat is capable of imaging a 510-km swath (see Appendix E), and offers the geographic coverage required to enable complete mapping of the Northern and Southern Hemisphere ice covers in 2 d (with a possible repeat only every 6 d, due to data volume). Only then will it be possible to track parcels of sea ice indefinitely in space and time, or indeed throughout an entire ice season. The capability of Radarsat to image expansive areas on a weekly basis goes a long way to solving most of the problems associated with applying SAR data to geophysical problems. An additional feature that allows variable viewing geometry using multiple beams will enable beam steering to overcome the difficulties of ice monitoring in problematic regions of ice contraflow or particularly dynamic ice margins.

24.3 Coupled Data and Model Approaches

Wind-Driven Ice Motion and the Momentum Balance

One important geophysical question is how sea ice responds to forcing from above and beneath by winds and currents, as the transfer of momentum to the sea ice is critical to understanding sea ice dynamics. This argument may be extended to the way ice motion responds to local winds causing discontinuous ice motion or divergence and the formation of leads. A major uncertainty in calculations of regional air-sea-ice heat exchange is the rate of ice divergence.

In this section pairs of Antarctic ERS-1 SAR images are coupled with data from Argos buoy data to derive the ice kinematics and divergence in response to the surface wind forcing. The current state of the art in routine postprocessing of image data allows tracking of sea-ice features. Geophysical data products generated at the Alaska SAR Facility through the Geophysical Processor System (GPS) are (1) ice-motion vector fields, made from pairs of images separated by a few days in time from successive passes over the same ice; and (2) ice classification images[7,8] (an example illustrating these products from the Beaufort Sea is shown in Figure 24.1. The motion field may be used simply to generate a deformation field in which the gridded relative displacements can be used to measure divergence or convergence in the ice together with the spatial derivatives in the x and y image directions. This approach is currently being extended at the Jet Propulsion Laboratory (JPL) to ERS-1 data products recorded for Antarctic ice. An example of a GPS ice-motion product is shown in Figure 24.2 for a pair of images located near to a drifting Argos buoy.

In the unenclosed Weddell Sea, Antarctica, ice motion is less affected by internal stresses generated by coastal contact. In situations away from the coast under largely divergent conditions, one may assume that the mean ice motion $\bar{\mathbf{v}}_i$ is steady and in balance with the mean current $\bar{\mathbf{c}}_i$ and the mean geostrophic wind $\bar{\mathbf{U}}_g$:

**DISPLACEMENT
FIELD**

**DEFORMATION
GRID**

50 km

ICE TYPE MAPS

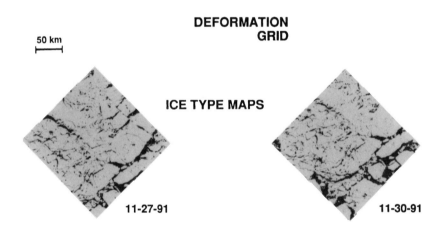

Figure 24.1 Example of geophysical products routinely generated at the Alaska SAR Facility from ERS-1 SAR data (courtesy Ron Kwok and Ben Holt, JPL). Images were acquired 3 d apart in the Beaufort Sea in November 1991. The ice-motion vector and deformation grids are shown in the center panels with the classified images shown in the lowermost panels.

$$\overline{\mathbf{v}}_i - \mathbf{A}\overline{\mathbf{U}}_g - \overline{\mathbf{c}}_i = 0 \qquad (1)$$

A simple relationship then exists between the instantaneous time-varying part of the ice velocity $\mathbf{v}'_i$ and the instantaneous geostrophic wind $\mathbf{U}'_g$:[1]

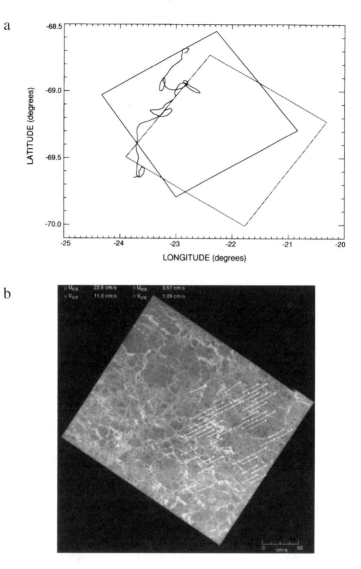

Figure 24.2   Illustration of ice drift and deformation extracted from Antarctic ERS-1 SAR images and an Argos drifter instrumented site in the central Weddell Sea, Antarctica. (a) Geolocated SAR image locations superimposed upon the July drift track of buoy 9364; (b) polar stereographically projected SAR image on July 14, 1992 (day 196) with superimposed vectors illustrating the ice-motion field over a 24-h period from day 195–196 (SAR image © ESA).

$$\mathbf{v}'_i = \mathbf{A}\mathbf{U}'_g - \mathbf{e} \qquad\qquad (2)$$

where the error term $\mathbf{e}$ contains all error sources including time-varying current and the divergence in the internal ice stress tensor. Importantly, the mean error term $\mathbf{e}$ is zero which accounts for its absence in Equation 1. Matrix $\mathbf{A}$ in each case may be considered as a complex scale factor ($|\mathbf{A}|$, δ) with the real part $|\mathbf{A}|$ giving the magnitude of the ratio of ice speed to wind speed, and the imaginary part δ the rotation or turning angle (positive = clockwise) between the ice and the wind vectors. Similar schemes have recently been used to explain the drift of sea ice in the Weddell Sea.[9,10]

Under divergent conditions, a free-drift form of the ice momentum balance can successfully be used to simulate eastern Weddell Sea ice drift, with some assumptions regarding ice-water drag and internal ice stress.[11] To date this approach has been used for two separate buoy drift experiments to derive the typical range of winter Antarctic 10-m neutral drag coefficients.[9,11] In 80–100% concentration ice they observed air-ice values in the range $1.45 \times 10^{-3} \le C_{dn10} \le 1.79 \times 10^{-3}$ with aerodynamic roughnesses spanning $0.27 \le z_o \le 0.47$ mm, and derived a value for the ice-ocean drag coefficient of 1.13×10^{-3}.

Data reported here characterize ice motion in the eastern-central Weddell gyre in an area of German Argos drifters deployed from the research vessel Polarstern in July 1992 during the Winter Weddell Gyre Study (WWGS '92). Figure 24.2 indicates selected results from these SAR images and six buoys placed in the Central Weddell Gyre. In Figure 24.2a the positions of two geolocated ERS-1 SAR images are shown with respect to the July portion of the drift track of buoy 6. In Figure 24.2b the corresponding SAR-tracked ice motion is shown for a 1-d time period between this pair of images acquired on July 13 and 14. In Figure 24.3a, the daily mean drift distance and 24-h smoothed velocity of buoy 6 indicates the response to pulses in the geostrophic wind $\mathbf{U}'_g$ (calculated from the two-dimensional buoy pressure-gradient field) as it drifts along the track depicted in Figure 24.2a. The corresponding response of the area encompassed by a triplet of Argos buoys (including buoy 6) to $\mathbf{U}'_g$ is equally sensitive. This area shown in Figure 24.3b increases consistently over the month of July in response to the wind stress, and the fractional area of new ice production in this period is 1100 km,[2] equating to a mean divergence of $0.67 \pm 0.45\%$/d. Sea ice quickly reacts to pulses in geostrophic wind, as shown in the 24-h smoothed divergence in Figure 24.3b. Each pulse of relatively stronger winds is interleaved by relaxation events where the ice pack converges and leads close under internal stresses. In contrast, divergence over the whole six-buoy array in July results in a mean value of $0.4 \pm 0.6\%$/d. Southeastern buoys tend to converge over the same July period, and spatial variability in the divergence results in a lower mean and larger variance than that represented in Figure 24.3b. This spatial variability in divergence on different scales makes it imperative that the satellite data are additionally employed in monitoring regional divergence and hence region-wide new ice production.

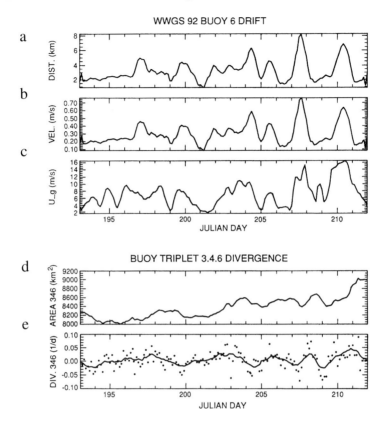

Figure 24.3 (a) Daily displacement and (b) drift speed of Antarctic Argos buoy 9364 in comparison with (c) geostrophic wind speed (U_g) in July 1992; (d) buoy triplet area change and (e) 3 hourly divergences (points) superimposed by the smoothed mean daily divergence.

The ice motion derived from the SAR motion field in Figure 24.2b indicates a mean ice-motion vector magnitude V_i of 0.255 m/sec, with a bearing θ_i of 17.466° (when the vectors are correctly referenced to the Southern Hemisphere SSM/I grid by a 45° rotation). The mean geostrophic wind $\overline{U}_g$ over the same period is 6.3 m/sec, on a bearing of 36.5°. Preliminary results comparing these buoy-derived winds with the first Antarctic ERS-1 SAR-derived ice motion give values for ($|A|$, δ) of (0.04 , −19°). The turning angle is consistent with a mean of $|A| = 0.03$ and δ = −20° (to the left of the wind) observed during periods of free drift during WWGS '86 and '89.[11]

Further processing of SAR ice-motion data acquired during WWGS '92 will form an Antarctic satellite ice-motion database for comparison with buoy statistics of velocity and divergence. The tested relationships between wind stress and divergence, under essentially free-drift conditions, then can also be extended to yield similar bulk drag coefficients through this simple formulation of the momentum balance. In addition, more accurate long-term measurements of divergence can be made in order to make estimates of the impact on heat exchange between the upper ocean and atmosphere.

SAR and Ice-Ocean Model Surface Flux Estimates

Coupling SAR Observations and Physical Ice-Growth Models

A simple extension can be made of the results shown in Figure 24.2 by coupling the regional SAR-derived divergence measurements with an ice-growth model. In Figure 24.4a the full energy balance is computed using longwave radiation (LW) budget and shortwave (SW) fluxes (from WWGS '92 shipborne measurements) and estimates of the turbulent fluxes of sensible (SH) and latent heat (LH) driven by the mean surface geostrophic wind speed values plotted in Figure 24.3. The energy balance is then used in a coupled two-dimensional ice growth model (extended from Reference 12) to simulate growth of an ice sheet (Figure 24.4b) and the conductive heat flux (F_{COND}) for the month described by the buoy drift and SAR motion data. Typical snow data are added in the model to reproduce observed natural conditions, and with uninterrupted growth the ice sheet reaches a thickness of 55 cm in only 16 d. Figure 24.4b indicates salinity and temperature profiles of this ice sheet at 10-cm growth-thickness intervals. Corresponding brine volume profiles are computed as a direct input to backscattering models that require vertical profiles of dielectric properties within the ice sheet. The direct relevance of this technique for oceanography, however, is shown in Figure 24.4c and d, where the impact on the upper ocean is monitored by calculating the flux of salt from ice to upper ocean as brine rejection and drainage take place. During early growth, when the air temperatures were close to $-30°C$, winds of between 4 and 10 m/sec combine to cause a net outgoing heat flux of around -100 W m^{-2}. With clear skies and relatively brisk winds the sensible heat flux drives a net heat flux (F_{NET}) that on day 205 reaches -300 W m^{-2}. Resulting ice growth is rapid with the first day growth rate peaking at 15 cm/d. After this early peak the rate slows to more typical values of 2 cm/d. A similar trend is seen in the salt flux data (Figure 24.4d) where at peak growth the salt flux exceeds 65 kg m^{-2} per month.

This brief example continues from the theme developed using the buoy and ice-motion data. Future SAR measurements of regional ice divergence and the fraction of open water, together with a physical ice-growth model such as that presented here will enable the regional heat and salt budgets of such experiment areas to be assessed.

Time-Series Backscatter Data and the Energy Budget

A further extension to our understanding of microwave backscattering response from sea ice can also be drawn from studying radar backscatter time series in conjunction with this ice-modeling and radiation balance approach. In this example a time series of surface C-band scatterometer measurements is shown during a 3-d Weddell Sea ice drift station at the end of the month of July when the net heat

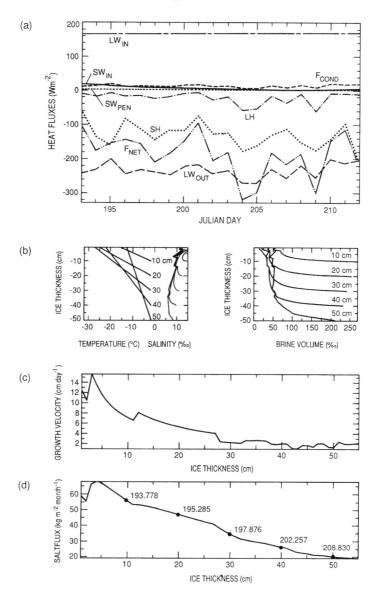

Figure 24.4 Radiation balance and new ice growth during July 1992, in the region around the buoy shown in Figure 24.2. (a) Radiation balance of a newly opened lead; (b) ice properties during 1 month of new ice growth starting on day 193; (c) ice growth rate; and (d) the resulting salt flux over this period.

flux budget shows a dramatic transition.[13] Field experiment data from WWGS '92 for a 1-m thick ice sheet in Figure 24.5 records 10-min averages of wind stress and turbulent sensible heat flux together with the net energy budget (courtesy of W. Frieden of Hannover University, Germany) during a C-band radar scatterometer

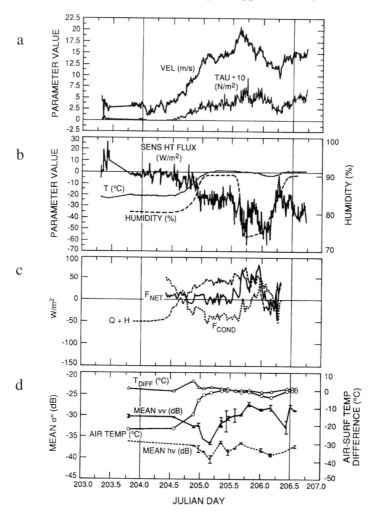

Figure 24.5 Shipborne C-band microwave radar scatterometer time-series comparison with surface fluxes of heat and momentum. (a) Turbulent flux of momentum; (b) turbulent flux of sensible heat with the variation in air temperature and humidity; (c) the energy balance and its major components; and (d) 45° incidence VV- and HV-polarized backscatter response with temperature and air-surface temperature variations.

time-series of measurements between days 203 and 206 (July 21–24, 1992). In the lowermost panel of Figure 24.5, the 45° incidence C-band signature is plotted starting at Julian day 203.79. The solid line with error bars represents the mean VV-polarized backscatter and the dashed line the corresponding HV-polarized back-scatter. A 10-dB variation in VV-polarized backscatter (σ°_{vv}) and 5-dB variation in HV-polarized backscatter (σ°_{hv}) is observed in response to the changes in the heat fluxes presented.

Figure 24.5a indicates a large variation in wind stress during this period, increasing rapidly to over 0.5 N m^{-2} when the winds peaked at 20 m/sec. The increase in wind took place with a sudden rise in air temperature (Figure 24.5b) from around −23 to 0°C during the passage of a warm frontal system. As a consequence of the overcast skies, warm temperatures, and the high wind, the sonic-anemometer-thermometer recorded a net negative (downward/incoming) flux of sensible heat that peaked during the strong winds at 60 W m^{-2}. A sharp decrease in the amount of incoming turbulent sensible heat occurs on the morning of July 24, due to the clear night sky.

A net energy budget (F_{NET}) is shown in Figure 24.5c as a solid line, using the conventional system of negative net outgoing flux of heat and positive net incoming heat flux. The largest measured components of this budget are indicated as: the net radiative flux (Q), the net sensible heat flux (H), and the conductive heat flux (F_{COND}). At the start of the period, the surface lost nearly 50 W m^{-2} in radiative and sensible heat, with this loss balanced largely by conducted heat. Some heat is supplied by the freezing at the base of the ice sheet, as indicated by ice core measurements and a thermistor chain frozen into the sea ice. Immediately the sky became overcast on day 204, temperatures rose, and light snowfall began. Several decibel variability in backscatter (Figure 24.5d) indicates several orders of magnitude change in backscattering over the conditions indicated during the following period. The change in heat flux regime is strongly correlated with a reduction in VV and HV backscatter as the net radiative and sensible heat flux (Q + H) swings from outgoing to incoming. At first, conduction takes care of the surface warming, carrying heat away from the snow and ice surface until the time at which the ice sheet becomes nearly isothermal (at day 205.5). Melting then begins to dampen the snow surface and the backscatter rapidly rises to its peak. Subsequently, high winds and a brief period of cloud-free night at Julian day 206.25 encourage evaporative cooling, a minimum in F_{NET}, and a local minimum in the values of VV- and HV-polarized backscatter.

Surface measurements in association with the C-band radar data indicate that the VV-polarized data clearly signal the change in heat flux environment when air temperatures rise above −20°C. While surface temperatures rose dramatically, snow-grain transformations began and equi-temperature metamorphism is first focused at the surface reducing the large angular grains to small rounded grains. As the warm temperatures penetrate down to the ice surface, the previously exaggerated snow temperature gradients become reduced. However, the wind speed continues to rise to gale force. The $\sigma°_{VV}$ reduction mirrors the rise in Q + H in Figure 24.2c, reaching a minimum below −35 dB. A reversal then occurs in $\sigma°_{VV}$ with a rise in backscatter values up to a level higher than that on day 203. During this period, no significant amount of surface wetness accumulated in the shallow 3-cm deep snow layer until the latter part of day 205. Instead, the wind speeds at the surface were large enough that sufficient heat is removed by surface sublimation to prevent visible melting and free-water appearance in the snow (as described in Reference 14). Until after day 205.5, it is suggested that in the case of these strong winds enough heat is removed by sublimation and conduction to preclude significant melting. The large change in thermal profile in the snow rapidly changed snow crystal characteristics under these conditions from hoar-style angular crystals (induced earlier by strong negative heat gradients before day 204) to rounded crystals. Together with the layering that developed in the snow, it is proposed that these snow changes play a large role, first by reducing and then subsequently by increasing volume scatter as rounded grains grow larger. The last significant change in σ°_{VV} takes place after cooling and overnight refreezing (day 206) of moisture in the snow layer on a brief reversal of the net heat budget from positive to negative. Diurnal cooling and the swing in the humidity during this period on day 206 result in a brief minima in σ°_{VV}. The rapid return to values above -30 dB coincides with the brief period of incoming short-wave energy during scattered clouds at midday and signifies the reappearance of moisture at the surface of the snow.

This winter example clearly indicates that the sea-ice surface properties respond to the balance of fluxes at the surface. Furthermore, C-band values of σ°_{VV} react equally sensitively to these changes. Though these temperature swings may be more commonly associated with spring conditions, this change in events may be recognized in time-series data to reflect transformations in the surface heat and vapor flux environment. It is proposed that with the aid of buoy data and weather analysis fields (for specifying boundary conditions), satellite SAR and surface data to be used together with physical models to understand how microwave data reflect key changes in the energy balance. The power of time-series measurements using SAR backscatter has barely been exploited. Coupling data analyses in this manner together with the tracking capability of SAR provides a powerful method for studying geophysical processes and surface changes.

24.4 Conclusions

SAR is now guiding us to geophysical variability previously unrecognized by way of the resolution and the time dimension often absent in sampling of polar geophysical media because of the logistical difficulty of routine observations. The goal of future satellite data sets therefore shall be to provide both a Lagrangian time series of the characteristics and dynamics of sea-ice floes tracked across the Arctic basin or Southern Ocean. Parcels of sea ice can already be followed through their evolution from thin new ice to multiyear ice using a Lagrangian style sea-ice tracker.[15] This requires observations to be frequent, continuous, and indefinite over a large geographic expanse. We are already close to this goal when Radarsat arrives with an existing mandate to map the complete sea-ice cover of both hemispheres once every week. Tracking a piece of ice throughout its lifetime negates the necessity to identify ice types — or indeed to use such ice categories as a proxy for sea-ice thickness. Having sufficient accompanying environmental information enables a physical ice-growth model to be coupled with the observed life cycle of the ice in simulating true thickness. In this way, contiguous parcels of ice over extensive regions may be integrated into a regional ice thickness distribution.

The main problem confronting scientists wishing to exploit SAR remote sensing of the polar ice pack is coping with the vast amounts of data being recovered. Automated algorithms that are tested and validated must become the accepted norm, and are a necessary step to digesting the spatial and temporal dimensions of the resulting data sets. The key to realizing geophysical applications of these data is recognizing the requirement to process vast amounts of data routinely with knowledge of the errors inherent in the postprocessing techniques used. Quantitative data can then be derived from these SAR products and used to reflect the changing surface characteristics of the sea-ice pack in response to the changing surface heat and salt flux environment.

Acknowledgments

This research was performed at the Jet Propulsion Laboratory, California Institute of Technology under contract to the National Aeronautics and Space Administration. MRD is grateful to Petra Heil who implemented the ice-growth model. Wolfgang Frieden and Rüdiger Brandt of the University of Hannover analyzed and kindly provided the long-station flux data set. Christoph Kottmeier is acknowledged for the July '92 Argos-buoy drift data, and Peter Lemke is thanked for his planning support during WWGS '92. The late Thomas Viehoff of the PIPOR Office, and ESA are gratefully appreciated for their support in Antarctic ERS-1 SAR acquisition scheduling.

References

1. Holt, B. M., Rothrock, D. A., and Kwok, R., Determination of Sea Ice Motion from Satellite Images, in Microwave Remote Sensing of Sea Ice, *Geophys. Monogr.*, *68*, American Geophysical Union, 1992, 343.
2. Martin, S., Holt, B., Cavalieri, D. J., and Squire, V., Shuttle imaging radar B (SIR-B) Weddell Sea ice observations: a comparison of SIR-B and scanning multichannel microwave radiometer sea ice concentrations, *J. Geophys. Res.*, 92(C7), 7173, 1987.
3. Carsey, F. D., Barry, R. G., and Weeks, W. F., Introduction, Microwave Remote Sensing of Sea Ice, *Geophys. Monogr.*, *68*, 1992, 1.
4. Massom, R., *Satellite Remote Sensing of Polar Regions*, CRC Press, Boca Raton, FL 1991, 307 pp.
5. ESA, ERS-1 system, ESA Spec. Publ., SP-1146, 1992, 87 pp.
6. Drinkwater, M. R. and Squire, V. A., C-band SAR observations of marginal ice zone rheology in the Labrador Sea, *IEEE Trans. Geosci. Remote Sensing*, 27(5), 522, 1989.
7. Kwok, R., Curlander, J. C., McConnell, R., and Pang, S. S., An ice-motion tracking system at the Alaska SAR facility, *IEEE Trans. Oceanic Eng.*, 15, 44, 1990.
8. Kwok, R., Rignot, E., and Holt, B., Identification of sea-ice types in spaceborne synthetic aperture radar data, *J. Geophys. Res.*, 97(C2), 2391, 1992.
9. Martinson, D. G. and Wamser, C., Ice drift and momentum exchange in winter Antarctic pack ice, *J. Geophys. Res.*, 95, 1741, 1990.

10. Kottmeier, C. and Engelbart, D., Generation and atmospheric heat exchange of coastal poynyas in the Weddell Sea, *Boundary Layer Meteorol.*, 60, 207, 1992.
11. Wamser, C. and Martinson, D. G., Drag coefficients for winter Antarctic pack ice, *J. Geophys. Res.*, 98(C7), 12,431, 1993.
12. Cox, G. F. N. and Weeks, W. F., Profile Properties of Undeformed First-Year Sea Ice, CRREL Rep., 88-13, 1988, 57 pp.
13. Drinkwater, M. R., Hosseinmostafa, R., and Dierking, W., Winter Microwave Radar Scatterometer Sea Ice Observations in the Weddell Sea, Antarctica, Proc. IGARSS '93, 2, 446, 1993, Tokyo, Japan.
14. Andreas, E. L. and Ackley, S. F., On the differences in ablation seasons of Arctic and Antarctic Sea ice, *J. Atmos. Sci.*, 39, 440, 1982.
15. Kwok, R., Rothrock, D. A., Stern, H. L., and Cunningham, G. F., Determination of the age distribution of sea ice from Lagrangian observations of ice motion, *IEEE Trans. Geosci. Remote Sensing*, 33(2), 392, 1995.

25

Study of Waves in Ice-Covered Oceans Using SAR

Anthony K. Liu

25.1 Introduction

Marginal ice zones (MIZs) are areas of transition between the ice cover and the open ocean. There is a complex interaction in these regions between winds, ocean currents, fronts, wave fields, and the characteristics of ice (such as thickness and roughness that affect the oceanic fluxes of heat, brine, mass, and momentum). These are dynamic regions both spatially and temporally, being zones of rapid ice growth and decay, with sharp oceanic gradients and air-sea-ice temperature gradients. Ocean surface waves from the open sea may penetrate into the MIZ and contribute to the breakup of floes and to other processes that modify the ice cover. Ice conditions are generally not uniform within the MIZ. Here strong mesoscale air-sea-ice interactive processes occur that control the advance and retreat of the ice margin. Wave-ice interaction is an important phenomenon in the MIZ because ice breakup due to the wave-induced flexural failure is the chief determinant of floe size distribution. The breakup of the ice floes effectively increases the melt rate because of the increased ice surface area in contact with the above freezing water.

Synthetic aperture radar (SAR) images of ocean surface from Seasat demonstrated that refracting surface waves, wind signatures, oceanic fronts, and mesoscale eddies can be detected via their influence on the short wind waves responsible for microwave backscatter.[1] Similar surface signatures of swell, wind, fronts, meanders, and eddies have also been observed near the ice edge by SAR from satellite (Seasat, SIR-B) and airborne radars.[2] The ability of a SAR to provide valuable information on the type, condition, and motion of the ice cover has been amply demonstrated. The processes that control oceanic and atmospheric transport of heat to and from the MIZ are of special significance because these ocean-ice interaction processes are primarily responsible for controlling ice edge location. It is an excellent opportunity to study the wave-ice interaction in the MIZ using the First European Remote Sensing Satellite (ERS-1) and the Japanese First Earth Resources Satellite (JERS-1), SAR data received at Alaska SAR Facility (ASF).

SAR imaging of surface gravity waves in the ocean has been demonstrated by many investigators, although the relative importance of imaging mechanisms such as hydrodynamic modulation, tilt modulation, and velocity bunching is still not that well understood.[3-4] Recently, extensive research has been conducted as reported by Beal[5] in the Symposium of Measuring Ocean Waves from Space, and by Shemdin[6] from the Tower Ocean Wave and Radar Dependence (TOWARD) experiment. It is generally accepted that the dominant mechanism that allows a SAR to image waves in ice is velocity bunching,[7,8] a natural consequence of coherently sensing scatterer motions at the wave orbital velocity. The waves-in-ice patterns observed in the SAR imagery and the corresponding SAR image spectra are further explained in terms of SAR wave-imaging models by Vachon et al.[9]

25.2 SAR Observations of Waves in Ice

Using the SAR imagery, we can estimate the spatial variability of the dominant wavelength through spectral analysis. Figure 25.1 shows a C-band SAR image of sea ice taken on March 23 off Newfoundland during the Labrador Ice Margin Experiment (LIMEX '87). An overview of LIMEX '87 ice observations has been reported by Carsey et al.[10] The image shows the swell penetrating into the MIZ and a periodic displacement of the ice-water boundary, apparently due to velocity-bunching effects associated with the wave orbital motion. Note the clearly discriminated shear line that also indicates the extent of swell penetration and action on ice rafting and compaction. On the basis of the wave spectrum from the SAR data and the concurrent wave frequency spectrum from the ocean buoy data and accelerometer data on the ice during LIMEX '87, the dispersion relation has been estimated and compared reasonably well with the observations.[11] Due to the change of wave dispersion relation in ice, wave refraction at an ice edge has been observed by SAR during LIMEX '87.[12] Waves in frazil and pancake ice and their detection in Seasat SAR imagery were studied by Wadhams and Holt[13] for a mass-loading effect on dispersion relation. They showed that wavelength changes in such an ice field can thus be used to calculate the thickness of the frazil slick.

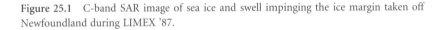

Figure 25.1 C-band SAR image of sea ice and swell impinging the ice margin taken off Newfoundland during LIMEX '87.

Wave attenuation in the marginal ice zone during LIMEX has been studied by Liu et al.[8] By extracting many SAR subscenes in the wave propagation direction and estimating the decreasing contrast for each wave number band, the wave attenuation coefficient can be obtained from SAR data as shown in Figure 25.2 for both LIMEX '87 and LIMEX '89 cases. Note that the viscous attenuation model (solid curve) developed by Liu and Mollo-Christensen[14] compares reasonably well with SAR data and predicts a rollover at high wave numbers, which was also observed by the ice-motion sensor data.[15] The wave-ice interaction model takes account of the physical characteristics of the ice cover and the wave parameters to predict wave dispersion, attenuation, and refraction, and possibly ice thickness.

Since the launch of ERS-1 satellite on July 16, 1991, its mission has been proceeding smoothly. Intensive campaign activities have been conducted to calibrate the

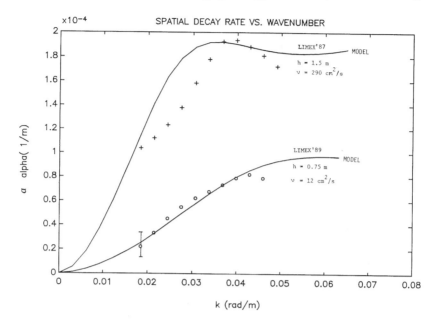

Figure 25.2   Model-data comparisons for wave attenuation coefficient as a function of wave number for LIMEX.

sensor and validate the geophysical data. This commissioning phase was initiated in early August and completed by mid-December 1991. The ERS-1 satellite has finished the ice phase with 3-d repeating cycle in April. The 3-d orbit has the advantage of frequent looks at the same SAR swaths so that dynamically changing phenomena can be monitored. European Space Agency (ESA) is now in the middle of data acquisition/processing for the multidisciplinary phase, lasting from mid-April until the end of 1993. The ASF is currently receiving on average 50 min of signal data per day from ERS-1. Excellent opportunities now exist to continue the wave-ice interaction work using SAR data streams from the ERS-1 and JERS-1.

25.3 Wave-Ice Interaction

Ocean surface waves from the open sea may penetrate into the marginal ice zone and contribute to the breakup of floes and to other processes that modified the ice cover. Mesoscale eddies are key features associated with the ice margin. Using SAR imagery, the spatial variability of the wave and current field can be estimated through spectral analysis and wave model. The surface effect models induced by wave-current interaction and atmospheric instability are used to interpret the mesoscale features of eddies and fronts from Seasat and ERS-1 SAR images.[16] Based on SAR observed wave, current and wind inputs, two-dimensional coupled ocean-ice interaction numerical models can be used to predict the ice edge and ocean dynamics.

During LIMEX '87, ice edge and swell penetration were observed in the aircraft SAR images in the MIZ. The wave effects on the ice edge meandering were observed by SAR photo mosaic coverages of southeast Newfoundland between March 21–23, and 26, 1987. On March 23, ocean swell was observed coming from the northeast and starting to penetrate the ice pack with relatively smooth ice edge. On March 26, because of strong easterly wind and swell, the ice was heavily compacted against the coast. Ocean swell was observed to have penetrated the entire pack to the shore. Ice edge meanderings with eddies are thought to be manifestations of the wind and wave actions.

A two-dimensional coupled ocean-ice model has been developed by Liu et al.[17] with the wind and wave effects included for the study of MIZ dynamics. The ice model is coupled to the reduced gravity ocean model (f-plane) through interfacial stresses.[18] The internal ice stresses are important at high ice concentration, and ice motion is affected by the wave radiation pressure. For typical MIZ ice concentrations the main dynamic balance in the ice medium is between the water-ice stress, wind stress, and wave radiation stresses. By considering the exchange of momentum between waves and ice pack through radiation stresses for decaying waves, a parametric study is performed to compare the effects of wave stress with wind stress on ice edge dynamics. The numerical results show significant effects from wave action. Upwelling at the ice edge and eddy formation can be enhanced due to the nonlinear effects of wave action; wave action sharpens the ice edge and produces ice meandering that enhances local Ekman pumping, pycnocline anomalies, and cyclonic eddy formation.

In order to know the locations of the MIZ during the ice phase, the standard ice analysis/forecast graphic products are routinely collected from the Navy/National Oceanic and Atmospheric Administration (NOAA) Joint Ice Center. The C-band ERS-1 SAR image in the MIZ on January 14 during the ice phase is shown in Figure 25.3. The SAR image is 100×100 km in size and the resolution is 25 m. The center coordinates for the SAR image in Figure 25.3 is (58.38° N, 165.49° W). The dark region near the ice edge at the north end of image is an upwelling region of cold water that stablizes the atmospheric boundary layer and thus has a smooth surface.

A swell of 230-m wavelength is clearly visible both in the open ocean and in the ice in Figure 25.3. Figure 25.4 shows the SAR wave spectra of wave in ice and ocean waves from the SAR data collected on January 17, 1992. From the SAR spectra, the swell system propagates northeasterly from the open ocean into the MIZ, and ocean waves of 90 m propagate westerly. It is evident that the wind waves attenuate in the ice and refract near the ice edge; thus wind waves do not penetrate deep into the ice cover. Based on the buoy data and weather map, the wind was from the northeast direction (off-ice) and the ice bands were evident near the ice edge perpendicular to the wind direction. After wind and wave forcing for 3 d, an upwelling with eddy formation was generated near the ice edge in the middle of SAR image collected on January 17, 1992.

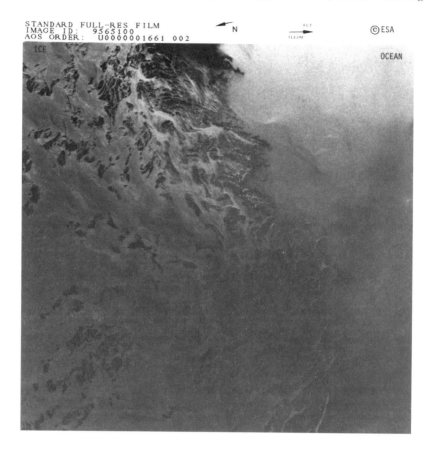

STANDARD FULL-RES FILM
IMAGE ID: 9565100
AOS ORDER: U0000001661 002

ICE

OCEAN

N   FLT   ILLUM   ©ESA

Figure 25.3   C-band ERS-1 SAR image of marginal ice zone in the Bering Sea on January 14, 1992 during the ice phase.

25.4 Discussion

The ability of a SAR to provide valuable information on the type, condition, and motion of the ice cover is well demonstrated. The wave parameters (e.g., wavelength, direction, and attenuation rate) can be extracted from SAR and buoy data as the inputs for the mesoscale ocean-ice model. To demonstrate the wave effects, a special case of incoming wave train and wind direction were considered as shown in ERS-1 SAR data. The numerical results show significant effects from wave action. The ice edge is sharper and ice edge meandering has begun to form in the MIZ due to forcing by wave action after a couple of days. Upwelling at the ice edge and eddy formation can be greatly enhanced due to the nonlinear effects of wave action; wave action sharpens the ice edge and produces ice meandering that enhances local Ekman pumping, pycnocline anomaly, and cyclonic eddy formation. The ice-ocean

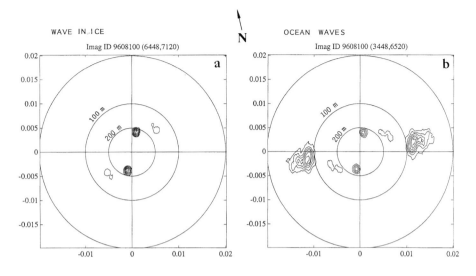

Figure 25.4 ERS-1 SAR spectra for (a) wave in ice; and (b) ocean waves from January 17, 1992 in the marginal ice zone of the Bering Sea.

model can be used for ERS-1 SAR data during the ice phase with 3-d repeating cycles to study the mesoscale processes in the MIZ.

These studies have demonstrated the value of SAR imagery for the study of wave-ice interaction in the MIZ. Great progress can therefore be expected using ERS-1 and JERS-1 with suitable ground validation.

Acknowledgment

The author wishes to thank Chich Peng, Sirpa Hakkinen, and Ben Holt for their valuable discussions and suggestions. Goddard Space Flight Center (GSFC) work was supported by the National Aeronautics and Space Administration and Office of Naval Research.

References

1. Fu, L.-L. and Holt, B., Seasat View Oceans and Sea Ice with Synthetic Aperture Radar, JPL Publication 81-120, Pasadena, CA, 1982.
2. Carsey, F., Holt, B., Martin, S., McNutt, L., Rothrock, D., Squire, V., and Weeks, W. F., Weddell-Scotia Sea marginal ice zone observations from space, October, 1984, *J. Geophys. Res.*, 91, 3920, 1986.
3. Alpers, W. R., Ross, D. B., and Rufenach, C. L., On the detectability of ocean surface waves by real and synthetic aperture radar, *J. Geophys. Res.*, 86, 6481, 1981.
4. Hasselmann, K., Raney, R. K., Plant, W. J., Alpers, W., Shuchman, R. A., Lyzenga, D. R., Rufenach, C. L., and Tucker, M. J., Theory of synthetic aperture radar ocean imaging: a MARSEN view, *J. Geophys. Res.*, 90, 4659, 1985.

5. Beal, R. C., Measuring ocean waves from space, *Johns Hopkins Tech. Dig.*, 8,(1), 1987.
6. Shemdin, O. H., Tower ocean wave and radar dependence experiment: an overview, *J. Geophys. Res.*, 93, 13829, 1988.
7. Lyzenga, D. R., Shuchman, R. A., Lyden, L. D., and Rufenach, C. L., SAR imaging of wave in water and ice: evidence for velocity bunching, *J. Geophys. Res.*, 90, 1031, 1985.
8. Liu, A. K., Vachon, P. W., Peng, C. Y., and Bhogal, A. S., Wave attenuation in the marginal ice zone during LIMEX, *Atmos.-Ocean*, 30, 192, 1992.
9. Vachon, P. W., Olsen, R. B., Krogstad, H. E., and Liu, A. K., Airborne synthetic aperture radar observations and simulations for waves-in-ice, *J. Geophys. Res.*, 98, 16,411, 1993.
10. Carsey, F., Digby-Argus, S., Collins, M., Holt, B., Livingstone, C., and Tang, C., Overview of LIMEX'87 ice observations, *IEEE J. Geosci. Remote Sensing*, 27, 468, 1989.
11. Liu, A. K., Holt, B., and Vachon, P. W., Wave propagation in the marginal ice zone: model predictions and comparisons with buoy and SAR data, *J. Geophys. Res.*, 96, 4605, 1991.
12. Liu, A. K., Vachon, P. W., and Peng, C. Y., Observation of wave refraction at an ice edge by SAR, *J. Geophys. Res.*, 96, 4803, 1991.
13. Wadhams, P. and Holt, B. Waves in frazil and pancake ice and their detection on Seasat SAR imagery, *J. Geophys. Res.*, 96, 8835, 1991.
14. Liu, A. K. and Mollo-Christensen, E., Wave propagation in a solid ice pack, *J. Phys. Oceanogr.*, 18, 1702, 1988.
15. Wadhams, P., Squire, V. A., Ewing, J. A., and Pascal, R. W., The effect of marginal ice zone on the directional wave spectrum of the ocean, *J. Phys. Oceanogr.*, 16, 358, 1988.
16. Liu, A. K., Peng, C. Y., and Schumacher, J. D., Wave-current interaction study in the Gulf of Alaska for eddy detection by SAR, *J. Geophys. Res.*, submitted.
17. Liu, A. K., Hakkinen, S., and Peng, C. Y., Wave Effects on ocean-ice interaction in the marginal ice zone, *J. Geophys. Res.*, 98, 10,025, 1993.
18. Hakkinen, S., Coupled ice-ocean dynamics in the marginal ice zones: upwelling/downwelling and eddy generation, *J. Geophys. Res.*, 91, 819, 1986.

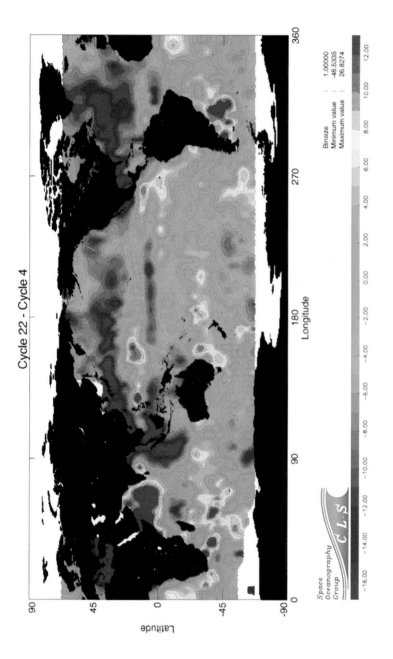

Figure 6.2 Difference between cycle 4 (October 22-November 1, 1992) and Topex/Poseidon cycle 22 (April 19-29, 1993). Units are in centimeters; no orbit error correction was applied. The figure shows the global annual cycle of sea level.

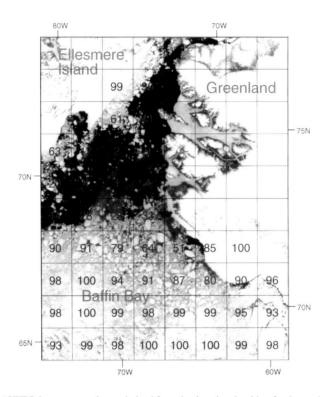

Figure 19.1 AVHRR ice concentration as derived from the tie point algorithm for the northern Baffin Bay area, May 30, 1993. The AVHRR LAC channel 1 image shows a subscene of 400 × 500 km with the Greenland ice sheet on the left and Ellesmere Island on the right. This area is also known as the North Water due to its reduced ice concentration caused by strong surface winds (ice divergence) and upwelling of warmer water along the Greenland coast. The scene is divided into 50 × 50 km grids (blue mesh) and the mean ice concentration is given for each grid cell in red color.

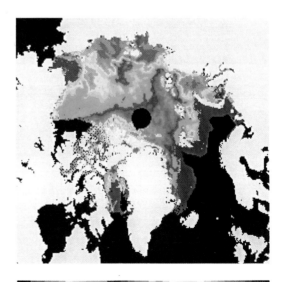

Figure 22.3 Depicted is σ° evaluated at 40° incidence angle over the Arctic Ocean in November 1992. (Color scale from 0.0 to 0.17; zones of open water in brown).

Part V

Appendices

A

AVHRR on NOAA

Sei-ichi Saitoh

A.1 NOAA Spacecraft

The Advanced TIROS-N satellites (ATN), which are labeled NOAA-A, -B, etc. before launch and NOAA-6, -7, etc. after successful launch, provide operational coverage of the entire earth four times per day by using two satellites (at present, NOAA-11 and NOAA-12). The launch date and names after launch of each satellite in the past is summarized in Table A.1. The sensor systems on the ATN satellites include: TIROS Operational Vertical Sounder (TOVS), Advanced Very High-Resolution Radiometer (AVHRR), Space Environment Monitor (SEM), ARGOS Data Collection and Platform Location System (DCS), Search and Rescue System (SARSAT). TOVS is a three-instrument system consisting of: High-resolution Infrared Radiation Sounder (HIRS/2), Stratospheric Sounding Unit (SSU), and Microwave Sounding Unit (MSU). Other sensors are Solar Backscatter Ultraviolet radiometer (SBUV) and Earth Radiation Budget Experiment (ERBE).[1,2,3]

The ATN is 4.2 m long, 1.9 m in diameter, and carries a 2.4- × 4.9-m solar array for power.[1-3] The nominal configuration of the ATN satellites is shown in Figure A.1. The ATN were designed to operate in a near-polar, sun-synchronous orbit. The orbital period is about 102 min, which produces 14.1 orbits per day. Two nominal altitudes, 833 and 870 km, have been chosen to permit concurrent operation of two satellites. The local solar time (LST) of equator crossing is usually 1400 or 1930 for ascending node and 0200 or 0730 for descending node.

Table A.1   Launch and Data Available Dates from NESDIS/NOAA for the Advanced TIROS-N Series Satellites

| Satellite Name | | Sensor | | |
After Launch	Before Launch	(Number of Channels)	Launch Date	Date Range
TIROS-N	TIROS-N	AVHRR/1 (4)	Oct. 13, 1978	Oct. 19, 1978–Jan. 30, 1980
NOAA-6	NOAA-A	AVHRR/1 (4)	June 27, 1979	June 27, 1979–Mar. 5, 1983
				July 3, 1984–Nov. 16, 1986
—	NOAA-B	AVHRR/1 (4)	May 29, 1980	Failed to achieve orbit
NOAA-7	NOAA-C	AVHRR/2 (5)	June 23, 1981	Aug. 19, 1981–June 7, 1986
NOAA-8	NOAA-E	AVHRR/1 (4)	Mar. 28, 1983	June 20, 1983–June 12, 1984
				July 1, 1985–Oct. 31, 1985
NOAA-9	NOAA-F	AVHRR/2 (5)	Dec. 12, 1984	Feb. 25, 1985–Nov. 7, 1988
NOAA-10	NOAA-G	AVHRR/1 (4)	Sept. 17, 1986	Nov. 17, 1986–present
NOAA-11	NOAA-H	AVHRR/2 (5)	Sept. 24, 1988	Nov. 8, 1988–present
NOAA-12	NOAA-D	AVHRR/2 (5)	May 14, 1991	Sept. 1, 1991–present

A.2 AVHRR Characteristics

AVHRR data are transmitted in real time to both automatic picture transmission (APT) and high-resolution picture transmission (HRPT) users. The APT system transmits data from any two of the AVHRR channels. The HRPT system transmits data from all AVHRR channels. The HRPT data format is found in the technical report.[4]

The AVHRR is a five-channel scanning instrument sensitive to visible, near-infrared, and infrared portion of the spectrum. Table A.2 lists the spectral characteristics of the AVHRR.[2,3] AVHRR/1 with four channels and AVHRR/2 with five channels are operational sensors. AVHRR/3 will be a future sensor on NOAA-K,L,M. The major changes from AVHRR/2 to AVHRR/3 are two points: a near-infrared channel will be a narrow band for vegetation research, and the sixth channel (1.6 μm; channel 3A in Table A.2) will be added for providing better discrimination between snow and clouds. The 3.7 μm will be transmitted during the night portion of each orbit, and the 1.6-μm channel will be transmitted during the daylight portion of the orbit.[5] Table A.3 is a listing of the basic AVHRR parameters.[2,3] The instantaneous field of view (IFOV) for all channels is specified to be 1.3 mrad. Spatial resolution is 1.1 km at the subpoint. A noise equivalent difference temperature (NEDT) is better than 0.12 K for a 300 K scene. Data of AVHRR consists of 10 b, and they are sent from the right column to the left column in each line that contains 2048 samples of data. The maximum scan angle is 55.4° and swath width becomes about 2900 km.

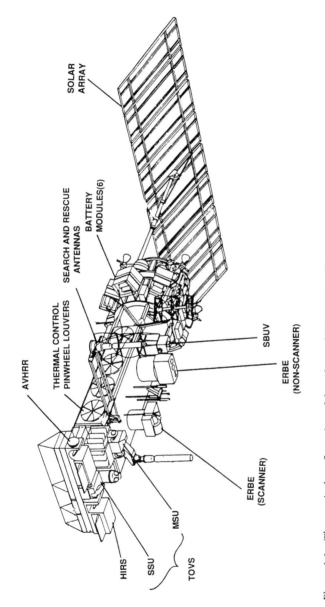

Figure A.1 The nominal configuration of the Advanced TIROS-N satellites.

Table A.2   Spectral Characteristics of the Advanced TIROS-N Series AVHRR Instruments

Channel No.	AVHRR/1		AVHRR/2	AVHRR/3
	TIROS-N	NOAA-6,8,10	NOAA-7,9,11,12,I,J	NOAA-K,L,M
1	0.55–0.90	0.55–0.68	0.55–0.68	0.55–0.68
2	0.72–1.10	0.72–1.10	0.72–1.10	0.84–0.87
3A	—	—	—	1.58–1.64
3/3B	3.55–3.93	3.55–3.93	3.55–3.93	3.55–3.93
4	10.5–11.5	10.5–11.5	10.3–11.3	10.3–11.3
5	—	—	11.5–12.5	11.5–12.5
Total channels	4	4	5	5

A.3 In-Flight Calibration

In order to work on AVHRR image data, there are three kinds of processing to correct the images. These are in-flight calibration, atmospheric correction, and geometric correction. It is necessary to prepare both instrument parameters for each AVHRR and orbiter parameters for each satellite. For the in-flight calibration, the prelaunch instrument parameters such as normalized response function, nonlinearity correction table, etc. are available from NESDIS/National Oceanic and Atmospheric Administration (NOAA) as a revised Appendix B of the technical report.[4]

Since characteristics of an infrared sensor vary with time, data of infrared channels need to be calibrated.[4-8] The AVHRR provides the following in-flight calibration data in every scan line:[4]

1. Output of four platinum resistance thermometers (PRTs) that measure temperature of the internal calibration target (ICT)
2. Observation data of ICT by AVHRR channels 3, 4, and 5
3. Observation data of space by all AVHRR channels

Table A.3   The Main Characteristics of AVHRR

Parameters	Characteristics
Instantaneous field of view (IFOV)	1.3 mrad (ca. 1.1 km on the ground)
Noise equivalent differential temperature (NEDT)	<0.12 K at 300 K
Scanning angle	−55.4° to + 55.4° (ca 2900 km on the ground)
Line rate	360 Lines per minute
Calibration	Visible, near IR: no calibration Thermal IR : deep space, blackbody
Quantization	10 b/pixel
Automatic picture transmission (APT)	2 Bands with 4 × 4 km resolution (visible and thermal IR)

PRTs are extracted from five consecutive AVHRR image scan lines and produce a reference value and four PRTs. Consequently, the five scan lines are a minimum data set for the in-flight calibration.

Radiometers are designed so that output X is proportional to radiation N:

$$N = G \cdot X + I \tag{1}$$

G and I represent characteristics of the sensor. (G is negative in the case of AVHRR channels 3, 4, and 5.) Time varying G and I are to be determined by observation of both space and ICT in the case of infrared sensors.

ICT is regarded as a blackbody. Planck's law relates radiation and temperature of a blackbody. Radiation density at a wave number v, and temperature T, is given by Equation 2:

$$\beta(v, T) = \frac{C_1 v^3}{\exp(C_2 v / T) - 1} \tag{2}$$

$C_1 = 1.1910659 \times 10^{-5} (\text{mW sr}^{-1} \text{ m}^{-2} \text{ cm}^{-4})$
$C_2 = 1.438833 \text{ cm K}$

Radiation N detected by a sensor is calculated by integration after multiplication by the response function. Response functions ϕ are defined at 60 discrete points:

$$N(T) = \sum_{i=1}^{60} \beta(v_i, T) \phi \Delta v \tag{3}$$

where the spacing between successive points is in Δv wave numbers. The expression (Equation 3) is used both to calculate the radiation of ICT and to obtain temperature from radiation determined from Equation 1. Inverse function of Equation 3 is required to calculate T from N. Since Equation 3 is a monotone function, linear interpolation can be used to calculate T.

Temperature PT measured by PRT is calculated from output PX by using the following equation:

$$PT = \sum_{j=0}^{4} a_{ij} PX_i^j \tag{4}$$

where PX_i is the mean count for PRT with $i = 0, 1, 2, 3, 4$; a_{ij} are the coefficients of conversion algorithm in each PRT.

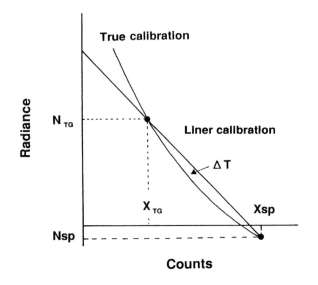

Figure A.2   Nonlinearity correction terms in the in-flight calibration processing.

The average of the internal blackbody temperature PT_i can be calculated as:

$$T_{TG} = \sum_{i=1}^{4} b_i PT_i \tag{5}$$

where b_i is the weighting factor of each PRT.

Let X_{TG} and Xsp be outputs of a sensor from ICT and space, respectively, and N_{TG} and Nsp be radiance of ICT and space, respectively. N_{TG} can be calculated by Equation 3 from T_{TG}, and Nsp is almost zero. G and I are given by Equations 6 and 7 (Figure A.2):

$$G = \frac{Nsp - N_{TG}}{Xsp - X_{TG}} \tag{6}$$

$$I = Nsp - G \cdot Xsp \tag{7}$$

Since the output of thermal infrared sensors is not usually proportional to the corresponding radiation, it is necessary to make nonlinearity correction (Figure A.2). The nonlinearity error(ΔT) that remains after linear calibration is given by the NESDIS nonlinearity correction table, which were generated from data obtained in the prelaunch experiments at ITT Aerospace/Optical Division.[6,8] The table is defined by matrix of internal target temperature and observing space temperature.

The in-flight calibration process is summarized as follows:

1. Determine Xsp, X_{TG}, and Nsp.
2. Get T_{TG} using Equation 5.
3. Get N_{TG} from T_{TG} using Equation 3.
4. Get G using Equation 6 and I using Equation 7 on a regression line through two points (X_{TG}, N_{TG}) and $(X$sp$, N$sp$)$.
5. Get No (radiance of observing pixels) from Xo (counts of observing pixels) using Equation 1.
6. Get To (temperature of observing pixels) from No using Equation 3.
7. Get ΔT using nonlinearity correction table with To and T_{TG}.
8. Get Tture (final corrected temperature) from To and ΔT.

A typical procedure of in-flight calibration is shown in Figure A.3. Five lines are a minimum consecutive data set for the procedure and n is a multiple coefficient. The minimum number of n is 1. When n is equal to 10, the processing is operated for 50 scan lines. In the correction of nonlinearity error, general method is employed with the procedure using conversion from radiance to temperature. Recent studies propose the conversion from radiance to radiance in nonlinearity correction process.[9,10] In the computer program, we employ the table lookup method for the procedure of count -> radiance -> temperature conversion.

A.4 Atmospheric Correction and Cloud Screening

There are two approaches for atmospheric correction to derive sea surface temperature using AVHRR: the empirical method and the radiative transfer method for determining the coefficients of the atmospheric correction equation.

According to the expression by Sobrino et al.,[11] the radiance $N_i(T_i)$ in channel i received by the AVHRR on NOAA can be written as:

$$N_i\left(T_i\right) = \varepsilon_i N_i\left(T_s\right)\tau_i + R_{uai} + \tau_i\left(1 - \varepsilon_i\right)R_{dai} \qquad (8)$$

In Equation 8, all quantities refer to a spectral integration over the band width of channel i. $N_i(T_s)$ is the radiance from sea surface, ε_i is the surface emissivity, τ_i is the total transmittance of the atmosphere, R_{uai} is the upwelling atmospheric radiance, and R_{dai} is the downwelling atmospheric radiance. The first term on the right-hand side is the radiation emitted by the sea surface. The second term is the radiation emitted by the atmosphere that depends on the water vapor and temperature profiles. The third term is the downwelling radiation emitted by atmosphere.

Hepplewhite[12] shows the results of analysis of the contributions to the deficit for the AVHRR. Atmospheric absorption of infrared radiation in spectral regions such as channels 3, 4, and 5 of AVHRR is caused primarily by water vapor and, particu-

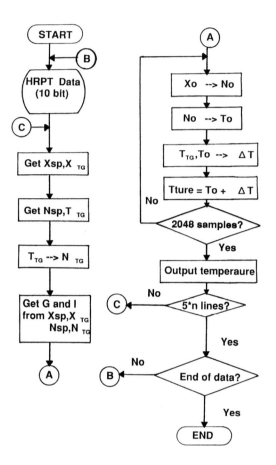

Figure A.3 Typical procedure of the in-flight calibration. The minimum number of *n* is 1. When *n* is equal to 10, the processing is operated for 50 scan lines.

larly, the water vapor continuum. The minor contribution factors are the uniformly mixed gases, such as CO_2, N_2O, and CH_4, and aerosols. The water vapor continuum absorption is most important in the window of channels 4 and 5 where line absorption is very weak. In the window of channel 3, dominant contributors are water vapor lines and aerosols. τ_i varies with these atmospheric absorption in Equation 8.

In general, seawater is radiatively black in the infrared with an emissivity of between 0.98 and 0.99. In the experience by Schluessel et al.[13] in the North Atlantic Ocean, spectrally averaged emissivities are from 0.886 to 0.891 when sea surface temperature varied between 280 and 300 K. If seawater is similar to a blackbody, ε_i is close to 1. When ε_i is close to 1, the first term of Equation 8 becomes relatively large and the third term becomes relatively small. That is an ideal condition, but the effects for decreasing ε_i value such as white caps on the sea surface must be considered for understanding of the variability of emissivity.

The theoretical development results in an algorithm of the form:

$$T_{SST} = T_i + \alpha\left(T_i - T_j\right) + \beta \tag{9}$$

where T_i and T_j are brightness temperature measurements at two different wavelengths in the thermal infrared region such as channels 4 and 5. The coefficients α and β are determined by regression either theoretically by using Equation 8 from a set of atmospheres, or empirically, by satellite data against *in situ* measurements by buoy and ship.[11]

A common empirical method is the multichannel sea surface temperature (MCSST) method[14] which is applied to several satellite SST data sets for the world ocean. The local precise parameters were proposed using the MCSST method.[15] NESDIS/NOAA has been applying this method to global area coverage (GAC) data collected by the TIROS/-N/NOAA series of satellites. Coefficients of the MCSST algorithm for each satellite are available from the NOAA users guide,[16] and coefficients for the operating AVHRRs are also available from on-line database NOAA.SAT.

Cloud detecting and screening are carried out in MCSST processing.[14] Cloud screening is very important for selecting a cloud-free ocean area since clouds are noises for oceanographic purposes.[17] This problem associated with clouds is discussed and reviewed by Darzi.[18] Various combinations of AVHRR visible and infrared channels are used to detect the presence of clouds. There are two kinds of operations on cloud detecting and screening for daytime and nighttime data. In daytime, it is possible to utilize visible and near-infrared channels for cloud detection, but the glitter area must be neglected where the ocean reflectance can exceed cloud reflectance. At nighttime, there are no visible and near-infrared data; differences in thermal channel responses (channels 3–4 and channels 4–5) are utilized for cloud detection. The recent AVHRR applications are summarized in several articles.[19-22]

A.5 Navigation and Mapping

It is necessary to correct the images for geometric distortions due to earth shape, earth rotation, variations in satellite orbits, and satellite altitude, etc. for overlaying and comparing with different times AVHRR image data. Emery et al.[23] reviewed these problems on navigation and mapping of AVHRR data. They pointed out that there are essentially two different approaches to the problem of polar orbiter image navigation. The first method assumes no orbital information, and thus relies on nominal orbital parameter values and image corrections computed by matching ground control points (GCPs) that apply to correct for errors in earth shape, scan geometry, satellite orbit, and satellite attitude. The second method uses high-quality satellite ephemeris data to make the correction with a limited number of GCPs. A GCP approach only works when land is present in the image, because there are very few GCPs in the open ocean. An alternative approach is to acquire accurate satellite ephemeris data and to use these data to allow image navigation with a fully elliptical model. With these corrections, it is possible to achieve image navigation accuracy of about 1.0–1.5 km that is equivalent to one pixel of AVHRR.

Following image navigation, remapping procedure can be separated into three steps:

Step 1. Transform geometrically from the output plane (desired map locations) to the input plane (remapped satellite image).
Step 2. Select image data with nearest scan line number and pixel number.
Step 3. Put image data to the matrix of output image.

This procedure can be generalized to simultaneously navigate all of the AVHRR channels at the same time rather than to do each channel separately, but the memory requirements for the transformation increase significantly to be able to store the scan line and pixel location arrays.[23]

A method with subpixel accuracy for geometric correction was proposed by Moreno and Melia.[24] This method allows the use of AVHRR data in a nearly local scale in combination with other high-resolution data, such as LANDSAT TM or SPOT.

The orbiter parameters are also available from NESDIS/NOAA as the TBUS information that is published by using the NOAA electronic bulletin board (NOAA.SAT) on Omnet/SCIENCEnet or SprintNet every day. Procedures for accessing NOAA.SAT must be referred to the NOAA users guide.[16]

References

1. Cotter, D. J., The United States operational polar and geostationary satellite, in *Weather Satellite: System, Data and Environmental Applications*, Rao, P. K., Holmes, S. J., Anderson, R. K., Winston, J. S., and Lehr, P. E., Eds., American Meteorological Society, Boston, 1990, chap. III-1.
2. Schwalb, A., The TIROS-N/NOAA satellite series, NOAA Tech. Memo, NESS 95, 1978.
3. Schwalb, A., ATN and GOES sensors, in *Weather Satellite: System, Data and Environmental Applications*, Rao, P. K., Holmes, S. J., Anderson, R. K., Winston, J. S., and Lehr, P. E., Eds., American Meteorological Society, Boston, 1990, chap. IV-2.
4. Lauritson, L., Nelson, G. J., and Porto, F. W., Data extraction and calibration of TIROS-N/NOAA radiometers, NOAA Tech. Memo, NESS 107, 1979.
5. Koczor, R. J. and Comeyne, G. J., Jr., Present and future uses of AVHRR multispectral data, *Proc. 19th Int. Symp. Remote Sensing Environ.*, Ann Arbor, MI, 1985, 739.
6. Brown, O. B., Brown, J. W., and Evans, R. H., Calibration of advanced very high resolution radiometer infrared observations, *J. Geophys. Res.*, 90(C6), 11667, 1985.
7. Tozawa, Y., Iisaka, J., Saitoh, S., Muneyama, K., and Sasaki, Y., SST Estimation by NOAA AVHRR and Its Application to Oceanic Front Extraction, Tokyo Scientific Center Rep., G 318-155, IBM Japan Ltd., 1981, 14 pp.
8. Weinreb, M. P., Hamilton, G., and Brown, S., Nonlinearity corrections in calibration of advanced very high resolution radiometer infrared channels, *J. Geophys. Res.*, 95(C5), 7381, 1990.
9. Steyn-Ross, D. A. and Steyn-Ross, M. L., Radiance calibration for advanced very high resolution radiometer infrared channels, *J. Geophys. Res.*, 97(C4), 55521, 1992.

10. Brown, J.W., Brown, O.B., and Evans, R.H., Calibration of Advanced Very High Resolution Radiometer infrared channels: a new approach to nonlinear correction, *J. Geophys. Res.*, 98(C10), 18257, 1993.

11. Sobrino, J. A., Li, Z.-L., and Stoll, M. P., Impact of the atmospheric transmittance and total water vapor content in the algorithms for estimating satellite sea surface temperatures, *IEEE Trans. Geosci. Remote Sensing*, 31, 946, 1993.

12. Hepplewhite, C. L., Remote observation of the sea surface and atmosphere and the oceanic skin effect, *Int. J. Remote Sensing*, 10, 801, 1989.

13. Schluessel, P., Emery, W. J., Grassl, H., and Mammen, T., On the bulk-skin temperature difference and its impact on satellite remote sensing of sea surface temperature, *J. Geophys. Res.*, 95(C8), 13341, 1990.

14. McClain, E. P., Pichel, W. G., and Walton, C. C., Comparative performance of AVHRR-based multi-channel sea surface temperature, *J. Geophys. Res.*, 90(C6), 11587, 1985.

15. Sakaida, F. and Kawamura, H., Estimation of sea surface temperature around Japan using the advanced very high resolution radiometer (AVHRR)/NOAA-11, *J. Oceanogr.*, 48, 179, 1992.

16. Kidwell, K. B., NOAA Polar Orbiter Data Users Guide (TIROS-N, NOAA-6, NOAA-7, NOAA-8, NOAA-9, NOAA-10, NOAA-11, and NOAA-12), NESDIS, NOAA, National Climate Data Center, Satellite Data Service Division, Washington, D.C., 1991.

17. Bernstein, R. L., Sea surface temperature estimation using the NOAA-6 satellite advanced very high resolution radiometer, *J. Geophys. Res.*, 87(C12), 9455, 1982.

18. Darzi, M., Cloud screening for polar orbiting visible and infrared satellite sensors, NASA Tech. Memo 104565, SeaWiFS Technical Rep. Ser., 7, 1992.

19. Saunders, R. W., An automated scheme for the removal of cloud contamination for AVHRR radiances over western Europe, *Int. J. Remote Sensing*, 7, 867, 1988.

20. Saunders, R. W. and Kriebel, K. T., An improved method for detection of clear sky and cloudy radiances from AVHRR data, *Int. J. Remote Sensing*, 9, 123, 1988.

21. Simpson, J. and Humphrey, C., An automated cloud screening algorithm for daytime advanced high resolution radiometer imagery, *J. Geophys. Res.*, 95(C8), 13459, 1990.

22. Gallegos, S. C., Hawkins, J. D., and Cheng, C. F., A new automated method of cloud masking for Advanced Very High Resolution Radiometer full-resolution data over the ocean, *J. Geophys. Res.*, 98(C5), 8505, 1993.

23. Emery, W. J., Brown, J., and Nowak, Z. P., AVHRR image navigation: summary and review, *Photogr. Eng. Remote Sensing*, 55, 1175, 1989.

24. Moreno, J. F. and Melia, J., A method for accurate geometric correction of NOAA AVHRR HRPT data, *IEEE Trans. Geosci. Remote Sensing*, 31, 204, 1993.

B

Coastal Zone Color Scanner on Nimbus-7

Marlon R. Lewis

B.1 Nimbus-7

The Nimbus-7 satellite was launched into polar orbit in October of 1978. The satellite was extraordinarily successful; it more than outlived the projected lifetime of the satellite and many of the instruments that it carried. The satellite orbited at an altitude of 955 km in a sun-synchronous mode with an ascending equator crossing time of near noon. One of the instruments carried onboard was the first satellite sensor specifically designed for the estimation of the pigment content of the surface ocean, the Coastal Zone Color Scanner (CZCS).

B.2 Coastal Zone Color Scanner

Sensor Characteristics

The Coastal Zone Color Scanner was a six-channel, visible and infrared scanning spectrometer with the primary purpose of observing of the color of the sea surface (see Chapter 11). The CZCS was designed as a proof of concept mission with a design life of only 1 year, a projected 10% duty cycle, and only 10% of the resulting data projected to be analyzed. The sensor in fact continued to operate for almost 8 years, until 1986, and returned data that has proved to be of immense value to

0-8493-4525-1/95/$0.00+$.50
© 1995 by CRC Press, Inc.

Table B.1   CZCS Sensor and Instrument Specifications[4]

Waveband (FWHM, nm)	Saturation Radiance (mW m^{-2} μm^{-1} sr^{-1})	SNR	Radiometric Accuracy (%)	Nadir Resolution (km)	Swath Width (km)
433–453	11.4	350	5	0.825	1800–1600
510–520	8.0	342	5	0.825	1800–1600
540–560	6.4	280	5	0.825	1800–1600
660–680	2.9	209	5	0.825	1800–1600
700–800	24.0	50	5	0.825	1800–1600
10,500–12,500	N/A, SST				

biological and physical oceanography. The resulting images have revolutionized the way oceanographers view the biological workings of the sea.

The main characteristics of the optical bands are given in Table B.1. CZCS had five bands in the visible and one band in the thermal infrared. Depolarization optics were used, along with a selectable gain for bands one through four. The sensor was capable of tilting fore and aft 20° along the velocity vector of the satellite to avoid sun-glint contamination.

Bands one through four proved to be the most useful for retrieval of the water-leaving radiances. The resulting 60,000 images (each covering approximately an area 970 × 1500 km) were processed to derive the pigment concentration and attenuation coefficient on a global scale[1] (also see Chapter 11).

The satellite operated in two transmission modes. First, high-resolution data (the so-called local area coverage [LAC]) represented a resolution at nadir of 0.825 km and was broadcast directly. To reduce the data volume stored onboard the satellite, the LAC data was subsampled to the so-called global area coverage (GAC) that represented approximately 4-km ground resolution.

Data Processing

After initial screening, the entire CZCS archive was processed in a herculean effort by the Goddard Space Flight Center (GSFC) and the University of Miami[1,2] to result in a unique data set now available to the scientific community. The processing consisted of several levels, with much care put into the quality control of the resulting data. A fuller description of the processing methodologies, the quality control, and the science objectives can be found elsewhere.[3]

References

1. Feldman, G., Kuring, N., Ng, C., Esaias, W., McClain, C., Elrod, J, Maynard, N., Endres, D., Evans, R., Brown, J., Walsh, S., Carle, M., and Podesta, G., Ocean color: availability of the global data set, *Eos*, 70, 634, 1989.
2. Esaias, W., Feldman, G., McClain, C.R., and Elrod, J., Satellite observations of oceanic primary productivity, *Eos*, 67, 835, 1986.

3. McClain, C., Feldman, G., and Esaias, W., A review of the Nimbus-7 Coastal Zone Color Scanner data set and remote sensing of biological oceanic productivity, in *Global Change Atlas*, Parkinson, C., Foster, J., and Gurney, R., Eds., Cambridge University Press, London, 1993.

4. Ball Aerospace Systems Division, Development of the Coastal Zone Color Scanner for Nimbus-7: Vol. 2 -Test and Performance Data, Ball Aerospace Systems Division, Boulder, CO, Final Rep. F78-11 Rev. A, 1979, 101 pp.

C

Microwave Radiometers

Kristina B. Katsaros and W. Timothy Liu

This appendix provides the main characteristics of Scanning Multichannel Micro-wave Radiometers (SMMRs) and Special Sensor Microwave/Imagers (SSM/Is).

0-8493-4525-1/95/$0.00+$.50
© 1995 by CRC Press, Inc.

Table C.1   Microwave Radiometers Launched Since the Late 1960s

Year	Platform	Sensor
1967	Mariner-2	
1968	Cosmos-243	
1970	Cosmos-384	
1971	Nimbus-5	ESMR
1973	Skylab	3-193
		S-194
1974	Meteor	
1975	Nimbus-6	ESMR
1978	Seasat	SMMR
1978	Nimbus-7	SMMR
1987	MOS-1	MSR
1987	DMSP F8	SSM/I
1990	DMSP F10	SSM/I
1991	DMSP F11	SSM/I
1991	ERS-1	ATSR/M
1992	Topex/Poseidon	TMR
1994	DMSP F12	SSM/I
1996	DMSP F13	SSM/I
1997	TRMM	TMI
1998	DMSP F14	SSM/I
1998	ADEOS-II	ATSR/M
2000	EOS PM	MIMR

Note: Sounders are not included

Acronyms:

ATSR/M	Along-Track Scanning Radiometer and Microwave Sensor.
EOS	Earth Observing System.
ERS	European Remote Sensing Satellite.
ESMR	Electronically Scanning Microwave Radiometer.
DMSP	Defense Meteorological Satellite Program.
MIMR	Multifrequency Imaging Microwave Radiometer.
MSR	Microwave Scanning Radiometer.
SMMR	Scanning Multichannel Microwave Radiometer.
SSM/I	Special Sensor Microwave/Imager.
TMI	TRMM Microwave Imager.
TMR	Tropospheric Moisture Radiometer.
TOPEX	Topographic Experiment.
TRMM	Tropical Rainfall Measuring Mission.

Table C.2   Characteristics of the Two Polar Orbiting
Radiometers: SMMR on Seasat and on Nimbus-7 and SSM/I
on the F8, F10, and F11 DMSP Satellites

SMMR		SSM/I	
Freq. GHz	Approx. Resolution (km)	Freq. GHz	Approx. Resolution (km)
6.6	150	—	—
10.7	100	—	—
18	65	19.35	55
21	60	22.235	50
37	35	37.	35
—	—	85.5	15

Swath Width: Swath Width   1400 km
  Seasat: 650 km
  Nimbus-7: 780 km
Period of Operation   Period of Operation
  Seasat: July–October 1978   July 1987–Present
  Nimbus-7: October 1978–
           Fall, 1987

D

Radar Altimeter Sensors and Methods

Lasse H. Pettersson, Paul Samuel, and Åsmund Drottning

D.1 Introduction

Active microwave remote sensing technology for spaceborne radar altimeter (RA) sensor systems has evolved over the last 20 years. Evaluation and applications of RA sensors started with the Skylab mission in 1973.[1] This was followed by the U.S. Geos-3 mission (1975–1978) (e.g., Reference 2) and the National Aeronautics and Space Administration (NASA) Seasat mission that was operational for 3 months in 1978.[3,4] The first long-term, global satellite RA mission was the Geosat launched by the U.S. Navy in 1985 and was operational until 1989.[5,6] The first 2 years of this mission were dedicated for mapping the global geoid with a spatial sampling interval down to 4 km; and the major part of the data from this phase is classified. In November 1986, Geosat commenced its Exact Repeat Mission (ERM), with a 17-day repeat cycle; and data from this phase has been freely available to the scientific community. In July 1991, the European Space Agency (ESA) launched its

Table D.1   A Summary of Some Key Information of the Present and Past Satellite Missions Equipped with RA Sensor Systems

	GEOS-3	Seasat	Geosat	ERS-1	TOPEX/Poseidon
Operational	1975–1978	1978	1985–1989 ERM[a] 1987–1989	1991–	1992–
Agency	NASA	NASA	US Navy	ESA	NASA CNES
Data distribution	NASA JPL	NASA JPL	NOAA NGS	ESA ESRIN	JPL/DAAC CLS/AVISO
Frequency (GHz)	13.9	13.5	13.5	13.8	T: 13.6 and 5.30 P: 13.65
Auxiliary sensors used in the processing of geophysical parameters	TRANET tracking network Laser reflector array (LRA)	TRANET LRA Multichannel microwave radiometer (SMMR)	TRANET	PRARE tracking system failed) LRA Along-track scanning radiometer (ATSR)	LRA DORIS tracking system Dual-frequency altimeter for ionospheric correction TOPEX microwave radiometer (TMR) GPS demonstration receiver
Inclination	115°	108°	108°	98.5°	66°
Orbital height (km)	843	800	800	780	1336
Repeat cycle of orbit (days)	Various	3	17 d during ERM	3, 35, or 168	10 T ≈ 90% P ≈ 10%

[a]   ERM: Exact repeat mission, an unclassified 17-d repeat cycle.

first RA sensor on the European Remote Sensing Satellite-1 (ERS-1).[7] So far ERS-1 has been operational in different orbit configurations with 3- and 35-d repeat cycles and will completed its geodetic phase with a 168-d repeat orbit from April 1994 to March 1995, giving coverage at a spatial resolution of about 16 km at the equator. A cooperative space venture between NASA (U.S.) and CNES (France) resulted in the launch of the TOPEX/Poseidon satellite, the first dedicated RA mission,[8,9] on August 10, 1992. TOPEX/Poseidon is equipped with two different RA sensors and dedicated auxiliary instrumentation for computing optimal correc-

Table D.2   Typical Amplitudes and Spatial Scales of the Various Components of the Altimeter Height Measurements

Height Component	Vertical Range	Typical Spatial Scale
Ocean dynamic topography		
Mesoscale eddies	Tens of cm	Tens to hundreds of km
Continental boundary	Eastern: ≈20 cm	≈500 km
Currents	Western: ≈100 cm	≈100km
Ocean gyres	≈100 cm	>1000 km
Equatorial currents	≈20 cm	>500 km
Sea-state effect (EM-bias)	≈10 cm	>20 km
Ocean tide	Up to several m	Tens of km
Inverse barometric effect	Up to 50 cm	Tens of km
Geoid	−100 m to +80 m	Spatial variation on all scales

tions to the altimeter-derived height data. The satellite has a 10-d repeat orbit and an expected lifetime of 3–5 years. A summary of some key information on the present and past RA satellite missions is given in Table D.1.

Several additional RA missions are planned for launch during the coming years, ensuring the continuity of the altimeter data set. ESA has scheduled a second launch of RA sensor system on ERS-2 in 1995 and is planning a third RA sensor on ENVISAT-1 in 1998. Both a Geosat follow-on (1996–2010) and a TOPEX/Poseidon follow-on (1998) are under consideration. RA sensors are also candidates for inclusion in the NASA EOS space platform payloads in the beginning of the next decade.

Analysis of the RA return signal can yield information on ocean circulation, significant wave height and near-surface winds, as discussed in Chapters 2, 4, 6, 8, 12, and 14. Here a brief description of RA measurement principles and data processing methods are given.

D.2 Radar Altimeter Measurement Principles

The RA sensor is a nadir-looking active microwave sensor usually in the 13–14 GHz frequency band. The signal pulse, transmitted vertically downward, interacts with the land or ocean surface and is reflected back to the altimeter antenna. The time taken for the round-trip is accurately clocked and is used, together with knowledge of the propagation speed of electromagnetic waves, to compute the altitude of the antenna above the reflecting surface. If the exact position of the sensor with respect to a fixed reference surface, such as the reference ellipsoid, is known from orbit computations, the instantaneous height of the sea surface above the reference surface can be computed. Repeated altimeter observations over the same location make it possible to resolve both the stationary and transient components of the ocean topography. For many of the geophysical components constituting the sea surface height signal, such as those due to ocean circulation, subdecimeter accuracy is required to resolve the typical scales present (see Table D.2). This requirement

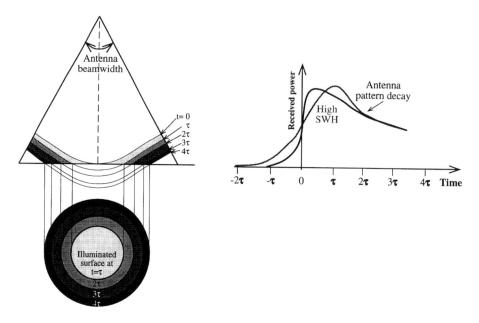

Figure D.1   A schematic view of the geometry of the pulse-limited RA and the signal interaction with the ocean surface. The shown return pulse is based on an average of a large number of reflected pulses.

determines the constraints to be placed on RA sensor performance and the accuracy of the auxiliary data sets used for correcting the RA measurements.

To measure the height (e.g., of ERS-1 at an altitude of approximately 780 km), the travel time and speed of the signal pulse should be determined to an accuracy of 1 in 10^7, which in turn requires a clock accuracy better than 0.5 nsec. Such high timing accuracy requires wide frequency bandwidths. The alternative is to use a relatively narrowband pulse and to fit the reflected pulse to a curve from which the arrival time can be determined. The latter technique requires that the received pulse has a high signal-to-noise ratio, which is realized by chirp frequency modulation and compression of the signal. To further increase the pulse duration and energy level, a number of pulses, typically 1000, are averaged.

For measurements over the ocean, pulse-limited altimeters, with an antenna beam surface footprint larger than the effective pulse surface footprint, are generally used. The pulse footprint radius is dependent on the satellite altitude, the pulse duration, and the significant wave height (SWH) (see Reference 10). In the case of the Geosat, typical values range from about 1.7 km for calm seas to 9.8 km at 20 m SWH. The geometric principles of the pulse-limited signal interaction with a flat surface are illustrated in Figure D.1. The pulse initially illuminates a point on the surface that gradually expands until the trailing edge of the pulse reaches the surface. After this, an annular area of increasing radius, but nearly constant area, will be illuminated. The reflected pulse will correspondingly show an initial linear increase in power after which it levels off (see Figure D.1). For a rough ocean surface, the slope of the initial rise will be smaller since the time over which the power increases will be longer, corresponding to the time between the arrival of the leading edge at the highest crest and that of the trailing edge at the lowest trough. Thus an analysis of the form of the return pulse can yield estimates of the sea state, or more precisely, SWH.[11]

Surface wind speed is estimated by its effect on the return pulse amplitude. With increasing wind, the sea surface becomes rougher so that the number of specularly reflecting faces decreases and more energy is scattered in off-nadir directions. This leads to a decrease in the amplitude of the mean backscattered power, which can thus be related to wind speed. In practice, the radar backscatter coefficient is computed from the automatic gain control (AGC) that is used to normalize the mean detected return pulses for acquisition and tracking. A number of algorithms have been proposed to retrieve wind speed from the altimeter return pulse.[12,14]

D.3 Geophysical Parameters

To increase the level of accuracy of the altimeter-derived geophysical parameters, the return pulses are typically averaged to provide one data point every second along the satellite ground track. At a satellite ground speed of 7 km/s, each altimeter data value represents an elongated area of about 7 km along track, or more according to the actual sea state condition. These 1-s average values are the basis for the derived geophysical parameters included on the Geophysical Data Records (GDR) from the different satellite missions which are distributed to the application science community.

A high level of accuracy is required for all the correction terms in the altimeter height budget depicted in Figure D.2. The error budget for the RA sensors used on the major altimeter missions is presented in Table D.3;[15] there the corrections and errors due to the RA instrument, atmospheric factors, satellite orbit uncertainty, as well as those due to the various components of the instantaneous sea level, are included.

As to the instrument, an exact determination of the spacecraft internal geometry including the location of the altimeter antenna and the actual center of mass of the spacecraft is required. As mentioned above, the pulse bandwidth limitation introduces an uncertainty due to the fact that each return pulse is an average of a large number of individual pulses with a variable shape. This introduces a noise in the ability of the sensor to estimate the exact arrival time of the reflected pulse, termed tracker bias. Current sensor technology has an instrument error at the 2-cm level. Another error arises from the fact that the troughs of ocean waves present a greater number of reflecting faces than the crests, leading to a bias toward the troughs when the mean sea surface is computed. This so-called electromagnetic bias (EM-bias) of the range estimate increases with SWH and is usually corrected by means of an empirical function related to the altimeter-derived SWH.

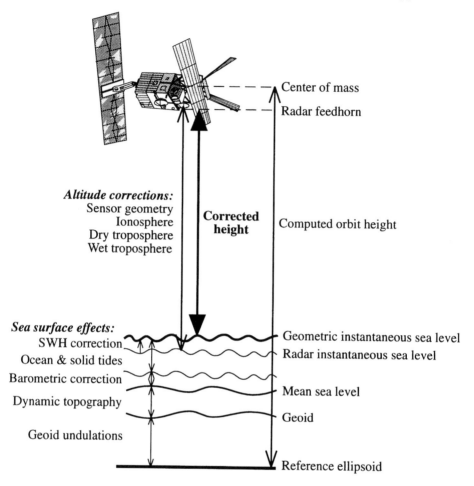

Figure D.2 A schematic view of the altimeter height budget and the various components and corrections that are relevant to determine the sea surface topography.

Although the atmosphere is quite transparent to microwave radiation, the propagation speed of the RA signal is affected by free electrons in the ionosphere and by air molecules and water vapor in the troposphere.

The ionospheric range correction, or the range delay of the RA pulse due to the ionosphere, is related to the total electron content (TEC). During periods of high solar activity the dynamic range of the ionospheric effect on the height estimates can be up to 30 cm. To meet the accuracy of 1–2 cm in range determination, the TEC needs to be determined at the level of 5% accuracy. For the earlier altimeter missions the ionospheric range determination was based on indirect observations of Faraday rotation (magnetic field observation) at ground stations, ionosodes, and global models of the TEC (e.g. Reference 16). The TOPEX altimeter, on the other

Table D.3   Residual Accuracies of Absolute Sea Level Measurements

Sources	GEOS-3	Seasat	Geosat	ERS-1	TOPEX/Poseidon
Altimeter	13.9 GHz	13.5 GHz	13.5 GHz	13.8 GHz	5.3 GHz/13.6 GHz (13.65 GHz)
Instrument noise	50 cm	10 cm	5 cm	3 cm	<2 cm
Bias uncertainty	—	7 cm	5 cm	3–5 cm	3 cm
Time bias	—	5 ms	3–5 msec	1–2 msec	<1 ms
Electromagnetic (EM) bias	10 cm	5 cm	2 cm	2 cm	<2 cm
Skewness	2 cm	1 cm	1 cm	1 cm	1 cm
Dry troposphere	2 cm	2 cm	1 cm	1 cm	1 cm
Wet troposphere	15 cm	3 cm	4 cm	1.2 cm	1.2 cm
Ionosphere	2–3 cm	2–3 cm	2–3 cm	2–3 cm	1.3 cm (2 cm)
Orbit	30–50 cm	30 cm	20 cm	18 cm	3.5 cm
Root-sum-squared error	67 cm	33 cm	22 cm	19 cm	<5 cm

Notes:

1. TOPEX altimeter is dual-frequency and Poseidon altimeter is single-frequency.
2. Three-channel radiometer for TOPEX/Poseidon.
3. ERS-1 assessment conducted using the Ocean Products Data (OPR02).
4. Altimeter data average over 1–3 s.
5. Root-sum-squared error excludes time bias error.
6. Seasat has tracker bias error included in the EM bias estimates.
7. Ionosphere error estimated for normal solar activities. Poseidon ionospheric correction to be improved using Doris measurements (error given in parenthesis).
8. TOVS/SMMI wet troposphere correction model for Geosat. Geos-3 altimeter is assumed to have no available wet tropospheric corrections.
9. Assumptions: significant wave height is 2 m, wave skewness is 0.1; and the pressure field is assumed to be accurate to +3 mbar for Geosat, ERS-1 and TOPEX/Poseidon.

After Shum, C. K., Ries, J. C., and Tapley, B. D., *Geophys. J. Int.*, 121, 321–336, 1995. With permission.

hand, makes use of the fact that the TEC range delay of radio signals is frequency dependent. By making range measurements at two frequencies, 13.6 and 5.3 GHz, the boresight range delay due to the TEC can be estimated. The same dual frequency technique is the basis for using the satellite tracking system for estimation of TEC. The Doppler Orbitography and Radiopositioning Integrated by Satellite (DORIS) tracking system of the TOPEX/Poseidon satellite measures the differential TEC along the path between the satellite and the various ground station beacons. With the current global network of 40 to 50 stations the spatial resolution of the TEC estimates seems to be sufficient and the range accuracy of 2 cm rms is achieved by the DORIS measurments.[17-19]

The mass of dry air molecules in the atmosphere causes a range delay termed the *dry tropospheric effect*. This effect is directly proportional to the sea level pressure, with an average magnitude of about 2.3 m. To perform a range correction at 1 cm accuracy the pressure needs to be determined to 4-mbar accuracy. Over most of the oceans this level of accuracy is realizable from the current meteorological observations and modeling efforts. Generally, this correction is computed from global

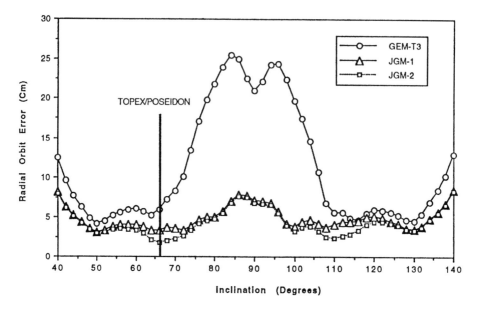

Figure D.3   The radial orbit error at the Topex/Poseidon altitude as a function of inclination, based on various gravity models. (After Shum, C.K., Ries, J.C., and Tapley, B.D., *Geophys. J. Int.*, 121, 321, 1995. With permission).

models such as those at the Fleet Numerical Oceanographic Center (FNOC) or the European Center for Medium Range Weather Forecasting (ECMWF).

The *wet tropospheric correction*, or the range delay due to atmospheric water vapor, varies considerably spatially and temporally. Typically, the range effect of this correction is between 5 and 30 cm. The variations are correlated with atmospheric frontal systems with spatial scales of tens to hundreds of kilometers and temporal scales down to a few hours. Auxiliary information sources are needed to estimate the water vapor content. These could be interpolated from meteorological models (FNOC, ECMWF, etc.) or passive microwave radiometer data (Scanning Multichannel Microwave Radiometer [SMMR], Special Sensor Microwave Imager [SSM/I], etc.), but several investigators have voiced concern that the resulting errors could lead to erroneous interpretation of the ocean signal (e.g., Reference 20). On the ERS-1 satellite, estimates for this correction are available from the Along-Track Scanning Radiometer (ATSR) and on the TOPEX/Poseidon, from the nadir-looking TOPEX Microwave Radiometer (TMR). Preliminary results from a comparison of the TMR estimates with those from ECMWF indicates a bias of 1.5 cm.[17]

As to the *orbit*, the satellite ephemerides can be estimated from a global gravity field model and knowledge of the various forces acting on the satellite. The largest uncertainty in orbit computations is due to errors in the gravity field model and depends on the orbital inclination (Figure D.3). Unmodeled variations in atmospheric drag, solar radiation pressure, and other forces working on the spacecraft add to the uncertainty. The precision of the orbit determination depends on *a priori* knowledge of these factors as well as on the regularity with which the satellite can be tracked. Such satellite tracking systems have been developed for RA satellites and examples of these are the Laser Reflector Arrays (LRA) carried by all the altimeter satellites; others are the Tracking Network (TRANET) Doppler tracking system on Seasat and Geosat, the Precise Range and Range-Rate Equipment (PRARE) system that failed on ERS-1 is mounted on ERS-2, the DORIS two-frequency Doppler tracking system on TOPEX/Poseidon and the Global Positioning System (GPS) demonstration receiver on TOPEX/Poseidon. For RA data, the radial component of the orbit error becomes particularly important. It can be divided into geographically uncorrelated and geographically correlated parts. The former varies in time and is different for ascending and descending tracks passing through a given point. The latter are concentrated at a set of special frequencies determined by the rotational period of the earth and the orbital period of the satellite.[21] It depends only on the geographic location and is therefore identical for all ascending and descending tracks passing through a given geographical point. Such errors can be reduced only by improving the earth gravity model or the tracking abilities, and are usually assumed to be small and uniform over regional scales. Radial orbit errors still remain as important obstacles for exploiting the full potential of RA data.

There are several sea level geophysical components. Depending on the actual use of the sea surface topography data, the various geophysical signals — the ocean tides, the inverse barometric effect due to atmospheric pressure fluctuations, the dynamic topography due to the ocean currents, and the geoid due to the gravity field — may be regarded as the signal itself or as error sources.

Ocean Tides.   Usually, independent global ocean tidal models are the source of the tidal elevation correction included on the GDRs. Such global tidal models may show large differences realtive to regional tidal models. For the Seasat GDRs in the Nordic seas, differences up to 25 cm were found, with an average bias of 8 cm.[22] Such discrepancies may lead to erroneous interpretation of the sea level signal, as several investigators have noted.[23,24]

Barometric Pressure.   The sea level signal due to atmospheric pressure fluctuations is called the inverse *barometric effect*, and instantaneous values for it can be computed indirectly from the dry troposphere correction. However, the timescale of the oceanic response to pressure fluctuations is not well documented, and this effect may be a significant source of error in RA data. Comparison of the inverse barometric correction computed by a regional sea level model over the Nordic seas and that derived from the FNOC dry tropospheric correction on the Seasat GDRs showed agreement on average, but differences of up to 13 cm were seen.[22]

Dynamic Topography.   This is the contribution due to ocean circulation. To a first approximation, the geostrophic balance may be assumed, which implies surface slopes proportional to the current velocity. Surface slopes due to circulation features range from over 10^{-5} (1 m/100 km) for intense western boundary currents to less than 10^{-7} (50 cm/5000 km) for equatorial currents. These signals also have a wide range of spatial and temporal scales, from tens to thousands of kilometers, and from a few days to many years, respectively.

The Geoid. The geoid is a level surface of earth gravity potential and can be regarded as time invariant for typical RA applications. Its height relative to the reference ellipsoid has a peak-to-peak amplitude of about 200 m, with regional slope variations up to 20 cm/km. The mean sea surface determined from the RA data is mainly made up of the contribution from the geoid, but also includes the average height contribution from the geostrophic ocean currents. To derive the geoid height accurately from altimetric data alone, one therefore needs estimates of the sea surface height contribution due to these currents. Independent gravity estimates may also be used to determine the actual geoid variation, thus providing an estimate of the dynamic topography. Regionally the geoid height can be derived from gravity measurements from ships, but accurate global estimates will require a dedicated satellite gravity field mission such as the proposed Aristoteles satellite.[25]

Except for the last two components of the height budget, the above parameters are included on the GDRs from the different missions. These estimates are commonly used for processing RA data to obtain the various geophysical parameters constituting the height signal. Some of the commonly used processing methods are presented below.

D.4 Processing Methods for Ocean Dynamics Applications

Repeat Track Analysis

Repeat track analysis, or collinear analysis, is the standard processing technique used for applications of RA data to study oceanic variability;[26] and it relies on the fact that almost all RA missions use a repeating orbit where the satellite overflies the same geographic location at regular intervals.

The geoid is one of the major constituents of the geometric sea surface height, exhibiting spatial variations at all scales with amplitudes much larger than that due to ocean dynamics. Currently available geoid models do not have the accuracy required for oceanographic applications, except in some local areas. However, since the geoid is time invariant, this does not represent an insurmountable obstacle to the study of time-varying currents, as long as RA measurements sample the sea surface at an interval sufficient to resolve the energetic scales.

Another important error source is that due to inaccuracies in the orbit computations. However, this error is mostly at very long wavelengths and can be separated from the mesoscale signal.

Repeat track analysis eliminates most of the error due to these two sources and also reduces the error due to some of the other sources. Each segment of the track within the selected study area is processed separately. For convenience, we will refer to these segments as tracks and heights along a track as repeat passes. Given a number of repeat passes along the same track, a mean pass is first computed as the average of all the repeat passes. Provided the sampling interval and duration are adequate to resolve significant scales, the mean pass will represent the sum of the geoid, the mean dynamic sea level, and the mean of the error terms, including orbit error. This mean pass is then subtracted from each of the repeat passes, to retrieve the varying part of the dynamic sea level (deviations) and the error terms. The orbit error, being a long-wavelength component can then be reduced by subtracting a linear or quadratic function fitted to the deviations, provided the track is at least twice as long as the longest wavelength that one is interested in retaining. The above two operations also reduce the contribution of the other error terms and the remaining signal represents the sea surface variations induced by mesoscale circulation features.

Instead of subtracting a mean pass, one could also subtract a selected repeat pass from all the other repeat passes, reduce orbit error, and then subtract the mean of the corrected repeat passes to retrieve the sea level deviations. This method is preferable when there are data gaps along the repeat passes since these could give rise to sharp discontinuities in the computed mean pass if the first method is used.

Another variation of the repeat track method uses along-track slope deviations instead of sea level deviations to study the mesoscale.[27] Along-track slopes are computed after subtraction of a mean pass or reference pass. The advantage of this method is that orbit error is automatically eliminated since numerical derivation is a high-pass filtering operation. At the same time, the mesoscale signal of interest is enhanced. Also, slopes of the sea surface are more easily associated with the current component normal to the track (v_n) through the geostrophic relationship:

$$v_n = \frac{g}{f} \cdot \frac{\delta h}{\delta s} \tag{1}$$

where g is the acceleration due to gravity, f is the Coriolis parameter, and $\frac{\delta h}{\delta s}$ is the along-track slope.

These variations of the repeat track analysis can be used to study individual mesoscale features such as eddies, fronts, etc. Regional distributions of the mesoscale energy field can also be mapped, either as the root mean square (rms) of the sea surface height variations, or as the eddy kinetic energy. For a fuller utilization of the information content of the data set, dynamic models have to be introduced for appropriate space-time interpolation. Altimeter data are also seen as among the most promising data sources for oceanic data assimilation schemes.[47]

Combination of Satellite RA Data and Sea-Gravity Data

For regional studies of the mean ocean circulation and the gravity field, satellite RA data may be used in conjunction with conventional sea-gravity data. As mentioned above, repeat track analysis provides a set of mean passes and estimates of the ocean current variability. The mean passes can be merged to describe the mean sea surface by reducing the remaining geographically uncorrelated radial orbit error in a crossover adjustment. Since this error is dominated by long-wavelength components, the along-track error can be approximated by linear or quadratic functions, or Fourier series.[28] The coefficients of these functions are estimated by minimizing the height differences between crossing mean passes and possibly fitting the data to a precomputed mean sea surface model.[29]

After the crossover adjustment, the data can be regarded simply as the sum of the geoid height and the dynamic topography due to the mean ocean circulation. A separation of the two is complicated by the fact that the sum of the two is determined with higher accuracy than either component can be estimated separately. Furthermore, the geoid height is a much larger quantity than the dynamic topography. The separation is essential for obtaining an improved description of both the mean ocean currents and the gravity field using satellite RA data.

This requires that independent information about either the geoid height or the mean dynamic topography is introduced. Additional gravity field data can be acquired from ships. They measure the gravity anomaly — the variations of earth gravity acceleration relative to gravity acceleration associated with the reference ellipsoid. Such data either may be used independently to estimate the geoid height or together with the RA data within an integrating mathematical framework, e.g., least squares collocation.[30] The former approach makes no use of gravity field information in the RA data and is limited by deficiencies in the ability to determine high-resolution gravimetric geoids. The errors of these may exceed the signal from the ocean currents.

Simultaneous use of RA data and sea-gravity data by least squares collocation allows for the exploitation of the full signal content of the data.[31] The method takes into account the statistical properties of the various signals by using covariance functions, and the modeling of these is vital for obtaining realistic results. In particular, the covariance function models associated with the ocean circulation[32] need further development and verification. The multidisciplinary aspects of satellite RA data underline the need for an integrated approach for data processing and interpretation.

D.5 Other Applications of Radar Altimeter Data

Satellite Altimetry in Geodesy/Gravimetry

RA data from satellites have caused a dramatic improvement in the ability to describe the marine gravity field worldwide, in particular, in the medium- and short-wavelength range. Many ocean areas that previously were covered by few or no conventional observations are now mapped in great detail directly from space.[33,34] A near global coverage of gravity field data is thus available, and this has facilitated the computation of high-degree and order geopotential models[35] with significantly higher resolution than their predecessors without RA data.

The mean sea surface determined from satellite RA data is a close approximation to the marine geoid, which is also an important reference surface for precise positioning at sea. The spatial variations of the geoid are greater than both the temporal and spatial variations of the other signals and errors in the RA data. The variations of the gravity field are mainly caused by the density distribution of the interior of the earth. Of major importance are the density contrasts associated with the transition zone between crust and mantle, geologic structures in the crust, or the bathymetry.

Standard methods allow the conversion of geoid heights into, e.g., gravity anomalies or deflections of the vertical, the deviation of the plumb line from the reference ellipsoid normal.[36,37] Then the short-wavelength characteristics of the gravity field are enhanced, and this makes RA data useful tools in, e.g., oil exploration and plate tectonic studies. RA data provide the flexibility of acquiring gravity field information in all ocean areas without being dependent on expensive shipborne surveys or being hampered by political obstacles.

Satellite Altimetry over Polar Ice Caps

Satellite RA data provide the only practical means of mapping the topography of polar ice caps[38] and monitoring their mass balance,[39-41] given the vast areas these cover. The ice caps of Greenland and Antarctica cover about 10% of the surface of the earth, and changes in their ice volumes are both indicators and causes of climate change.

By the turn of the century a 15-year time series of ice elevation changes will be available, reflecting the changes in ice volume. These constitute a significant uncertainty in the mass budget of the oceans and should be considered in the context of sea level changes. In particular, the coupled ocean-atmosphere models used in the International Panel on Climate Change scientific assessment project[42,43] predict increased precipitation in the high latitudes of the Northern Hemisphere. Over the same region, the surface air temperature is also expected to increase.

RA return signals become more complex and variable over ice caps than over open ocean, causing an increased uncertainty in the range determination. The full waveform of the RA data should be used in a complete retracking of the reflection points.[44] When the ice surface is flat, the range is measured at the nadir point of the satellite. Otherwise, the reflection point is moved upward to the point on the ice surface closest to the satellite; therefore, the data need to be corrected for the slopes of the ice surface.[45] RA data over ice caps can only be acquired where the ice surface is relatively smooth, and require special processing to optimize the range determination.

References

1. McGoogan, J. T., Miller, L. S., Brown, G. S., and Hayne, G. S., The S-194 radar altimeter experiment, *IEEE Trans.*, 1974.
2. Stanley, R. H., The GEOS-3 project, *J. Geophys. Res.*, 84, 1979.
3. Seasat Special Issue I, Geophysical evaluation, *J. Geophys. Res.*, 87(C5), 1982.
4. Seasat Special Issue II, Scientific results, *J. Geophys. Res.*, 88(C3), 1983.
5. Geosat Special Issue I, Sea level from space, *J. Geophys. Res.*, 95(C3), 1990.
6. Geosat Special Issue II, Sea level from space, *J. Geophys. Res.*, 95(C10), 1990.
7. Francis, R., Graf, G., Edwards, P. G., McCraig, M., McCarthy, C., Dubock, P., Lefebvre, A., Pieper, B., Pouvreau, P.-Y., Wall, R., Wechsler, F., Louet J., and Zohl, R., The ERS-1 Spacecraft and Its Payload, European Space Agency, Bulletin No. 65, 1991.
8. Fu, L. L., Lefebvre, M., and Patzert, W., TOPEX/Poseidon science plan overview, *EOS Trans. AGU*, No. 73, American Geophysical Union Fall Meet., San Francisco, CA, 1992.
9. TOPEX/Poseidon Special Issue I, Geophysical validation, *J. Geophys. Res. Oceans*, 99(C12), 1994.
10. Walsh, E. J., Problems inherent in using aircraft for radio oceanographic studies, *IEEE Trans. Antennas Propag.*, AP 25(1), 145, 1977.
11. Fedor, L. S., Godbey, T. W., Gower, J. F. R., Guptill, R., Hayne, G. S., Rufenach, C. L., and Walsh, E. J., Satellite altimeter measurements of sea state — an algorithm comparison, *J. Geophys. Res.*, 84(B8), 3991, 1979.
12. Brown, G. S., Estimate of surface wind speeds using satellite-borne radar measurements at normal incidence, *J. Geophys. Res.*, 84, 3974, 1979.
13. Mognard, N. M. and Lago, B., The computation of wind speed and wave height from Geos-3 data, *J. Geophys. Res.*, 84, 3979, 1979.
14. Dobson, E., Monaldo, F., and Goldhirsh, J., Validation of Geosat altimeter-derived wind speeds and significant wave heights using buoy data, *J. Geophys. Res.*, 92(C10), 10719, 1987.
15. Shum, C. K., Ries, J. C., and Tapley, B. D., The accuracy and applications of satellite altimetry, *Geophys. J. Int.*, 121, 321, 1995.
16. Klobuchar, J., Ionospheric time-delay algorithm for single-frequency GPS users, *IEEE Trans. Aerosp. Electron. Systs.*, AES(23), 325, 1987.
17. Vincent, P. and Menard, Y., Evaluating the TOPEX/Poseidon system, *AVISO Newsl.*, CNES(2), May, 1993.
18. Menard, Y., Jeansou, E., and Vincent, P., Calibration of the TOPEX/Poseidon altimeters over Lampedusa. Additional results over Harvest, TOPEX/Poseidon Special Issue I: geophysical validataion, *J. Geophys. Res. Oceans*, 99(C12), 24,487, 1994.
19. Christensen, E. J. et al., Calibration of TOPEX/Poseidon at platform Harvest and GPS precise tracking of TOPEX/Poseidon: results and implications, TOPEX/Poseidon Special Issue I: geophysical validation, *J. Geophys. Res. Oceans*, 99(C12), 24,465, 1994.
20. May, D. A. and Hawkins, J. D., Impact of the Geosat altimeter wet troposphere range correction in the Greenland-Iceland-Norwegian Sea, *J. Geophys. Res.*, 96(C4), 7237, 1991.
21. Schrama, E. J. O., The role of orbit errors in processing of satellite altimeter data, Rep. No. 33, Netherlands Geodetic Commission, Publications on Geodesy, New Series, 1989.
22. Johansen, T. A., Samuel, P., and Pettersson, L. H., A comparison of results from a regional and a global model for computing tide and atmospheric corrections to altimeter data, NRSC Technical Rep. No. 30, Bergen, 1990, 22 pp.
23. Thomas, J. and Woodworth, P. L., The influence of ocean tide model corrections on Geosat mesoscale variability maps of the Northeast Atlantic, *Geophys. Res. Lett.*, 17(12), 2389, 1990.
24. Muench, R. D. and Overland, J. E., Eds., *Polar Oceans and their Role in Shaping the Global Environment*, American Geophysical Union Monograph, 85, 1994.
25. Rummel, R. and Schrama, E. J. O., Two complementary systems on-board 'Aristoteles': Gradio and GPS, *ESA J.*, 15, 135, 1991.

26. Cheney, R. E., Marsh, J. G., and Beckley, B. D., Global mesoscale variability from collinear tracks of Seasat altimeter data, *J. Geophys. Res.*, 88(C7), 4343, 1983.

27. Sandwell, D. T. and Zhang, B., Global mesoscale variability from the Geosat exact repeat mission: correlation with ocean depth, *J. Geohys. Res.*, 94 (C12), 17971, 1989.

28. Tai, C. K. GEOSAT crossover analysis in the tropical Pacific. I. Constrained sinusoidal crossover adjustment, *J. Geophys. Res.*, 93(C9), 10621, 1988.

29. Knudsen, P. and Brovelli, M., Collinear and cross-over adjustment of Geosat ERM and Seasat altimeter data in the Mediterranean sea, *Surv. Geophys.*, 14(4–5), 449, 1991.

30. Moritz, H., *Advanced Physical Geodesy*, Herbert Wichmann Verlag Karlsruhe, Abacus Press, Tunbridge Wells, Kent, 1980.

31. Drottning, Å., Gravity Field and Dynamic Sea Surface Topography Estimation in the Norwegian Sea Using Satellite Altimeter Data and Sea-Gravity Data, Ph.D. thesis, University of Bergen, Norway, 1993.

32. Knudsen, P., Simultaneous estimation of the gravity field and sea surface topography from satellite altimeter data by least-squares collocation, *Geophys. J. Int.*, 104, 307, 1991.

33. Rapp, R. H., Gravity anomalies and sea surface heights derived from a combined GEOS 3/Seasat altimeter data set, *J. Geophys. Res.*, 91(B5), 4867, 1986.

34. Knudsen, P., Andersen, O. B., and Tscherning, C. C., Altimetric gravity anomalies in the Norwegian-Greenland Sea — Preliminary results from the ERS-1 35 Day Repeat Mission, *Geophys. Res. Lett.*, 19(17), 1795, 1992.

35. Rapp, R. H. and Pavlis, N. K., The development and analysis of geopotential coefficient models to spherical harmonic degree 360, *J. Geophys. Res.*, 95(B13), 21885, 1990.

36. Tscherning, C. C., Comparison of some methods for the detailed representation of the Earth's gravity field, *Rev. of Geophys. Space Phys.*, 19(1), 213, 1981.

37. Schwarz, K. P., Sideris, M. G., and Forsberg, R., The use of FFT techniques in physical geodesy, *Geophys. J. Int.*, 100, 485, 1990.

38. Zwally, H. J., Bindschadler, R. A., Brenner, A. C., Martin, T. V., and Thomas, R. H., Surface elevation contours of Greenland and Antarctic ice sheets, *J. Geophys. Res.*, 88(C3), 1589, 1983.

39. Zwally, H. J., Growth of Greenland ice sheet: interpretation, *Science*, 246, 1589, 1989.

40. Douglas, B. C., Cheney, R. E., Miller, L., Agreen, R. W., Carter, W. E., and Robertson, D. S., Greenland ice sheet: is it growing or shrinking?, *Science*, 248, 288, 1990.

41. Partington, K. C., Cudlip, W., and Rapley, C. G., An assessment of the capability of the radar altimeter for measuring ice sheet topographic change, *Int. J. Remote Sensing*, 12(3), 585, 1991.

42. Houghton, J., Jenkins, G., and Ephraums, J., The supplementary report to The IPCC Scientific Assessment, Cambridge University Press, 1992.

43. Hougton, J., Jenkins, G., and Ephraums, J., Climate Change: the IPCC Scientific Assessment, Cambridge University Press, 1990.

44. Martin, T. V., Zwally, H. J., Brenner, A. C., and Bindschadler, R. A., Analysis and retracking of continental ice sheet radar altimeter waveforms, *J. Geophys. Res.*, 88(C3), 1608, 1983.

45. Brenner, A. C., Bindschadler, R. A., Thomas, R. H., and Zwally, H. J., Slope-induced errors in radar altimetry over continental ice sheets, *J. Geophys. Res.*, 88(C3), 1617, 1983.

46. Fu, L. L., Christensen, E. J., Yamarone, C. A., Lefbvre, M., Menard, Y., Dorrer, M., and Escudier, P., TOPEX/Poseidon mission overview, in TOPEX/Poseidon Special Issue I: Geophysical validation, *J. Geophys. Res.*, 99(C12), 24,369, 1994.

47. Evensen, G. and van Leeuwen, P. J., Assimilation of Geosat altimeter data for the Agulhas current using the ensemble Kalman filter with a quasi-geostrophic model, *Monthly Weather Rev.*, May 30, 1995.

E

SAR Systems*

R. Keith Raney

A Synthetic Aperture Radar (SAR) is an imaging system whose primary output product is a two-dimensional mapping of the reflectivity of the illuminated scene. This Appendix summarizes SAR imaging principles,[1] reviews ocean reflectivity at microwave frequencies, and tabulates selected parameters of SAR systems relevant to remote sensing of the ocean.

E.1 Principles of Imaging Radar

Resolution, or image sharpness, is the hallmark of a SAR. As with all imaging systems, resolution is measured in two spatial dimensions. For a side-looking radar, the two image dimensions are known as range (distance from the radar flight path), and azimuth (distance along the radar flight path), respectively. Slant range is directly proportional to echo time delay, measured along the line of sight (LOS) from the radar to each reflector. Slant range is scaled to ground range through $1/\sin\theta_i$, where the incidence angle θ_i is defined with respect to vertical at the corresponding range in the scene. Most SAR image products are presented in nominal ground range. Range image distortion from terrain slope and altitude is not a significant issue in ocean applications.

SAR systems differ from other radars primarily by the way in which the azimuth resolution is achieved. For SAR systems, the data from an area are collected and placed into computer memory, line by line, in time sequence. The memory contents are processed later to form the image, which may have a resolution several hundred times sharper than that of the original data. The mathematics of image formation is identical for SAR and many other focused systems such as an optical lens or large radio telescope antenna. It follows that SAR imaging is equivalent to wave front reconstruction, for which the fundamental process is coherent: it depends on the phase structure of all constituent signals. Coherent integration leads to a single-look azimuth resolution as small as half of the azimuth length of the SAR antenna.

For reflections from many distributed reflectors, the random phases of the individual elements cause mutual interference, leading to brightness modulation of the imaged field. This is known as speckle. Speckle may be reduced by averaging, which implies a trade-off with image resolution. Looks describes the number of statistically independent subimages used to form the averaged image. The number of looks, divided by the product of the resolution parameters (in range and azimuth), is a fundamental parameter that describes the Nyquist information capacity of the system.[1] This may be expressed as $Q_{SAR} = N_L/(r_A r_R)$. Usually, SAR range resolution is equal to the inverse range bandwidth (scaled to ground range), and azimuth resolution is approximated by half of the azimuth aperture length D_{Az} multiplied by the number of statistically independent looks. Larger Q_{SAR} is better.

In oceanographic applications, the scene of interest is in motion, which has serious consequences on SAR system performance.[2,3] Wave translations larger than image resolution may be observable. This effect becomes more evident as the exposure time of the radar is increased, in certain ways analogous to the motion sensitivity of a conventional camera. SAR exposure time $\Delta T \approx (\lambda/D_{Az})(R/V)$, which is proportional to radar wavelength. The R/V factor, known as the range/velocity ratio, is a time/space scaling parameter applicable to ocean wave imaging by SAR. In general, smaller R/V is better. Values for this ratio and for exposure time are included in the tables of this Appendix, based on range at the minimum incidence angle available. The SAR azimuth image, formed by integration over the time sequence of recorded signals, may be successfully focused only to the extent that the phase structures of the reflections are preserved. This implies that scatterer movements as small as a fraction of the illuminating wavelength, during radar exposure time, can degrade SAR image resolution for oceanic scenes.

E.2 Radar Reflectivity from the Sea

For a distributed scene such as the sea, radar reflectivity is described by sigma naught (σ^0), which is defined by the ratio of average powers of backscattered to incoming illumination, per unit area in the horizontal plane. Sigma naught is a function of both instrumentation and of geophysical parameters. The instrumentation factors include polarization (pp) and wavelength (λ) of the radar illumination, and the incidence angle (θ_i) at which the scene is observed. Geophysical parameters include surface roughness and surface conductivity, and environmental factors that affect surface properties. For radar wavelengths, reflectivity of the ocean tends to decrease with increasing incidence angle,[4,5] as shown in Figure E-1. Radar wavelengths and their band designations may be found in the tables below.

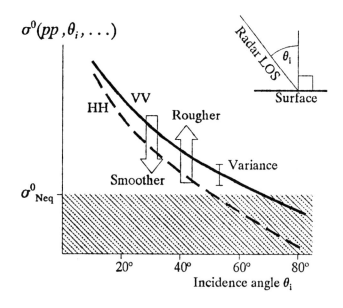

Figure E.1 Typical radar reflectivity σ^0 as a function of incidence angle.

Polarization of an emitted electromagnetic field is determined by the geometry of the antenna. The observed reflectivity depends on the polarization state of the receive antenna as well as of the transmitted field, where vertical (V) or horizontal (H) linear polarizations are the most commonly used for SAR. Most systems use the same antenna for both functions, leading to like-polarized σ^0 estimates. Ocean reflectivity is usually larger for VV polarization than for HH polarization. For example, σ^0_{VV} is larger than σ^0_{HH} by about 2.5 dB at 30° and C-band. The cross-polarized component σ^0_{VH} is weaker by as much as 10 dB. The set of possibilities spans all elliptical polarization states, including the special cases of circular and linear polarizations. It is possible to synthesize any combination of transmit and receive polarizations by transformation of the scattering matrix[1] of the scene. The complete scattering matrix is obtained through a (calibrated) quadrature polarimetric radar,[6,7] known as quad-pol for short. This requires that the radar must: (1) transmit on two orthogonal polarizations; (2) receive the like and the cross-polarized components of the reflected fields; and (3) maintain relative phase coherence between all channels. Radars that lack one or more of these features may yield polarization diversity, but the data cannot be manipulated to replicate other polarization states.

All radar observations have inherent statistical variance, dominated by speckle, described above. Furthermore, all radar observations of scene reflectivity are in competition with additive radar noise, for which level may be described by an equivalent reflectivity number. This is known as noise equivalent sigma naught,

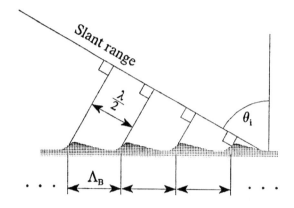

Figure E.2   Bragg scattering.

σ^0_{Neq}. The level of scene reflectivity should be at least 5 dB above the noise for useful applications, and a 10 dB signal-to-noise ratio is better.

For a wide range of incidence angles, the dominant cause of radar reflectivity from the ocean is Bragg scattering (Figure E.2), first proposed by Wright.[8] Subsequently, substantial theoretical and experimental evidence has accumulated in support of Bragg scattering. The most convincing proof may be derived through polarimetric analysis.[6] Bragg scattering is due to a systematic phase fit between the illuminating half wavelength, $\lambda/2$, projected onto the sea surface; and a matching periodic ocean wave component, Λ_B. (The factor of one half arises because of the two-way path traveled by radar illumination.) The Bragg condition is $\Lambda_B = \lambda/(2\sin\theta_i)$. For the wavelengths and incidence angles of typical ocean remote sensing radars, Bragg waves range from long capillaries (~2 cm) to short gravity waves (~50 cm). Roughness in the microwave remote sensing context implies this special Bragg regime of surface wave components. These waves, which have relatively short lifetimes, are locally generated, and are modulated and advected by longer and more energetic seas. It follows that the larger wave structure is observable by microwave systems primarily through patterns expressed in the Bragg waves. This is known as the two-scale model of microwave ocean scattering.[2,8]

Roughness at Bragg wavelengths is set up by local wind stress. For a given incidence angle and radar wavelength, there is a first-order power law dependence between radar reflectivity and wind speed,[5] as suggested in Figure E.3. The exponent γ is a function of wavelength, among other parameters. For ERS-1[9] that has $\lambda = 5.7$ cm, VV polarization, and $\theta_i = 23°$, the exponent is $\gamma = 1.1 \pm 0.1$; on the other hand, at the same incidence angle, for Seasat[10] ($\lambda = 23$ cm at HH polarization), $\gamma = 0.5 \pm 0.1$. The σ^0 power law has a systematic aspect angle dependence that may be exploited to estimate wind direction as well as wind speed.[11] Within limits, the power law is robust. The upper limit is approached as the radar reflectivity tends to saturate with increasing wind speed u_s. The saturation wind speed u_{sat} is above 25 m/sec for European Remote Sensing Satellite-1 (ERS-1). The lower limit is more abrupt. Bragg wave formation occurs only for wind speeds above the lower thresh-

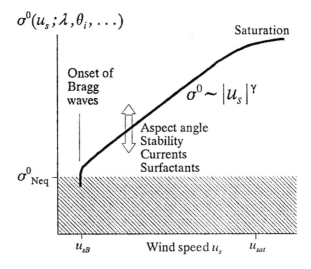

Figure E.3 Typical radar reflectivity σ^0 as a function of wind speed.

old u_{sB}, which is about 2.5 m/sec for ERS-1. Both the onset of Bragg waves, and the level of σ^0 are functions of environmental conditions, such as the relative air-sea temperature, surface currents, and surfactants. If σ^0_{Neq} is sufficiently low that the threshold is visible, then the environmental effects may lead to dramatic oceanic features, often observed in ERS-1 SAR imagery. In contrast, for the Japanese Earth Resources Satellite-1 (JERS-1) the signal-to-noise ratio is only about 3 dB at $u_s = 13$ m/sec, which means that ocean imagery from this system for lesser wind speeds is predominantly noise, and image contrast at the Bragg threshold is lost.

E.3 SAR System Parameters

Orbital SAR 1978–1991

Virtually all remote sensing SAR satellite systems owe both their inspiration and baseline performance standards to Seasat, and by implication to its design and science teams. Principal parameters of Seasat[12] are listed in Table E.1. The design lifetime of Seasat was 2 years. Unfortunately, the spacecraft failed after 3 months of SAR operation due to a massive short circuit in the slip ring assembly of the solar panel primary power system. Data provided by Seasat have proved to be of high quality and immense interest to the ocean applications communities. Precedents for certain image parameters, particularly resolution and number of looks, were firmly established by Seasat. Unless one is willing to give up other image quality aspects such as swath width, there is rather little flexibility available.

Table E.1   Orbital SAR 1978–1991

	Seasat	SIR-A	SIR-B	(Kosmos 1870)	Almaz
General					
Country	U.S.	U.S.	U.S.	U.S.S.R.	U.S.S.R./Russia
Agency	NASA	NASA	NASA	Glavkosmos	Glavkosmos
Spacecraft	Seasat	Shuttle	Shuttle	Salyut	Salyut
Launch date	June 26, '78	Nov. 12 , '81	Oct. 5, '84	Jul. 25, '87	Mar., 31 '91
Lifetime	3 mos	2.5 days	8 days	2 years	1.5 years
Radar					
Band (wavelength [cm])	L (23.5)	L (23.5)	L (23.5)	S (10)	S (10)
Frequency (GHz)	1.275	1.278	1.282	3.0	3.0
Antenna	Corporate	Corporate	Corporate	Waveguide	Waveguide
Length × height (m)	10.7 × 2.16	9.4 × 2.16	10.7 × 2.16	15 × 1.5 (2)	15 × 1.5 (1)
Polarization	HH	HH	HH	HH	HH
Incidence angle (°)	23	50	15–64	30–60	30–60
Range resolution (m)	25	40	25	(~30)	15–30
Azimuth resolution (m)	25	40	58–17	(~30)	15
Looks	4	6	4	(4)	>4
Swath width (km)	100	50	10–60	20–45	20–45
Recorder on board?	N	Y	Y	Y	Y
Processing	O, D	O	O, D	O	D
(optical, digital)					
Noise equivalent σ^0 (dB)	−24	−32	(−28)		—
Mission					
Nominal altitude (km)	800	260	350, 225	250–280	300
Inclination (°)	108	38	57	71.5	73
Sun synchronous?	N	N	N	N	N
Downlink rate (MB/sec)	110	(None)	30	Analog	—
Repeat cycle (d)	17, 3	Nil	Nil	Variable	Nil
Operation per orbit	10	—	—	3 min/tape	3 min/tape
(min)					
Nominal R/V (s)	115	50	35	50	60
Exposure time (s)	2.5	1.25	0.8	0.35	0.4

The Shuttle SAR missions[13] extended the foundations established by Seasat in the dimension of incidence angle. Both SIR-A and SIR-B provided valuable results in spite of technical limitations. The R/V ratio for the shuttle radars is much more favorable for oceanographic observations than that of higher altitude satellites.

The U.S.S.R. operated the Kosmos 1870 radar[14,15] during 1987–1989. At the time it was classified, but its images were made public after the end of the mission. Kosmos 1870 was one of three satellite SARs of the late 1970s built by NPO Machinostroenye. (The other two are known at present by the name of Almaz, discussed below.) The Soviets had a series of well-known real aperture radars in space (the Kosmos 1500 series), but Kosmos 1870, in addition to Venera 16 and 17 to Venus, was its first publicized orbital SAR.

Almaz,[13,14,16] or diamond in the rough, was very similar to Kosmos 1870, but was upgraded in several regards. Data were stored onboard in four parallel video recorders with capacity limited to the length of each data take. The downlink was analog for both real-time and recorded data. The most significant improvements in

Table E.2   Orbital SAR 1991–1994

	ERS-1	JERS-1
General		
Country	Europe	Japan
Agency	ESA	MITI/NASDA
Spacecraft	ERS-1	JERS-1
Launch date	Jul. 16, '91	Feb. 11, '92
Lifetime (design)	2–3 years	2 years
Radar		
Band (wavelength [cm])	C (5.7)	L (23.5)
Frequency (GHz)	5.3	1.275
Antenna	Waveguide	Corporate
Size (m), length × height	10 × 1	11.9 × 2.4
Polarization	VV	HH
Incidence angle (°)	23	39
Range resolution (m)	26	18
Azimuth resolution (m)	28	18
Looks	6	3
Swath width (km)	100	75
Recorder on board?	N	Y
Processing (optical, digital)	D	D
Noise equivalent σ^0 (dB)	–24	–14.5
Mission		
Nominal altitude (km)	780	568
Inclination (°)	98.5	97.7
Sun synchronous?	Y	Y
Downlink data rate (MB/sec)	105	30 (x 2)
Repeat cycle (d)	3, 35, 176	44
Operation time per orbit (min)	10	20
Nominal R/V (s)	115	100
Exposure time (s)	0.7	2

the Almaz signal chain followed from conversion to digital data handling and processing. Another Almaz launch is planned to occur after 1995. From an applications point of view, the Almaz series is interesting, in that it is the only orbital SAR system to use S-band. The smaller incidence angles are best for ocean observation.

Orbital SAR 1991–1994

In contrast to the almost total absence of (civilian) space radar capability in the 10 years following Seasat, the 1990s began with a substantial increase in orbital SAR activity. After extensive studies and preparations from about 1975, the European Space Agency launched ERS-1 in 1991. The primary payload[17,18] onboard ERS-1 is the Active Microwave Instrument (AMI), the heart of which is a SAR with parameters that are listed in Table E.2. The orbit period may be changed in response to mission requirements. The first several months used a 3-d repeat orbit for system verification and data validation. Normal earth observation operations are based on a 35-d repeat orbit. A 176-d repeat orbit is available to support altimetric

Table E.3 Orbital SAR 1994–1999

	SIR-C/X-SAR	ERS-2	Radarsat	ENVISAT
General				
Country	U.S./Germany/Italy	Europe	Canada	Europe
Agency	NASA/DLR/DARA	ESA	CSA/U.S.	ESA
Spacecraft	Shuttle	ERS-2	(Dedicated)	(Envisat)
Launch date	1994, 1994	1995	1995	1999
Design lifetime	10 d	2–3 years	5 years	5 years
Radar				
Band	L, C; X	C	C	C
Wavelength (cm)	24, 5.7, 3.1	5.7	5.7	5.7
Frequency (GHz)	1.25, 5.3, 9.6	5.3	5.3	5.3
Antenna	Arrays + WG (X)	Waveguide	WG array	
Size (m), length × height	12 × (3, 0.75, 0.4)	10 × 1	15 × 1.5	
Polarization	Quad-pol L + C; X_{VV}	VV	HH	VV + HH
Incidence angle (°)	15–55	23	<20–50	<20–50
Range resolution (m)	10–30	26	10–100	~25
Azimuth resolution (m)	30	28	9–100	~25
Looks	~4	6	1–8	~4
Swath width (km)	15–60	100	10–500	100 (500)
Recorder on board?	Y (+D/L)	N	Y	Y
Processing (optical, digital)	D	D	D	D
Noise equivalent σ^0 (dB)	−40 < −28	−24	−23	
Mission				
Nominal altitude (km)	225	780	800	700
Inclination (°)	57	98.5	98.6	98.5
Sun synchronous?	N	Y	Y	Y
Downlink data rate (MB/sec)	45 (TDRS)	105	74–105	~105
Repeat cycle (d)	Nil	TBD	24	
Operation time (min/orbit)	(60 h, total)	10	20	10+
Nominal R/V (s)	33	115	115	100
Exposure time (s)	0.7, 0.16, 0.085	0.7	0.5	

experiments. The AMI includes a microwave scatterometer[17] for observation of wind stress over extended areas of the oceans.

The JERS-1[19] was launched in 1992. The SAR antenna was successfully deployed after nearly 2 months of delay due to a stubborn mechanism. The transmitter has not been able to achieve full radiated power. As a result, the operational σ^0_{Neq} listed in Table E.2 is 6 dB larger than its design value. At an incidence angle of 39°, the ocean HH reflectivity is rather small. Weak reflectivity combined with reduced radiated power implies that JERS-1 ocean imagery is unsatisfactory in general.

Orbital SAR 1994–1999

The SIR-C/X-SAR missions[13,20] (Table E.3), represent a major milestone in space-based radar. SIR-C/X-SAR is a joint venture between the U.S., NASA/Jet Propulsion Laboratory (JPL), and a European consortium of Deutsche Forschungsan-stalt für Luft-und Raumfahrt e.V. (the German Aerospace Research Establishment known

Table E.4   Radarsat Imaging Modes

Mode	Resolution (R[a] × A, m)	Looks[b]	Width (km)	Incidence (degrees)
Standard	25 × 28	4	100	20–49
Wide (1)	48–30 × 28	4	165	20–31
Wide (2)	32–25 × 28	4	150	31–39
Fine resolution	11–9 × 9	1	45	37–48
ScanSAR (N)	50 × 50	2–4	305	20–40
ScanSAR (W)	100 × 100	4–8	510	20–49
Extended (H)	22–19 × 28	4	75	50–60
Extended (L)	63–28 × 28	4	170	10–23

[a]   Ground range resolution varies with range.
[b]   Range and processor dependent.

as DLR) and the Italian Agenzia Spaziale Italiana (ASI). JPL is responsible for the C- and L-band systems, and DLR/ASI is responsible for the X-band system. Antennas for both the C- and L-band radars use active phased array technology. The H- and V-phased arrays, backed up by parallel receiver and data recording chains, may be cycled to achieve reception of the complex scattering matrix of the scene, which is designed to support the quadrature polarization technique. SIR-C/X-SAR will be the first time that this capability is available from space.

Radarsat, Canada's first earth resources remote sensing satellite[21] is designed for 5 years of service in orbit. The only imaging instrument is a SAR. The baseline configuration of the spacecraft has the SAR pointing to the north side of the orbital plane. This makes possible regular coverage of the Arctic up to the pole, but prevents coverage of the central Antarctic region. For two periods during the first 2 years of the mission, the satellite will be rotated 180° about its yaw axis so as to direct the radar antenna beam to the south side of the orbital plane. The purpose of this maneuver is to obtain a complete SAR map of Antarctica during seasons of maximum and minimum ice cover. Several imaging modes are available (Table E.4), covering a wide span of incidence angles. For the first time, a special wide swath mode[22,23] known as ScanSAR is available from a satellite radar.

Airborne SAR Systems

Since most civilian airborne remote sensing SARs are meant for technology development as well as applications experiments, their hardware is frequently changed. The most widely deployed airborne SAR, and the one having the most modes, is that of NASA, which carries the AirSAR radar[7,13] of the JPL. The quadrature polarimetric capability at the three frequencies of this radar offers a unique and very rich data source that is made available by NASA to investigators around the world. AirSAR also offers two other special modes[1] that merit attention, as noted in Table E.5. InSAR (JPL terminology), or ΔR interferometry, is a mode created by using data from two antennas, one mounted above the other on the side of the aircraft.

Table E.5 Status of Selected Airborne SAR Systems (1994)

	AirSAR	C/X SAR	E–SAR	KRAS	STAR-1
General					
Country	U.S.	Canada	Germany	Denmark	Canada
Agency	JPL/NASA	CCRS	DLR	TUD	Intera
Aircraft	DC–8	CV–580	DO–228	Gulfstream	Cessna
Nominal altitude (km)	8	6	3.5	12.5	10
Nominal airspeed (m/sec)	300	130	70	300	145
Purpose (experimental or operational)	E	E	E	E	O
Radar					
Band	C, L, P	X, C	X, C, L	C	X
wavelength (cm)	5.7, 24, 68	3.2, 5.7	3.1, 5.7, 23	5.7	3.1
Frequency (GHz)	5.3, 1.25, 0.44	9.3, 5.3	9.6, 5.3, 1.3	5.3	9.6
Antenna length (m)	1.3, 1.6, 1.8	~1.2	0.15, 0.24, 0.85	1.2	1.2
Antenna motion controller ?	N	Y	N	Y	Y
Polarization diversity	(Quad)	HH, VV, HV	HH, VV	VV	HH
Quadrature polarization	C, L, P	C	N	N	N
Incidence angle (°)	20–60	0–85	15–60	20–80	45–80
Range resolution (m) (slant)	7.5 (S)	6–20	2 (S)	2, 4, 8	6, 12
Range cells		4096	2048	8192	4096
Azimuth resolution (m)	2	<1–10	2	2, 4, 8	6
Looks	4	1–7	1–8	2–16	7
Swath width (km) (slant)	7–13	18–63	3 (S)	9–48 (S)	40, 60 (S)
Range gain function ?		Y	Y	Y	N
Processor on board ?	Y (1 ch)	Y	Y (QL)	(Y)	Y
Noise equivalent σ⁰ (dB) (for each band)		–30, –40	–40, –30, –35	–42	–30
Interferometry modes	Δt, ΔR	ΔR			
Nominal h/V (s)	25	45	50	40	60
Minimum exposure (s)	1.1, 3.7, 9.5	1.2, 2	10, 12, 14	1.9	1.5

This configuration is able to measure the range vector (distance and elevation angle) of each reflection. The interference pattern between the two received signals may be used to deduce terrain height information. The technique has been used in studies of ice shelf dynamics.[24] The other special mode also requires two separate antennas, spaced along the line of flight, leading to time difference, or Δt interferometry. Interferometric measurements with these two antennas may be used to observe dynamic phenomena in the scene, such as currents on the ocean surface. Direct estimation of ocean backscatter coherence time is possible through Δt interferometry.[25]

The SAR flown by the Canada Center for Remote Sensing (CCRS)[26] is on a Convair-580. Both X- and C-bands are fully supported by onboard real-time digital processors, and have a variety of modes and data combinations available. The standard image products are produced at seven looks by the real-time onboard processor. Signal data are recorded so that ground processing may be used for specific experimental purposes, such as investigations requiring access to separate looks for optimized oceanographic SAR wave imagery. The system has been modified to incorporate quadrature polarimetry on C-band, and may be operated in a topographic interferometric mode. The CCRS SAR system includes time difference interferometry.

The airborne radar of DLR[27-29] continues to be upgraded with new modes and capabilities. It is designed primarily for high-resolution and technology development; hence it has a narrower swath width than do the previous two systems. Its extension to X-band is in support of DLR vested interest in the SIR-C/X-SAR mission. The DLR radar includes a quick-look (QL) onboard processor having 50 × 50-m resolution.

The radar[30,31] of the Technical University of Denmark (TUD) has been designed to offer a variety of incidence and image parameter values. Within the limits set by their respective imaging geometries, the radar is matched to the ERS-1 SAR, aided by the use of a G-3 Gulfstream aircraft capable of high speed and high altitude.

STAR-1 is a civilian SAR[32,33] owned and operated by a Canadian company, and used for large-area surveys. This X-band system has performed more than 75% of all of the airborne radar mapping done for commercial clients worldwide since 1986, and has seen dedicated service in real-time ice surveillance in the Arctic. Data products from this system are digitally rectified to map accuracy standards. With the recent use of the Global Positioning System (GPS), accurate mapping is possible with no need for surveyed control points. Residual geometric distortion is less than 3 pixels, plus terrain relief effects.

There are several other airborne SARs that deserve mention. The Netherlands supports the advanced system known as PHARS.[34] This system is a C-band SAR mounted in a Swaeringen Meteor II, a twin-engined business plane. Nominal resolution is about 5 m with about 6 looks. The antenna is VV polarized to support experiments with ERS-1. The program is committed to implementing a full quadrature polarimetric version. One of the pioneers in the field of SAR is the Environmental Research Institute of Michigan (ERIM), for which its civilian experimental SAR[35] is managed by the U.S. Navy Air Development Center (NADC), and is mounted on a P-3 aircraft. The radar operates at X-, C-, and L-bands, and is fully polarimetric. Nominal resolution is in the 3-m range. The French airborne SAR[36] is known as VARAN-S. Design resolution is on the order of 5 m, with four looks. The Chinese have developed and expanded their X-band CASSAR,[37] with ocean observation as one of its application areas.

References

1. Raney, R. K., Radar fundamentals: technical perspective, *Manual of Remote Sensing,* 3rd ed., American Society of Photogrammetry and Remote Sensing, in press, chap. 2.
2. Hasselmann, K., Raney, R. K., Plant, W. J., Alpers, W., Shuchman, R. A., Lyzenga, D. R., Rufenach, C. L., and Tucker, M. J., Theory of SAR ocean wave imaging: a MARSEN view, *J. Geophys. Res.,* 86, 6481, 1981.
3. Vachon, P. W. and Krogstad, H., Synthetic Aperture Radar imagery of ocean waves, in *Oceanographic Applications of Remote Sensing,* Ikeda, M. and Dobson, F. W., Eds., CRC Press, Boca Raton, FL, 1995, chap. 15.

4. Long, M. W., *Radar Reflectivity of Land and Sea*, Lexington Books, D. C. Heath, Toronto/London, 1975.

5. Ulaby, F. T., Moore, R. K., and Fung, A. K., *Microwave Remote Sensing: Active and Passive*, Vol. 2, Addison-Wesley, Reading, MA, 1982, chap. 11.

6. Zebker, H. A., van Zyl, J. J., and Held, D. N., Imaging radar polarimetry from wave synthesis, *J. Geophys. Res.*, 92, 683, 1987.

7. van Zyl, J. J., AIRSAR Reference Manual, Jet Propulsion Laboratory, California Institute of Technology, Pasadena, CA, 1991.

8. Wright, J. W., A new model for sea clutter, *IEEE Trans. Antennas Propag.*, 16, 217, 1968.

9. Snoeij, P., Swart, P. J. F., and Unal, C. M. H., Study on the Response of the Radar Echo from the Ocean Surface Wind Vector at Frequencies between 1 and 18 GHz, Final Rep. to the European Space Agency, Delft University of Technology, Laboratory for Telecommunication and Remote Sensing Technology, Delft, The Netherlands, 1992.

10. Alpers, W. and Brümmer, B., Imaging of atmospheric boundary layer rolls by the synthetic aperture radar aboard the European ERS-1 satellite, *Proc. Geosci. Remote Sensing Symp.*, *IGARSS'93*, Tokyo, Japan, August 18–21, 1993, 540.

11. Topliss, B. J. and Guymer, T. H., Scatterometers, in *Oceanographic Applications of Remote Sensing*, Ikeda, M. and Dobson, F. W., Eds., CRC Press, Boca Raton, FL, 1995, chap. 13.

12. Jordan, R. L., The Seasat-A synthetic aperture radar system, *IEEE J. Oceanic Eng.*, 5, 154, 1980.

13. Way, J. and Smith, E. A., The Evolution of synthetic aperture radar systems and their progression to the EOS SAR, *IEEE Trans. Geosci. Remote Sensing*, 29, 962, 1991.

14. Chenard, S., Soviet earth observation gets less remote, *Interavia Space Markets*, 6, 17, 1990.

15. Wirin, W. B. and Williamson, R. A., Satellite remote sensing in the USSR: past, present, and future, *Remote Sensing Yearbook 1990*, Burgess Scientific Press, Basingstoke, U.K., chap. 3.

16. Li, F. L. and Raney, R. K., Prolog to special section on spaceborne radars for earth and planetary observations, *Proc. IEEE*, 79, 773, 1991.

17. Attema, E. P. W., The Active microwave instrument on-board the ERS-1 satellite, *Proc. IEEE*, 79, 791, 1991.

18. ERS-1 Special Issue, *ESA Bull.*, European Space Agency, 65, 1991.

19. Nemoto, Y., Nishino, H., Ono, M., Mizutamari, H., Nishikawa, K., and Tanaka, K., Japanese Earth Resources Satellite-1 synthetic aperture radar, *Proc. IEEE*, 79, 800, 1991.

20. Jordan, R. L., Huneycutt, B. L., and Werner, M., The SIR-C/X-SAR synthetic aperture radar system, *Proc. IEEE*, 79, 827, 1991.

21. Raney, R. K., Luscombe, A. P., Langham, E. J., and Ahmed, S., RADARSAT, *Proc. IEEE*, 79, 839, 1991.

22. Moore, R. K., Claasen, J. P., and Lin, Y. H., Scanning spaceborne synthetic aperture radar with integrated radiometer, *IEEE Trans. Aerospace Electron. Syst.*, 17, 410, 1981.

23. Tomiyasu, K., Conceptual performance of a satellite borne, wide swath synthetic aperture radar, *IEEE Trans. Geosci. Remote Sensing*, 19, 108, 1981.

24. Goldstein, R. M., Englehardt, H., Kamb, B., and Frolich, R. M., Satellite radar interferometry for monitoring ice sheet motion: application to an Antarctic ice stream, *Science*, 262, 1525, 1993.

25. Carande, R. E., Estimating ocean coherence time using dual-baseline interferometric synthetic aperture radar, *IEEE Trans. Geosci. Remote Sensing*, submitted.

26. Livingstone, C. E., Gray, A. L., Hawkins, R. K., and Olsen, R. B., CCRS C/X- airborne synthetic aperture radar: an R and D tool for the ERS-1 time frame, *IEEE AES Mag.*, 11, 1988.

27. Horn, R., C-band SAR Results Obtained by an Experimental Airborne SAR System, *Proc. Int. Geosci. Remote Sensing Symp.*, *IGARSS'89*, Vancouver, Canada, 1989, 2213.

28. Horn, R., Werner, M., and Mayr, B., Extension of the DLR Airborne Synthetic Aperture Radar, E-SAR, to X-band, *Proc. Int. Geosci. Remote Sensing Symp.*, *IGARSS'90*, Washington, D.C., 1990, 2047.

29. Dahme, C., Horn, R., Hounam, D., Öttl, H., and Schmid, R., Recent achievements of DLR's airborne experimental SAR system and image processing equipment, *Proc. Int. Geosci. Remote Sensing Symp., IGARSS'91*, Espoo, Finland, 1991, 245.

30. Dall, J., Jørgensen, J. H., Christensen, E. L., and Madsen, S. N., A Real-Time Processor for the Danish C-Band SAR, *Proc. Int. Geosci. Remote Sensing Symp., IGARSS'91*, Espoo, Finland, 1991, 279.

31. Madsen, S. N., Christensen, E. L., Skou, N., and Dall, J., The Danish SAR system: design and initial tests, *IEEE Trans. Geosci. Remote Sensing*, 29, 417, 1991.

32. Nichols, A. D., Wilhelm, J. W., Gaffield, T. W., Inkster, D. R., and Leung, S. K., A SAR for real-time ice reconnaissance, *IEEE Trans. Geosci. Remote Sensing*, 24, 383, 1986.

33. Lowry, R. T., Collette, P., and Tennant, J. K., Testing and validation of an operational SAR for open skies inspection, *Proc. Int. Symp. on Operationalization of Remote Sensing*, Vol. 5, ITC, Enschede, The Netherlands, 1993, 1.

34. Hoogeboom, P., Koomen, P. J., Pouwels, H., and Snoeij, P., First Results from The Netherlands SAR Testbed "PHARS", *Proc. Int. Geosci. Remote Sensing Symp., IGARSS'91*, Espoo, Finland, 1991, 241.

35. Kozma, A. D., Nichols, A. D., Rawson, R. F., Shackman, S. J., Haney, C. W., and Shanne, J. J., Jr., Multi-frequency, -polarization SAR for remote sensing, *Proc. Int. Geosci. Remote Sensing Symp., IGARSS'86, Zurich, Switzerland*, ESA Publication SP-254, 1986, 715.

36. Albrizio, R., Blonda, P., Mazzone, A., Pasquali, F., Pasquariello, G., Posa, F., and Veneziana, N., Digital Processing of X-band VARAN-S Airborne SAR Images, *Proc. Int. Geosci. Remote Sensing Symp., IGARSS'89*, Vancouver, Canada, 1989, 2203.

37. Yiming, S., Huiying, L., and Jiajun, S., Development and application of the airborne side looking synthetic aperture radar system in China, *Proc. Int. Symp. Operationalization of Remote Sensing*, Vol. 5, ITC, Enschede, The Netherlands, April 1993, 55.

F

The AMI-Wind Scatterometer

Alain Cavanié and Francis Gohin

The Active Microwave Instrument in wind measurement mode (AMI-Wind) flown on European Remote Sensing Satellite-1 (ERS-1) and to be flown on ERS-2 is a C-band (5.3 Ghz) VV-polarized radar functioning either in Synthetic Aperture Radar (SAR) image or wind scatterometer (AMI-Wind) mode. A detailed description of the instrument is given in the references that can be obtained from the European Space Agency (ESA). Here we will only give the most important information concerning the scatterometer mode, main descriptive features of the instrument, characteristics of its data, and points of contact in order to access the data.

F.1 The Instrument

The AMI-Wind is a three-beam scatterometer: fore, mid, and aft beams pointing 45°, 90°, and 135°, respectively, to the right of the satellite ground track. Yaw steering of ERS satellites maintains these directions constant during their orbits. The instrument swath begins 200 km to the right of the ground track and is 500 km wide. It is to be noted that, the orbit being inclined 98.5° to the equator, the extension of the swath reaches 87.5°N but only 75.2°S, which makes the instrument better to observe the Arctic Ocean than the Antarctic continent.

The precise determination of winds requires accurate backscattering measurements over the water surface. Speckle noise, the dominant source of the error, is reduced by using 256 measurement pulses per 50 km of ground track, in 32 pulse sequences for each of the beams successively.

Analysis of the return echoes, corrected for ambient and instrument noise as well as Doppler effects, and application of a spatial filter lead to the evaluation of backscattering coefficient σ^0 of all three beams over a regular grid, nominally 25 km in step size, along and across track.

Specifications for the instrument are the following:

Spatial Resolution $\geqslant$45 km (along and across track)

Radiometric resolution

    At 4 m/sec $\leqslant$ 8.5% (mid) $\leqslant$9.7% (fore/aft)

    At 24 m/sec $\leqslant$ 6.5% (mid) $\leqslant$7.0% (fore/aft)

Radiometric stability

    Common mode instrument $\leqslant$0.57 dB

    Interberam mode instrument stability $\leqslant$0.46 dB

Localization accuracy ~5 km (along and across track)

In practice, standard deviations of σ^0, given in the AMI-Wind products, range between 4 and 6%, somewhat better than initially planned. These values can be verified over sea ice in a simple way, using the fact that fore and aft beams are at the same incidence angles and backscattering over sea ice is isotropic in azimuth. The standard deviation of the ratio $(\sigma_1 - \sigma_3)/(\sigma_1 + \sigma_3)$, where σ_1 and σ_3 are the fore and aft beam backscattering coefficients, is therefore 0.52 (2 times the standard deviation of the individual σ^0, neglecting second-order terms. This has proved to be the case over a very large sample of values.

To extract wind fields from the backscatter data, a backscattering model including incidence and azimuth angles, wind speed, and direction must be used. These models have evolved much since prelaunch models were developed using airborne scatterometer data.

ESA is presently using a model developed at the European Center for Medium-Range Weather Forecasts (ECMWF) and calibrated with respect to model winds. A model based on National Oceanic and Atmospheric Administration (NOAA) and ESA buoy network wind data, developed at IFREMER, gives comparable results. Although they certainly can be improved, they both meet the ESA initial requirement that winds be extracted in the wind speed range of 4–24 m/sec with an accuracy direction of ~203, and in speed of 2 m/sec or 10%.

F.2 Processing and Products

Processing of AMI-Wind data is done at ESA ground stations and produces so-called fast delivery products (FDPs), which are transmitted in near-real time to the fast delivery nominated centers in Europe and in Canada. Off-lineprecision processing is conducted at the French Processing and Archiving Facility (F-PAF), starting from the backscattering coefficients and complementary parameters.

Because constraints are different, FDP and precision processing vary somewhat, but the basic algorithms are quite similar. Maximum likelihood techniques at each node of the scatterometer grid lead to several possible solution vector winds, which for a three-beam scatterometer are classed by probability. Moreover the two best-ranked solutions are nominally 1803 apart in directions; this allows the construction, by continuity, of two wind fields of approximately opposite directions, of which one is chosen by comparison with a metorologic model surface wind field.

FDP and precision products contain the same essential information in a 19×19 point grid, corresponding to a 500-km length of scatterometer swath:

- Geophysical positions of the points
- Backscattering coefficients and associated standard deviations
- Incidence and azimuth angles
- Computed surface wind vectors as well as general information (time of measurements, satellite track, open ocean/land/sea ice indicators, etc.)

F.3 Access to the Data

Access to the data is fully described in the ERS-1 "User Handbook", which can be obtained from ESA on request. Users requiring data in near-real time must contact the appropriate Fast Delivery Nominated Center.

Users interested in data already stored may consult the ERS-1 Central Catalog via VT-200 compatible terminals. Without network access, the Central Catalog may be consulted through the Help Desk for which the address is:

> ERS-1 Help Desk
> EECF
> ESA/Earthnet Program Office
> ESRIN C.P. 64
> I-00044 Frascati, Italy
> and which can also be reached by telephone:
> (39-6) 94180600 or fax (39-6) 94180510

Ordering of products is done through different Order Desks as follows. For users in Europe, North Africa, and the Middle East:

EURIMAGE ERS-1 Order Desk	Tel: (39-6)94180478
ESRIN C.P. 64	Fax: (39-6)9426285
I-00044 Frascati, Italy	Telex: 610637ESRINI

For users in Canada and the U.S.:

Radarsat International ERS-1 Order Desk	Tel: (613-238)6413
275 Slater Street, Suite 1203	Fax: (613-238)5425
Ottawa, Ontario, Canada	Telex: 053-3589

Users in other countries should go through the:

SPOT-Image ERS-1 Order Desk	Tel: (33-61)62194040
16 bis Avenue Edouard Belin	Fax: (33-61)281354
BP 4359, Paris, France	Telex: 532079FSPOTIM

These Order Desks will furnish order forms, price quotations, and complementary information on data support.

References

ERS-1 System, ESA SP-1146, September 1992, 1–86.
ESA ERS-1 Product Specification, ESA SP-1149, June 1992, 31–36.
ERS-1 User Handbook, ESA SP-1148, May 1992, 1–158.
ERS-1 Wind and Wave Calibration Workshop, ESA SP-262, September 1986.
ERS-1 Geophysical Validation, ESA WPP-336, August 1992, 1–205.

G

The Along-Track Scanning Radiometer

Peter J., Minnett

G.1 Introduction

There are many pressing needs for accurate measurements of sea surface temperature (SST) on regional and global scales for studies of climate perturbation, numerical weather forecasting, and oceanography; and infrared radiometers on earth
observation satellites are important sources of such data because they facilitate the
derivation of consistent global SST fields. Since the amplitudes of SST anomalies
are rarely greater than a few degrees, the useful accuracy of the satellite-derived
fields must be at least an order of magnitude smaller than this. The three most
important limits on the remote measurement of SST using infrared radiometers are
inadequacies in the prelaunch calibration and error characterization of the instrument; the deficiencies of the in-flight calibration procedures; and the failure to
correctly compensate for the effects of the intervening atmosphere. The additional
problem of obscuration of the sea surface by cloud is a further constraint on the SST

0-8493-4525-1/95/$0.00+$.50

derivation, but is a limitation of the technique imposed by the natural system; and has to be dealt with as best possible in the data processing algorithms. The issue of skin SST vs. bulk SST is not addressed here because that is associated with the choice of SST retrieval algorithm rather than the measurement principle; although from a physical standpoint, the satellite measurement is more related to the skin temperature.

In the design of the Along-Track Scanning Radiometer (ATSR) great attention has been paid to minimizing, as far as is practical, the residual errors from the three sources, and to providing enough information in the data stream to apply effective cloud screening algorithms without recourse to external data sources.

The ATSR is an experimental scientific instrument on the First Remote Sensing Satellite (ERS-1) of the European Space Agency (ESA). It was proposed in response to an Announcement of Opportunity issued by ESA and was selected on a competitive basis. The ATSR complements the other instruments on ERS-1 that are active microwave devices (radars).

ERS-1 is a polar-orbiting earth-observation satellite carrying a range of instruments designed for studying the ocean-ice-climate system. It was launched by an Ariane rocket from the Kourou Space Center in French Guiana on July 17, 1991, into a sun-synchronous, circular orbit at an altitude of about 780 km with an inclination of about 97.5° and a nodal period of about 100 min. The local solar time of the descending node is about 10:30. Initially the orbit had a 3-day repeat cycle, with a ground-track repeatability to within ±1 km; however, the spacecraft was subsequently maneuvered into a 35-d repeat orbit to give the possibility of contiguous coverage by the narrow-swath SAR images. This orbit transition was completed on April 2, 1992. ERS-1 returned to a 3-d repeat cycle on January 1, 1994, and entered a 176-d repeat orbit on April 1, 1994, where it is expected to stay for the rest of its lifetime. The sensor payload consists of the Active Microwave Instrument (AMI) that functions in two exclusive modes: a Synthetic Aperture Radar (SAR) and a Wind Scatterometer; a radar altimeter, an experimental high-accuracy satellite location system called the Precision Range and Range Rate Experiment (PRARE), which failed soon after launch; and the ATSR. Fuller details of ERS-1 are given elsewhere.[1,2]

The ATSR was provided to ESA by a consortium consisting in the U.K. of the Rutherford Appleton Laboratory, responsible for the overall design and construction of the instrument; Oxford University Department of Atmospheric, Oceanic, and Planetary Physics, responsible for the conceptual design and the scientific calibration of the assembled radiometer; Mullard Space Science Laboratory of University College London, principally responsible for the onboard blackbody calibration targets; and the U.K. Meteorological Office, responsible for the focal plane assembly containing the detectors. In addition there was significant contribution from the Commonwealth Scientific and Industrial Research Organization in Australia, supplier of the digital electronics package, as well as some of the simulation studies. The infrared radiometer is complemented by a microwave radiometer provided by the Centre de Recherches en Physique de l'Environment Terrestre et Planétaire in France.

The microwave radiometer of ATSR is a dual-channel Dicke-type instrument with two channels at 23.8 and 36.5 GHz. These are nadir-pointing and have a spatial resolution of ~20 km. In-flight calibration is achieved using a hot, internal load and a cold sky horn. The microwave measurements give a precise determination of the total atmospheric water vapor content to provide the correction for the effects of water vapor on the ERS-1 altimeter range measurement. They also can be used to furnish estimates of cloud liquid water, detection and characterization of surface ice cover, and detection of rain. In principle they also can improve the infrared SST retrieval, because water vapor is the main atmospheric constituent contributing to the contamination of the radiation originating at the sea surface, as it propagates through the atmosphere.

G.2 Instrument Characteristics

There are several design innovations incorporated into the ATSR (Figure G.1) to improve the absolute accuracy of the SST measurements. These include a conical scan pattern that facilitates the dual-angle measurement through atmospheric paths of different lengths, two onboard blackbody calibration targets in thermally controlled environments, a closed-cycle refrigerator to cool the detectors to a temperature close to that of liquid nitrogen, the digitization of the data stream to 12 bit, and the incorporation of a microwave radiometer to measure the atmospheric water vapor content. In addition, great care has been taken over the design of the optical path through the instrument to ensure good coregistration of the pixels from each channel,[3] over the mechanical design of the instrument to assure rigidity of the internal optical bench and its lack of dependence on temperature changes, over the thermal design to minimize the temperature changes around the orbit, and in the prelaunch calibration of the thermal infrared channels and characterization of the radiometer as a whole in both thermal equilibrium and nonequilibrium conditions.[4,5] Fuller details of the instrument are given by Edwards et al.[6]

Dual-Scan Angle

The dual-angle measurement[7] is achieved using a conical scan in which the cone traced out by the vector from the scan mirror to the instantaneous field of view (IFOV) is inclined at the half angle of the cone (Figure G.2a). A single revolution of the scan mirror directs onto the detectors the incoming radiation from a swath symmetrically placed about the subsatellite point (the nadir view) and from a swath ahead of the subsatellite point (the forward view). The cone half angle is 23.45°; and, with the satellite height of ~780 km, this leads to a local zenith angle to the spacecraft from the center of the forward view of ~55°. Thus the radiation from the center of the forward view passes through nearly twice as much atmosphere as the radiation from the center of the nadir swath. This is the essence of measurement of the atmospheric effect. The center of the forward view swath is about 900 km ahead

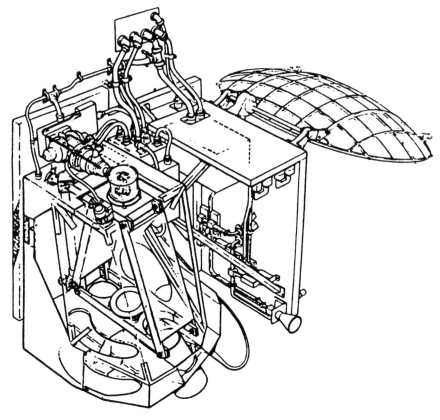

Figure G.1   A cutaway diagram of the ATSR, with the infrared radiometer on the left in front of the microwave radiometer. The flight direction is toward the top right and the direction to the earth surface is downwards. The curved forward-view and nadir-view orifices in the base of the infrared radiometer are fore and aft of the internal blackbody calibration targets and the off-axis parabolic mirror that directs the incoming radiation into the focal plane assembly in the cylindrical enclosure at the top of the instrument. (From Minnett, P. J., Satellite infrared scanning radiometers — AVHRR and ATSR/M, in *Microwave Remote Sensing for Oceanographic and Marine Weather Forecast Models*, Vaughan, R. A., Ed., Kluwer Academic Publishers, Norwell, MA, 1990. With permission.)

of the subsatellite point, and the time that elapses before the center of the nadir swath overlays the center of the forward view is ~135 s. The width of the swaths is ~500 km, with the forward view being slightly wider to accommodate the effects of the yaw steering of the satellite. The curvature of the swaths away from their centers leads to a smaller distance between the corresponding pixels from the nadir and forward views, and a shorter time between their superposition. It also reduces the difference in their atmospheric path lengths, and this must be taken into account in the atmospheric correction algorithm.

The swaths are viewed through two large curved orifices in the radiometer enclosure, and these are encircled by baffles to prevent direct sunlight from entering the instrument. In the space between the ends of the orifices are the two blackbody calibration targets. Thus each revolution of the scan mirror includes samples from both calibration targets between the two swaths of the surface of the earth (Figure G.2b).

Multichannel Characteristics

In addition to the dual-angle scan, the multiple channels of the ATSR permit a conventional atmospheric correction that exploits the spectral gradient of the atmospheric effect. This is the approach that has been used for over a decade with measurements of the AVHRR. The thermal channels of ATSR, in the spectral interval between 10 and 13 μm, match very closely those of the AVHRR; and by using the measurements from only the nadir swath, an AVHRR-type retrieval can be mimicked. Numerical simulations of the ATSR performance have shown that the incorporation of the multispectral information with the dual-angle measurements leads to an atmospheric correction that is more accurate than the multispectral or multiangle techniques alone.

A third channel, also in common with the AVHRR, is situated in the midinfrared range of the spectrum at ~3.7 μm, where the influence of water vapor is less than at the longer wavelength atmospheric window, and thereby provides additional information to improve the atmospheric correction. In addition to the emitted radiation from the sea surface, during the day this channel also contains some reflected solar radiation, which limits its use in SST determinations. Further use for data from this channel is identification of clouds in the IFOV during the night, because the spectral signatures of clouds are often different at 3.7 μm than at longer wavelengths.

Measurements from the fourth channel of the ATSR, at 1.62 μm, consist entirely of reflected insolation; and thus can be used during the day to identify coastlines, to check for the accuracy of the image geolocation, and for cloud detection because these have a high albedo in contrast to the dark sea surface. This wavelength was chosen over the shorter ones used on AVHRR because it gives better discrimination between clouds and underlying snow or ice. The next generation of AVHRR will also include a channel at this wavelength.

Cooled Detectors

The radiation from the IFOV is focused by an off-axis paraboloid mirror on a field stop, 0.695 mm square, which is the entrance aperture of the Focal Plane Assembly (FPA). This consists of the beam splitters, optical components, and detectors. The detectors for the two longwave channels are photoconductive cadmium mercury telluride, while those for the shorter wavelength channels are photovoltaic indium antimonide. The channel separation is achieved using a set of dichroic beam

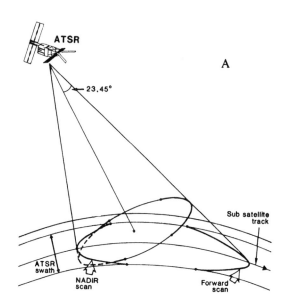

A

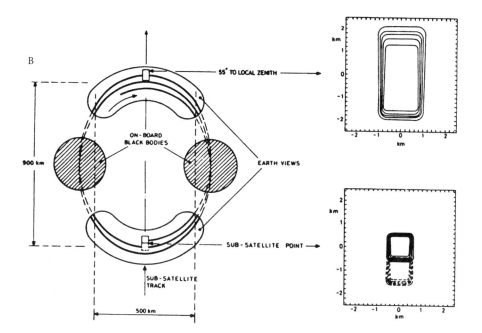

B

Figure G.2

splitters, and the channel spectral characteristics are determined by multilayer interference filters mounted in front of the detectors. To reduce the self-generated noise of the detectors the whole FPA is cooled to about 80 K by a Stirling cycle cooler that uses helium as the working fluid. This temperature is well below that which can be achieved using passive coolers, which, for example, are used to cool the AVHRR detectors to about 105 K. The resultant noise level, expressed as a noise equivalent temperature difference (NEDT), is given in Table G.1.

Onboard Calibration Targets

Located between the nadir and forward view orifices each of the two onboard blackbody targets is sampled for 16 pixels on each revolution of the scan mirror. To function well as calibration targets, these assemblies must have surface emissivities, ε, very close to unity; and be at temperatures that are well known, stable, and uniform. To achieve an uncertainty in the onboard calibration process of <0.1 K, the design goals for the calibration targets were $\varepsilon > 0.999$ for the 10–12 μm and >0.998 for the 3.7 μm spectral intervals; temperature differences across the viewed area of the target, <0.03 K; measurement uncertainty, <0.03 K; and temperature stability, 0.01 K/min.

The targets,[8] designed and built at the Mullard Space Science Laboratory of the University College, London, are constructed as cylindrical aluminum alloy cavities 140 mm in diameter, open at the top, with a high emissivity internal surface treatment. The resultant emissivity has been measured to be ~0.9995 and 0.9985 for the 10–13-μm and the 3.7-μm spectral intervals, respectively. Careful thermal design, involving isolation of the cavity from thermal disturbances (Figure G.3), has led to the viewed base of the cavity exceeding the design goals for temperature uniformity and stability. The temperature of each target is measured by six miniature platinum resistance thermometers, calibrated to 0.01 K, with an estimated overall readout accuracy within the design goal for the duration of the mission (Figure G.4).

Figure G.2 (A) The scan geometry of the ATSR on ERS-1. The satellite height is ~850 km above the earth surface and the center of the nadir scan is at the subsatellite point. (After Prata, A. J. F., Cechet, R. P., Barton I. J., and Llewellyn-Jones, D. T., The Along-Track Scanning Radiometer for ERS-1 — scan geometry and data simulation, *IEEE Trans. Geosci. Remote Sensing*, 28, 1990 (© 1990 IEEE). With permission.) (B) The configuration of the scan of the infrared radiometer showing the relative positions of the forward and nadir scan and the internal blackbody calibration targets. The relative sizes of the pixels at the centers of the forward and nadir swaths are shown, as contours of the spatial distribution of the IFOV at the 0.1, 0.3, 0.5, 0.7, and 0.9 levels of the maximum value. (From Minnett, P.J., Satellite infrared scanning radiometers — AVHRR and ATSR/M, in *Microwave Remote Sensing for Oceanographic and Marine Weather Forecast Models*, Vaughan, R. A., Ed., Kluwer Academic Publishers, Norwell, MA, 1990. With permission.)

Table G.1   Channel Noise Characteristics of the ERS-1 ATSR

Wavelength (μm)	Gain (mW/count)	NEΔT[a] (K)	Digitization (K/count)
Blackbody Temperature = 261.44 K (500 Samples)			
10.8	0.03451	0.026	0.030
12.0	0.03814	0.036	0.031
3.7	0.00022	0.065	0.107
Blackbody Temperature = 298.03 K (500 Samples)			
10.8	0.03451	0.019	0.021
12.0	0.03814	0.028	0.022
3.7	0.00022	0.025	0.020

[a]   Noise equivalent temperature difference.

Digitization and Data Transmission

The signal from each detector is integrated for 75 μsec before being digitized with 12-bit resolution. The scan mirror rotates at 400 rpm resulting in 2000 pixels per revolution. In normal circumstances 960 of these, constituting the nadir and forward view swaths and the blackbody measurements, are passed to the spacecraft data handling unit for transmission to the ground.

Unlike the AVHRR on the NOAA spacecraft, there is no direct broadcast of the ATSR signal; and the measurements are first stored by tape recorders before being dumped, orbit by orbit, over designated ground receiving stations. These are at Salmijärvi in Finland, near Kiruna in Sweden, and at Gatineau and Prince Albert in Canada. In addition, data from the Mediterranean area and North Africa are received in real time at Fucino in Italy and Maspalomas in the Canary Islands, Spain.

G.3 Prelaunch Calibration

An essential part of the instrument preparation is the characterization of performance of the radiometer as a whole. This prelaunch calibration took place in a specially constructed thermal-vacuum chamber at the University of Oxford, in which the conditions the instrument experiences in space can be duplicated in the laboratory. The process and results are fully described by Armitage et al.[4] and Mason.[5]

Objectives

There were three main objectives to the exercise: to ascertain the precise fields of view of the channels; to confirm numerical modeling predictions of the thermal environment in and around the ATSR in orbit; and to calibrate radiometrically the infrared channels. In addition, a simulation of the dynamic thermal environment around several complete orbits was undertaken to determine the behavior of the instrument in realistically changing, nonequilibrium thermal conditions.

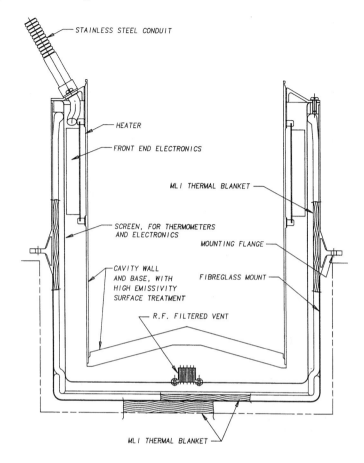

Figure G.3 Details of the design of the ATSR internal blackbody calibration targets. The targets have cylindrical symmetry.[8a]

Procedure

The ATSR thermal-vacuum facility consists of a large cylindrical chamber that can be evacuated to pressures of $<10^{-7}$ hPa, in which the thermal conditions of the instrument in orbit can be reproduced. This is done by enclosing the instrument in three separate copper shrouds, the surfaces of which were covered by a high emissivity black coating, and the temperatures of which could be controlled independently. One of the shrouds was a replica of the baseplate that attaches the ATSR to the ERS-1 spacecraft, and another was the Earthshine plate reproducing earth radiation that would fall onto the radiometer in orbit. The final shroud was a cold box that represented cold space and that completed the thermal enclosure of the ATSR.

The axes of both the ATSR scan mirror and the circular Earthshine plate were aligned along the axis of the cylindrical vacuum chamber. This allowed the two

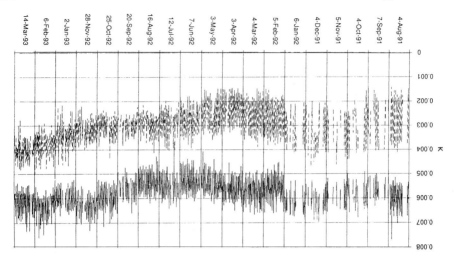

Figure G.4   The standard deviation of the temperatures measured by the six thermometers on the ATSR internal blackbody calibration targets, measured for 19 separate days over a 19-month period. The hot target shows a larger spread due to a radial change of ~0.02 K across the base, but in both cases relative drifts with time are negligible. Complementary measurements on duplicate blackbodies kept in identical thermal and vacuum conditions in the laboratory also show absolute calibration drifts to be <0.01 K over the same period.[8a]

external blackbody calibration targets that were embedded in the Earthshine plate to be brought into the field of view of the ATSR for any position of the rotation of the scan mirror, permitting a radiometric calibration of the infrared channels across the entire nadir and forward views. The emissivities of the external blackbody targets were >0.999, and their temperatures could be set in the range of 240–310 K. The temperatures were monitored using precision platinum resistance thermometers accurate to 0.02 K and traceable to a national standard.

The IFOVs of the ATSR channels were measured by scanning an image of an infrared point source (70-μm diameter) across the instrument field stop. This was to confirm the coregistration of the four channels and to map the spatial responses of the detector elements. The tests were also designed to reveal any distortion of the instruments optical bench, constructed out of carbon fiber, that may result from outgassing under vacuum.

All measurements were conducted under two thermal regimes, one simulating the spacecraft at the "beginning of life", and the other at the "end of life" when ambient spacecraft temperatures are expected to be somewhat warmer.

Results

The suite of measurements in the thermal-vacuum chamber revealed that the design targets of the instrument had indeed been met, and the ATSR would perform without unacceptable degradation up to the specified extremes of the spacecraft thermal environment. The infrared channels were found to be capable of measuring radiances with an accuracy corresponding to temperature uncertainties of <0.1 K, and the expected nonlinear responses of the longwave channels at high radiances were found and defined. An unexpected nonlinear response of the 3.7-µm channel was also identified. The measured NEDT for each channel was found to be in the range of 0.02–0.04 K (Table G.1). The spatial response of the 3.7-µm channel was found to be very uniform across the detector surface. The longwave channels were found to be distinctly wedge shaped in their response across the detector surface, and this is apparently due to minority carrier sweepout, a known characteristic of cadmium mercury telluride detectors. No significant degradation of the ATSR performance was detected due to distortion of the carbon fiber structure, due to outgassing, due to the nonequilibrium thermal effects around an orbit, or due to the gradual warming of the spacecraft expected during its lifetime.

G.4 In-Flight Performance

The postlaunch performance of the instrument has been excellent, with the only apparent degradation of the instrument after the launch being a slight pickup of the FPA cooler drive waveform in the 12-µm channel. This interference is at the level <~0.1 K and causes no significant degradation of the data quality.[9] However, on May 27, 1992 the 3.7-µm channel failed, and no data from this detector have been available since. This has had an impact on the nighttime SST retrievals in which this channel makes a major contribution to effective cloud screening. However, new cloud identification tests relying on the comparison between the nadir and forward views seem to have overcome this effect of the loss of this channel.[10]

G.5 Data Access

As with other measurements from the ERS-1 satellite, ATSR data and products may be accessed through the Earthnet ERS Central Facility,[11] which is situated at the European Space Research Institute (ESRIN), Via Galileo Galilei, 00044 Frascati, Italy. The numbers for the ERS-1 Help Desk are: Tel. (+39-6) 94180600, Fax (+39-6) 94180510, and Telex 610637 esrin i.

Current information on the status of the satellite and instruments, and any other material likely to be of interest to the ERS user community is available to users of the World Wide Web at http://services/esrin.esa.it. Similarly specific information about ATSR, and some example images, are available over the World Wide Web at http://atsrw3.ag.rl.ac.uk:80.

G.6 Recent Developments

A second ATSR (ATSR-2) for the ERS-2 spacecraft was launched on April 21, 1995. This is derived from the design of the ERS-1 ATSR, but incorporates precautions to avoid the premature loss of the 3.7-μm channel. ATSR-2 has additional channels in the visible part of the spectrum centered at 0.555, 0.659 and 0.865 μm, to provide data for the land-use research community. The signal-to-noise ratio of these channels is designed to be 20:1 at 0.5% albedo.

Acknowledgments

The author is indebted to members of the ATSR project team, especially C. T. Mutlow, A. M. Závody, and I. M. Mason, for recent information and figures included in this chapter. This was prepared under the auspices of the U.S. Department of Energy under Contract No. DE-AC02-76CH00016, with partial support from the National Oceanic and Atmospheric Administration, Grant NA26GP0266-01.

References

1. Duchossois, G., The ERS-1 mission objectives, *ESA Bull.*, 65, 16, 1991.
2. Louet, J. and Zobl, R., The ERS-1 satellite — Approaching two years in orbit, *ESA Bull.*, 74, 9, 1993.
3. Gray, P. F., The optical system of the Along Track Scanning Radiometer (ATSR), *Proc. Soc. Photo-Optical Eng. (SPIE)*, 589, 121, 1985.
4. Armitage, S., Corney, D., Mason, G., Taylor, W. H., Watkins, R. E. J., Williamson, E. J., Cragg, D., Mellor, G., Tinkler, D., Holmes, A. R., Linford W., and Spry, F., Test and calibration of the Along Track Scanning Radiometer (ATSR), *Proc. Int. Symp. Environ. Testing Space Programmes — Test, Facilities and Methods*, ESA SP-304, 1990, 559.
5. Mason, G., Test and Calibration of the Along Track Scanning Radiometer, a Satellite Borne Infrared Radiometer Designed to Measure Sea-Surface Temperature, D.Phil. thesis, Department of Atmospheric, Oceanic and Planetary Physics, University of Oxford, 1991.
6. Edwards, T., Browning, R., Delderfield, J., Lee, D. J., Lidiard, K. A., Milborrow, R. S., McPherson, P. H., Peskett, S. C., Topliss, G. M., Taylor, H. S., Mason, I., Mason, G., Smith A., and Stringer, S., The Along-Track Scanning Radiometer — Measurement of sea surface temperature from ERS-1, *J. Br. Interplanet. Soc.*, 43, 160, 1990.
7. Prata, A. J. F., Cechet, R. P., Barton I. J., and Llewellyn-Jones, D. T., The Along-Track Scanning Radiometer for ERS-1 — scan geometry and data simulation, *IEEE Trans. Geosci. Remote Sensing*, 28, 1990.
8. Mason, I. M., Sheather, P. H., Bowles J. A., and Davies, G., Black Body calibration sources of high accuracy for a spaceborne infrared instrument, the Along-Track Radiometer, *Applied Optics*, in review.
8a. Mason, I. M., personal communication, 1993.
9. Mutlow, C. T., personal communication, 1993.
10. Závody, A. M., personal communication, 1993.
11. Fea, M., The ERS ground segment, *ESA Bull.*, 65, 49, 1991.

H

OCTS on ADEOS

Sei-ichi Saitoh

H.1 ADEOS Spacecraft

The Advanced Earth Observing Satellite (ADEOS) has been conceived by National Space Development Agency of Japan (NASDA) to establish earth observation platform technology for future spacecraft, as well as to develop communication technology for the transmission of earth observation data. The main objectives of the ADEOS mission are to acquire data on global environmental changes such as the greenhouse effect, ozone layer depletion, deforestation of tropical rain forests, and abnormal climatic conditions.[1-3]

ADEOS is equipped with two NASDA core sensors and six Announcement Opportunity (AO) sensors for monitoring oceanic primary production, ocean surface winds, land utilization, greenhouse gases characteristics, aerosol distribution, and albedo distribution (Table H.1, Figure H.1).[3]

ADEOS is a large satellite with a mass of approx. 3500 kg and a power generation capability of more than 4500 W. It is made up of thermally, electrically, and mechanically independent units, which faciliate their integration and testing, including the Communication and Data Handling Subsystem (C&DH), the Electrical Power Subsystem (EPS), and the Altitude and Orbital Control Subsystem (AOCS). The satellite features an automatic, autonomous operation function capable of operating a large number of mission instruments, and interorbital communication equipment to transmit observation data via data relay satellite (Figure H.1, Table H.2).[3]

Table H.1 Main Specifications of Core Sensors and AO Sensors on ADEOS[3]

Sensor Name	Country	Organization	Objectives and Major Specifications
Core Sensors			
OCTS	Japan	NASDA	Measurement of ocean primary producion and sea surface temperature
AVNIR (Advanced Visible and Near Infrared Radiometer)	Japan	NASDA	Land utilization research, plant distribution research IFOV: 16 m (multiband) 8 m (panchromatic-band)
AO Sensors			
NSCAT (NASA Scatter meter)	U.S.	NASA/JPL	Measurements of surface wind speed and direction over the global oceans Accuracy speed: 2 m/sec direction: 20°
TOMS (Total Ozone Mapping Spectrometer)	U.S.	NASA/GSFC	Observation of total ozone changes
POLDER Earth's (Polarization and Directionality of the Reflectances)	France	CNES/LERTS	Observation of bidirectionality and polarization of the solar radiation reflected by the atmosphere
IMG (Interferometric Monitor for Greenhouse Gases)	Japan	MITI	Observation of greenhouse gases
ILAS (Improved Limb Atmospheric Spectrometer)	Japan	Environment Agency	Observation of the limb atmospheric microingredient over high latitude area
RIS (Retroreflector in Space)	Japan	Environment Agency	Observation of ozone, fluorocarbon, carbon dioxide, etc.

From NASDA Report 17, 1990, 8. With permission of the National Space Development Agency of Japan.

In early 1996, ADEOS will be launched into a sun-synchronous subrecurrent orbit of an altitude of about 797 km by the H-II launch vehicle from the Tanegashima Space Center.

H.2 OCTS Characteristics

The Ocean Color and Temperature Sensor (OCTS) is an optical radiometer devoted to the frequent global measurement of ocean color and sea surface temperature. The main characteristics of OCTS are provided in Table H.3. With a swath of 1400 km in each scan, OCTS will achieve global observation of the world oceans in 3 d.

OCTS has eight bands in the visible and near-infrared region and four bands in the thermal infrared region; spectral band characteristics are shown in Table H.4. The spatial resolution is about 700 m. The signal-to-noise ratio will be from 450 to 500 in the visible and near-infrared region. The noise equivalent differential temperature (NEDT) will be from 0.15 to 0.2 K at a target temperature of 300 K in the

Table H.2 Main Characteristics for ADEOS[3]

Parameter	Specifications
Shape	Main body (4 × 4 × 5 m)
	Solar array paddle (3 × 24 m)
Weight	Main body (2300 kg)
	Mission instruments (1300 kg)
Attitude control	Three-axis stabilized (zero-momentum)
Design life	3 years
Launch	
Vehicle	H-II Rocket
Site	Tanegashima Space Center
Period	Early 1996
Orbit	
Category	Sun-synchronous subrecurrent orbit
Local sun time	10:15–10:45 at descending node
Recurrent period	41 d
Altitude	ca. 797 km
Inclination	ca. 98.6°
Period	ca. 101 min
Data transmission	X-band and UHF-band
Data recorder	MDR × 3 (2 simultaneous operations)
	Recording (3/6/60 Mb/sec)
	Playback (60 Mb/sec)

From NASDA Report 17, 8, 1990. With permission of the National Space Development Agency of Japan.

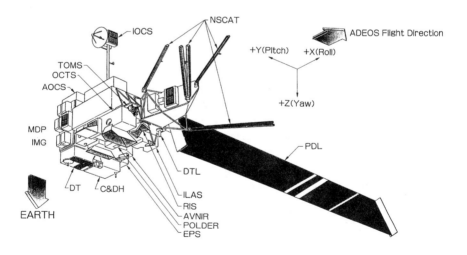

Figure H.1 On-orbit satellite overview of ADEOS. (From NASDA Report 17, 1990, 8. With permission of the National Space Development Agency of Japan.)

Table H.3   The Main Characteristics of OCTS

Parameters	Characteristics
Instantaneous field of view (IFOV)	0.85 mrad (approx. 700 m on the ground)
Scanning angle	−40 to + 40 degree (approx. 1400 km on the ground)
MTF	0.35
Polarization sensitivity	Band 1, less than 5%
	Bands 2–8, less than 2%
Tilting angle (in orbit)	−20°, 0°, +20° (3 steps)
Calibration	Visible, near IR: solar,
	Internal light source
	Thermal IR: deep space, blackbody
Quantization	10 b/pixel
Low data rate transmission	4 bands with 6- × 6-km resolution
(DTL)	(443 nm, 565 nm, 670 nm, 11.0 μm)

From NASDA Report 17, 1990, 8. With permission of the National Space Development Agency of Japan.

thermal region. The OCTS is calibrated in orbit using both solar light and a halogen lamp as the calibration sources. On-orbit calibration of OCTS will include observation of sunlight through relay optics; a stable halogen lamp for visible wavelength calibrations will also be used. Deep space and an internal blackbody will be used for calibration of the thermal infrared detectors.

OCTS has two data transmission modes. All raw pixel data are transmitted at X-band in a high data rate transmission mode. One pixel will be sampled from every 6×6 km area, and reduced resolution data is transmitted at UHF band in a low data rate transmission mode that is similar to the National Oceanic and Atmospheric Administration (NOAA) automatic picture transmission (APT). This subsampled data is called direct transmission for local users (DTL).

OCTS consists of scanning radiometer unit, which contains the optical system, the detector module, and the electronics. OCTS uses a catoptric optical system and a mechanical rotating, scanning method using a mirror. This is done because OCTS covers a wide range of wavelengths and includes scanning angles to ±40°. OCTS can tilt its line of sight along the track to prevent the sun glitter at the sea surface from contaminating the observations. For high sensitivity, each band has 10 pixels aligned to the track. The OCTS observation concept is shown in Figure H.2. The infrared detectors are cooled at 100 K by a large radiant cooler facing deep space.

H.3 Data Processing Plan and Derived Standard Products

The ADEOS Ground Data System will be located at the Earth Observation Center (EOC) in Hatoyama, Japan. In order to acquire the global OCTS data, recorded

Table H.4 The Observation Spectral Wavelength of OCTS

Band	Wavelength (μm)	Radiance ($Wm^{-2}\,sr^{-1}\,\mu m^{-1}$)	S/N
1	0.402–0.422	145	450
2	0.433–0.453	150	500
3	0.480–0.500	130	500
4	0.510–0.530	120	500
5	0.555–0.575	90	500
6	0.660–0.680	60	500
7	0.745–0.785	40	500
8	0.845–0.885	20	450
		Target Temp. (K)	**NEDT (K)**
9	3.55–3.88	300	0.15
10	8.25–8.80	300	0.15
11	10.3–11.4	300	0.15
12	11.4–12.5	300	0.20

From NASDA Report 17, 1990, 8. With permission of the National Space Development Agency of Japan.

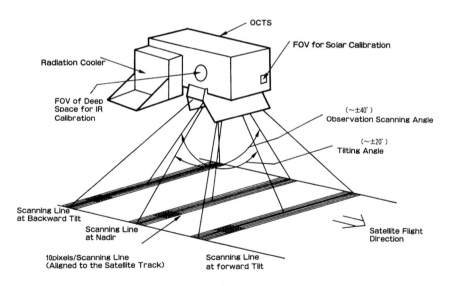

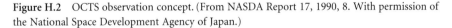

Figure H.2 OCTS observation concept. (From NASDA Report 17, 1990, 8. With permission of the National Space Development Agency of Japan.)

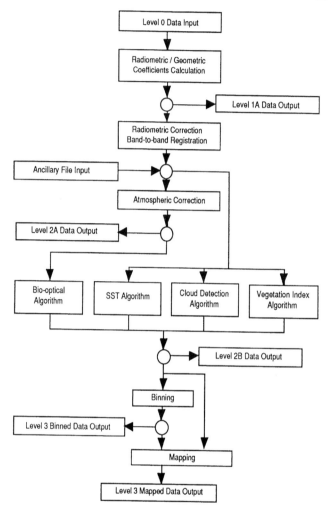

Figure H.3 Data processing scheme of OCTS. (From NASDA Report 17, 1990, 8. With permission of the National Space Development Agency of Japan.)

data on mission data recorders (MDRs) will be relayed to three ground stations: one at EOC and the other two at NASA's Alaska SAR Facility (ASF) in Fairbanks, Alaska, and at NASA's Wallops Flight Facility (WFF).

Data processing flow is shown in Figure H.3.[5] Level 2 products contain two kinds of data, level 2B will be generated after the processing by using bio-optical algorithm, SST algorithm and cloud-masking algorithm.

Level 0–3 products will be generated at EOC for OCTS. The standard products planned for OCTS are summarized in Table H.5.[4,5] Each data set will trace its

Table H.5   ADEOS/OCTS Data Products[4]

Level	Product Description
Level 0 (no distribution)	Individual files written
	for each pass of real-time data
	for each segment of recorded data
	No processing or reformating of minor frames
	Orbit data, time code and QQC information appended
Level 1A	Files written for individual scene (constant title angle)
	RTC: real-time data received at EOO
	LAC: full resolution for real-time and selected recorded data
	GAC: 4 km by 4 km sub-sampled from Level 0
	Records written by individual scans
	Radiometric and geometric coefficients appended (not applied)
Level 1B (no distribution)	Radiometrically corrected and band-to-band registrated
Level 2	Ocean Color Product;
	normalized water leaving radiance (412, 443, 490, 520, 565 nm)
	aerosol radiance (665, 765, 865 nm)
	epsilon map
	pigment concentration
	Chlorophyll a (TBD)
	diffuse attenuation coefficient (490 nm)
	quality flag
	SST Product:
	sea surface temperature
	NVI Product (TBD)
	vegetation indices
Level 3	Binned
	Spatial and time binning from Level 2 (GAC)
	10 km spatial bins (approx. 5.4E6 bins for global image)
	1 day, 1 week, 1 month and 1 year bins
	Statistical data for each parameter
Level 3	Map
	Parameters resampled on a map
	level 1B, 2 data (RTC, LAC)
Level 3	Binned Map
	Level 3 binned data mapped onto uniform latitude/longitude grid (2048 by 4096)
Browse Data (TBD)	Mapped parameters (pigment, SST) subsampled and compressed from:
	level 3′ Map (RTC)
	level 2 (GAC)
	level 3 Binned Map

From NASDA-EA-MITI JRA-93-001, 1993. With permission.

heritage from the Coastal Zone Color Scanner (CZCS), Sea-Viewing Wide Field-of-View Sensor (SeaWiFS), and Advanced Very High-Resolution Radiometer (AVHRR).

References

1. Iwasaki, N., Sensor operation of Advanced Earth Observing Satellite (ADEOS), *Proc. Asia-Pacific ISY Conf.*, 2, 47, 1992.
2. National Space Development Agency of Japan, Development of ADEOS and its sensors, NASDA Rep., 17, 1990, 8.
3. National Space Development Agency of Japan, Advanced Earth Observing Satellite, 1992, 24 pp.
4. National Space Development Agency of Japan, Japan Environmental Agency and Ministry of International Trade and Industry, Research Announcement Advanced Earth Observing Satellite (ADEOS) CAL/VAL and SCIENCE, NASDA-EA-MITI JRA-93-001, 1993.
5. Mitchell, B. G. and Saino, T., Eds., Assessment of global ocean productivity by satellite remote sensing, 1993, 49 pp.

Index

Index

A

AMI (Advanced Microwave Imager), 297, 301

AMI (Active Microwave Instrument), 360, 449–450, 457–458, 462

AVHRR (Advanced Very High Resolution Radiometer), 3–4, 29–31, 40–41, 98–99, 101–104, 106–107, 134, 146–147, 155, 163, 212, 263, 266, 278, 280, 307–311, 368–370, 372, 374–375, 377, 388, 407–408, 410, 413, 415–416, 464–466, 478

  albedo, 315

  channel, 309

  ice concentration, 309

  ice concentration accuracy, 311

  ice surface temperature, 314–315

  threshold technique, 310

  tie point algorithm, 310

Aerosol

  Mt. Pinatubo aerosol, 142

  statospheric aerosol, 133

Alaska SAR Facility, 377, 379, 398, 400

Algorithm

  bio-optical algorithm, 476

  geophysical algorithm, 206

  geophysical scatterometer algorithm, 297

  sea ice algorithm, 319, 326, 335, 379

  SST algorithm, 476

  sensor algorithm, 206

  split window algorithm, 146

  wind speed algorithm, 186–187, 189, 296–298

Aliasing, 47, 54

ATSR (Along-Track Scanning Radiometer), 131–142, 361, 434, 461–472

  black body calibration target, 467–470

  cooled detector, 465–466

  data access, 472

  inflight calibration target, 467–470

  inflight performance, 471

  measurement geometry, 463–466

  mechanical design, 463–468

  microwave radiometer, 463

  SST retrieval coefficient, 139

  stirling cycle cooler, 467

  thermal vacuum test, 468–471

ATSR-2, 472

  visible channel, 472

Altimeter, 16–19, 21, 23, 27, 45–54, 185–187, 189, 194, 199, 223, 227, 295, 297–299, 427–439

  dual frequency, 120

  error, 119, 124

  ERS-1, 117

  GFO (Geosat follow-on altimeter), 234

  sampling, 46–47

  sea level, 113

  Seasat, 186, 196

  sub-Nyquist sampling, 47, 49

  validation, 114, 117